Atmospheric Measurement Techniques

Atmospheric Measurement Techniques

Edited by **Dorothy Rambola**

SYRAWOOD
PUBLISHING HOUSE

New York

Published by Syrawood Publishing House,
750 Third Avenue, 9th Floor,
New York, NY 10017, USA
www.syrawoodpublishinghouse.com

Atmospheric Measurement Techniques
Edited by Dorothy Rambola

International Standard Book Number: 978-1-68286-014-4 (Hardback)

Printed in the United States of America.

Contents

Permissions

List of Contributors

Preface

The world is advancing at a fast pace like never before. Therefore, the need is to keep up with the latest developments. This book was an idea that came to fruition when the specialists in the area realized the need to coordinate together and document essential themes in the subject. That's when I was requested to be the editor. Editing this book has been an honour as it brings together diverse authors researching on different streams of the field. The book collates essential materials contributed by veterans in the area which can be utilized by students and researchers alike.

The study of the earth's atmosphere is a complex phenomena. It includes not just weather but also the diverse processes and interactions that occur within the atmosphere as well as with other ecological systems. The aim of this book is to present and elaborate discussion on the current and proposed atmospheric measurement techniques. It brings forth the latest innovations in this field and their applications in understanding atmospheric processes. This book will serve as a vital tool in facilitating the progress of this field.

Each chapter is a sole-standing publication that reflects each author's interpretation. Thus, the book displays a multi-facetted picture of our current understanding of applications and diverse aspects of the field. I would like to thank the contributors of this book and my family for their endless support.

Editor

Aerosol properties over the western Mediterranean basin: temporal and spatial variability

H. Lyamani[1,2], A. Valenzuela[1,2], D. Perez-Ramirez[3,4], C. Toledano[5], M. J. Granados-Muñoz[1,2], F. J. Olmo[1,2], and L. Alados-Arboledas[1,2]

[1]Andalusian Institute for Earth System Research (IISTA-CEAMA), 18006, Granada, Spain
[2]Department of Applied Physic, University of Granada, 18071, Granada, Spain
[3]Mesoscale Atmospheric Processes Laboratory, NASA Goddard Space Flight Center, Greenbelt, Maryland 20771, USA
[4]Goddard Earth Sciences Technology and Research, Universities Space Research Association (GESTAR/USRA), Columbia, Maryland, USA
[5]Atmospheric Optics Group (GOA), University of Valladolid (UVA), 47071, Valladolid, Spain

Correspondence to: H. Lyamani (hlyamani@ugr.es)

Abstract. This study focuses on the analysis of Aerosol Robotic Network (AERONET) aerosol data obtained over Alborán Island (35.90° N, 3.03° W, 15 m a.s.l.) in the western Mediterranean from July 2011 to January 2012. Additional aerosol data from the three nearest AERONET stations (Málaga, Oujda and Palma de Mallorca) and the Maritime Aerosol Network (MAN) were also analyzed in order to investigate the temporal and spatial variations of aerosol over this scarcely explored region. High aerosol loads over Alborán were mainly associated with desert dust transport from North Africa and occasional advection of anthropogenic fine particles from central European urban-industrial areas. The fine particle load observed over Alborán was surprisingly similar to that obtained over the other three nearest AERONET stations, suggesting homogeneous spatial distribution of fine particle loads over the four studied sites in spite of the large differences in local sources. The results from MAN acquired over the Mediterranean Sea, Black Sea and Atlantic Ocean from July to November 2011 revealed a pronounced predominance of fine particles during the cruise period.

1 Introduction

Atmospheric aerosol particles play an important role in the atmosphere because they can affect the Earth's radiation budget directly by the scattering and absorption of solar and terrestrial radiation (e.g., Haywood and Shine, 1997), and indirectly by modifying cloud properties (e.g., Kaufman et al., 2005), and hence have important climate implications. Understanding the influence of atmospheric aerosols on radiative transfer in the atmosphere requires accurate knowledge of their columnar properties, such as the spectral aerosol optical depth, a property related to aerosol amount in atmospheric column (Haywood and Boucher, 2000; Dubovik et al., 2002). Global measurements of columnar aerosol properties including spectral aerosol optical depth can be assessed from satellite platforms (e.g., Kaufman et al., 1997). However, satellite aerosol retrievals suffer from large errors due to uncertainties in surface reflectivity. Currently, the ground sun photometric technique is considered the most accurate one for the retrieval of aerosol properties in the atmospheric column (e.g., Estellés et al., 2012). Thus, many ground-based observation networks have been established in order to understand the optical and radiative properties of aerosols and indirectly evaluate their effect on the radiation budget and climate (e.g., AERONET (Aerosol Robotic Network)). However, the quantification of aerosol effects is very difficult because of the high spatial and temporal variability of physical and optical properties of aerosol (Forster et al., 2007). This high aerosol variability is due to their short atmospheric lifetime, aerosol transformations, aerosol dynamics, different meteorological characteristics, and the wide variety of aerosol sources (Haywood and Boucher, 2000; Dubovik et al., 2002). In this sense, Forster et al. (2007) highlighted the large uncertainties on the aerosol impact on radiation budget. Therefore, monitoring of aerosol properties at different areas in the world can contribute to reduce these uncertainties.

Most of the planet is covered by oceans and seas, and thus the study of marine aerosol is a topic of ongoing interest (e.g., Smirnov et al., 2002). Particularly, many efforts are being made to characterize this aerosol type from ground-based measurements, leading to the creation of the Maritime Aerosol Network (MAN) as part of the AERONET network (Smirnov et al., 2009). However, MAN lacks of continuous temporal measurements, and thus measurements from remote islands in the oceans and seas are required. Particularly, in the Mediterranean basin aerosol properties are characterized by a great complexity, due to the presence of different types of aerosols such as maritime aerosols from the Mediterranean Sea itself, biomass burning aerosols from forest fires, anthropogenic aerosols transported from European and North African urban areas, mineral dust originated from north African arid areas, and anthropogenic particles emitted from the intense ship traffic in the Mediterranean Sea (e.g., Lelieveld et al., 2002; Barnaba and Gobbi, 2004; Lyamani et al., 2005, 2006a, b; Papadimas et al., 2008; Viana et al., 2009; Pandolfi et al., 2011; Alados-Arboledas et al., 2011; Becagli et al., 2012; Valenzuela et al., 2012a, b; Mallet et al., 2013). Past studies revealed that the aerosol load and the aerosol direct radiative effect over the Mediterranean are among the highest in the world, especially in summer (e.g., Lelieveld et al., 2002; Markowicz et al., 2002; Papadimas et al., 2012; Antón et al., 2012).

In this framework, the characterization of aerosol over the Mediterranean has received great scientific interest. To date, a large number of studies has been done focusing on the eastern and central regions (e.g., Formenti et al., 1998; Balis et al., 2003; Gerasopoulos et al., 2003; Di Iorio et al., 2003; Kubilay et al., 2003; Pace et al., 2005, 2006; Fotiadi et al., 2006; Meloni et al., 2007, 2008; Di Sarra et al., 2008; Di Biagio et al., 2010; Boselli et al., 2012). However, few studies have been done in the western Mediterranean Basin (Horvath et al., 2002; Alados-Arboledas et al., 2003; Mallet et al., 2003; Estellés et al., 2007; Saha et al., 2008; Pérez-Ramírez et al., 2012, Foyo-Moreno et al., 2014). The majority of these studies have been performed in coastal Mediterranean urban sites largely influenced by local pollution emissions, except those carried out at Crete and Lampedusa islands in the eastern and central Mediterranean Sea regions. In general, columnar aerosol data are scarce over the Mediterranean Sea and almost absent over the western Mediterranean Sea. Thus, measurements of the aerosol properties over the western Mediterranean Sea are needed in order to evaluate the

aerosol regimes over this scarcely explored region (Smirnov et al., 2009). In order to fill this gap and provide columnar aerosol properties over the western Mediterranean Sea, the Atmospheric Physic Group of the University of Granada, Spain, in collaboration with Royal Institute and Observatory of the Spanish Navy (ROA), has installed a sun photometer at Alborán, a very small island in the westernmost part of Mediterranean Sea located midway between the African and European continents. Currently, this station is part of AERONET network (http://aeronet.gsfc.nasa.gov).

This study focuses on the characterization of aerosol load and aerosol types as well as on their temporal variability over Alborán Island in the western Mediterranean from 1 July 2011 to 23 January 2012. In addition, special attention is given to the conditions responsible for large aerosol loads over this island, and much attention is paid to identify the potential aerosol sources affecting Alborán. Furthermore, additional aerosol properties from three AERONET stations (Málaga, Oujda and Palma de Mallorca) surrounding Alborán Island and from a MAN cruise over the Mediterranean Sea, Black Sea and Atlantic Ocean (Fig. 1) are analyzed here to investigate the spatial aerosol variation over the Mediterranean basin.

The work is structured as follows. In Sect. 2 we describe the instrumentation used and the experimental sites. Section 3 is devoted to the main results, where we analyze the aerosol optical properties at Alborán Island and the spatial variability of aerosol properties in the Mediterranean. Finally, in Sect. 4 we present the summary and conclusions.

2   Instrumentation and study sites

2.1   AERONET measurements

Columnar aerosol properties were measured by a CIMEL CE-318-4 sun photometer, which is the standard automated sun photometer used in the AERONET network (Holben et al., 1998). This instrument has a full view angle of $1.2°$ and makes direct sun measurements at 340, 380, 440, 500, 670, 870, 940 and 1020 nm (nominal wavelengths). The direct sun measurements are then used to retrieve the aerosol optical depth at each wavelength, $\delta_a(\lambda)$, except for 940 nm which is used to compute precipitable water vapor (Holben et al.,1998). Detailed information about the CIMEL sun photometer can be found in Holben et al., 1998. The total estimated uncertainty in $\delta_a(\lambda)$ provided by AERONET is of ± 0.01 for $\lambda > 440$ nm and ± 0.02 for shorter wavelengths (Holben et al., 1998). Furthermore the spectral dependency of the $\delta_a(\lambda)$ has been considered through the Ångström exponent, $\alpha(440–870)$, calculated in the range 440–870 nm. The Ångström exponent provides an indication of the particle size (e.g., Dubovik et al., 2002). Small values of the Ångström coefficient ($\alpha(440–870) < 0.5$) suggest a predominance of coarse particles, such as sea salt or dust, while

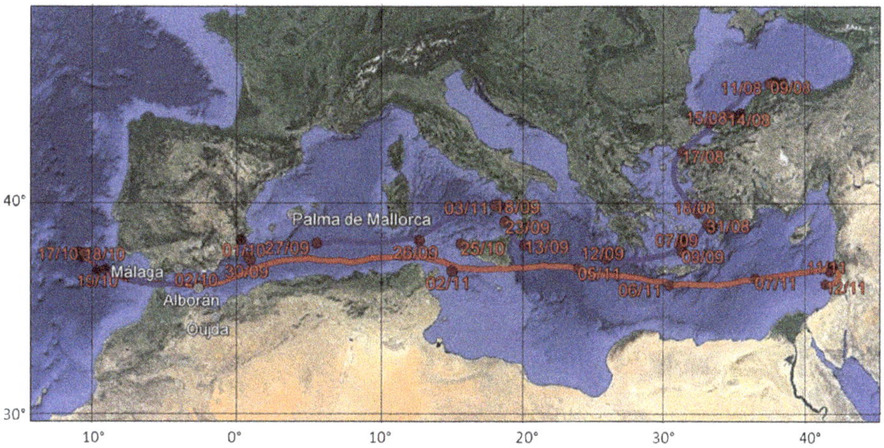

Figure 1. Map of Mediterranean basin showing the location of Alborán Island, Málaga, Oujda and Palma de Mallorca and a MAN cruise track over the Mediterranean Sea, Black Sea and Atlantic Ocean during 26 July–13 November 2011.

$\alpha(440\text{--}870) > 1.5$ indicates a predominance of small particles such as sulphate, nitrate and biomass burning particles. Also included in the analysis are aerosol optical depths at 500 nm for fine mode ($\delta_F(500\,\text{nm})$) and for coarse mode ($\delta_C(500\,\text{nm})$) as well as the fine mode fraction (FMF) (ratio of $\delta_F(500\,\text{nm})$ to $\delta_a(500\,\text{nm})$), determined using the spectral de-convolution algorithm method developed by O'Neill et al. (2003). In this study, the level 2 AERONET aerosol data are used.

2.2 AERONET stations

This study focuses on the AERONET sun photometer measurements acquired at the Alborán Island (35.90° N, 3.03° W, 15 m a.s.l.), in the western Mediterranean Sea, from 1 July 2011 to 23 January 2012. Alborán is a small island with an approximate surface of 7 ha, located ∼ 50 km north of the Moroccan coast and 90 km south of the Spanish coast (Fig. 1). Currently, only 12 members of a small Spanish Army garrison live on the island. The island and its surrounding area are declared a natural park and marine reserve. There is no significant local anthropogenic emission source at Alborán; however, the island is just south of an important shipping route (www.marinetraffic.com). Due to its location, Alborán Island is expected to be affected, depending on regional circulation, by anthropogenic pollutants originated in urban and industrial European areas, anthropogenic particles emitted from the ship traffic, desert dust transported from North African arid regions and maritime aerosols from the Mediterranean Sea. The climate of the region depends strongly on the Azores anticyclone. Winter is mainly characterized by low pressure systems passing over the Iberian Peninsula, resulting in the prevalence of westerly winds and enhanced rainfall. In this season, the weather is unstable, wet and windy. In summer, the well-established Azores high pressure produces dry and mild weather with easterly winds

that combine with sea/land breezes created by the aridity of the coastal mountains (Sumner et al., 2001).

In addition, to investigate the spatial variation of aerosol properties over the western Mediterranean, we used AERONET data obtained from 1 July 2011 to 23 January 2012 over three AERONET stations surrounding Alborán Island; Oujda, Málaga and Palma de Mallorca (see Fig. 1). These sites cover different environments including, urban, coastal and island sites, respectively, and have different background aerosol characteristics. Palma de Mallorca (39.35° N, 2.39° E, 13 m a.s.l.), the capital of the Balearic Islands, is the largest city in the Mallorca Island with a population of around 400 000. It is located in the western Mediterranean Sea, about 250 km from the African continent and 190 km from the Spanish coast. Málaga (36.72° N, 4.5° W, 40 m a.s.l.), with a population of around 600 000 is the major coastal city in southeast Spain on the Mediterranean coast. Oujda city (34.65° N, 1.89° W, 450 m a.g.l.) is located in eastern Morocco, 60 km south of the Mediterranean Sea, with an estimated population of 450 000.

2.3 Maritime Aerosol Network measurements

Furthermore, we used shipborne sun photometer measurements collected onboard the *Nautilus 11* on the Mediterranean Sea, Atlantic Ocean and Black Sea during the period 26 July–13 November 2011. These measurements were made in the framework of the Maritime Aerosol Network (MAN), a component of AERONET (Smirnov et al., 2011). More detailed information about the *Nautilus 11* cruise track can be found at http://aeronet.gsfc.nasa.gov/new_web/cruises_new/ Nautilus_11.html. MAN uses Microtops II hand-held sun photometers and utilizes calibrations and data processing procedures of AERONET network. The Microtops II sun photometer used in this cruise acquires direct sun measurements at 440, 500, 675 and 870 nm. The estimated uncertainty of the optical depth in each channel is around ± 0.02 (Knobelspiesse et al., 2004). Level 2 MAN data are used in this study.

2.4 Air mass trajectories

To characterize the transport pathways and the origins of air masses arriving at our studied AERONET sites, 5-day backward trajectories ending at 12:00 UTC at these sites for 500, 1500, 2500, 3500, 4500 and 5000 m above ground level were calculated using the HYSPLIT model for days with AERONET measurements (Draxler and Rolph, 2003). In addition, backward trajectories ending at the different points of MAN cruise for 500, 1500, 2500, 3500, 4500 and 5000 m above ground level were also performed for days with MAN observations. The HYSPLIT model version employed uses GDAS meteorological data and includes vertical wind.

3 Results and discussion

3.1 Temporal evolution of aerosol properties over Alborán Island

Figure 2 shows the temporal evolutions of daily mean values of aerosol optical depths at 500 and 1020 nm and $\alpha(440-870)$ measured at Alborán Island in the western Mediterranean from 1 July 2011 to 23 January 2012. There are some gaps in $\delta_a(\lambda)$ and $\alpha(440-870)$ data series due to some technical problems and the presence of clouds (invalid data). Table 1 presents a statistical summary of daily average values of all the analyzed aerosol properties. One of the main features observed is the large variability of $\delta_a(\lambda)$ (for example, $\delta_a(500\,nm)$ ranged from 0.03 to 0.54) that is primarily related to changes in the air masses affecting the study area, as can be seen hereafter. The coefficient of variation (COV), defined as the standard deviation divided by the mean value, can be used to compare the variability of different data sets. As shown in Table 1, the $\delta_a(\lambda)$ at 1020 nm (with COV of 91 %) showed much greater variability than at 340 nm (with COV of 60 %). It is well known that $\delta_a(\lambda)$ at higher wavelengths is more affected by naturally produced coarse particles (radius above $0.5\,\mu m$) like dust and sea salt particles, while $\delta_a(\lambda)$ at smaller wavelengths is more sensitive to the fine particles (radius below $0.5\,\mu m$) such as those from anthropogenic activities or biomass-burning. Thus, the higher variability of $\delta_a(\lambda)$ for larger wavelengths indicates strong variability in the coarse particle load (dust or sea salt) over Alborán Island. This result is also supported by the larger COV of $\delta_C(500\,nm)$ as compared to $\delta_F(500\,nm)$ (Table 1). Aerosol salt emission variations due to the wind speed variation and the changes in the frequency and intensity of dust intrusions over the island may explain the large variability in the coarse particle component and hence the large $\delta_a(\lambda)$ variability for large wavelengths. Moreover, coarse particles have shorter residence time in the atmosphere in comparison with fine particles, which could

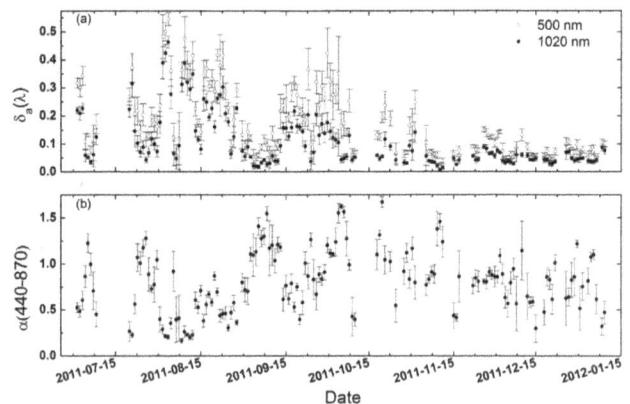

Figure 2. Temporal evolution of the daily mean values of (a) aerosol optical depth at 500 and 1020 nm and (b) the Ångström exponent calculated in the range 440–870 nm, measured at Alborán Island in the western Mediterranean from 1 July 2011 to 23 January 2012. The error bars are standard deviations.

explain also the large $\delta_C(\lambda)$ variability. On the other hand, $\alpha(440-870)$ values also show large variability and vary from 0.2 to 1.7 with mean value of 0.8 ± 0.5, indicating different atmospheric conditions dominated by different aerosol types (coarse particles, fine aerosols and/or different mixtures of both coarse and fine particles). It is noted that on 70 % of the analyzed days, the values of $\alpha(400-870)$ were lower than 1, suggesting that coarse particles dominated the aerosol population over the Alborán Island for most of the analyzed days. This is further supported by the analysis of fine mode fraction which ranged from 0.20 to 0.90 (mean value of 0.47 ± 0.15), with daily mean values less than 0.5 on 65 % of the analyzed days.

The observed mean $\delta_a(500\,nm)$ value over Alborán Island was significantly higher (by factor of 2) than that reported by Smirnov et al. (2002) ($\delta_a(500\,nm)$ in the range 0.06–0.08) for open oceanic areas in the absence of long-range transport influences. Moreover, the mean $\delta_a(500\,nm)$ and $\alpha(440-870)$ values obtained in this study were larger than the global mean $\delta_a(500\,nm)$ value of 0.11 and $\alpha(440-870)$ of 0.6 reported for maritime aerosols by Smirnov et al. (2009). On the other hand, average aerosol optical depths at 495.7 nm of 0.24 ± 0.14 and $\alpha(415-868)$ of 0.86 ± 0.63 were obtained from multi filter rotating shadowband radiometer at Lampedusa Island (in the central Mediterranean Sea) during July 2001–September 2003 (Pace et al., 2006). Using AERONET data measured in Crete Island (eastern Mediterranean Sea) during 2003–2004, Fotiadi et al. (2006) reported mean $\delta_a(500\,nm)$ value of 0.21 and $\alpha(440-870)$ of 1.1. The differences between aerosol properties observed over the islands of Alborán, Lampedusa and Crete could be explained in terms of differences in the period and duration of the measurements, in air mass circulation and in the methodologies employed. Later we compare the results obtained over Alborán to those observed over three nearby AERONET sta-

Table 1. Statistical summary of daily mean values of spectral aerosol optical depth at 1020, 500 and 340 nm, Ångström exponent, α(440–870), fine and coarse mode aerosol optical depths at 500 nm, δ_F(500 nm) and δ_C(500 nm), and fine mode fraction, FMF, observed over Alborán Island in the western Mediterranean during 1 July 2011–23 January 2012; SD is the standard deviation and COV is the coefficient of variation.

	Mean	SD	Minimum	Maximum	COV(%)
δ_a(1020 nm)	0.11	0.10	0.01	0.46	91
δ_a(500 nm)	0.17	0.12	0.03	0.54	70
δ_a(340 nm)	0.25	0.15	0.05	0.65	60
α(440–870)	0.8	0.4	0.2	1.7	50
δ_F(500 nm)	0.08	0.05	0.01	0.30	63
δ_C(500 nm)	0.10	0.09	0.01	0.4	90
FMF	0.47	0.15	0.20	0.94	32

tions (during the same period and using the same type of instruments).

According to the Smirnov et al. (2003) criterion, pure maritime situations can be generally found when δ_a(500 nm) < 0.15 and α(440–870) is less than 1. Considering this criterion, pure maritime situations were observed over Alborán Island on 40 % of the analyzed days. According to back trajectory analysis, almost all these days were characterized by advection of clean Atlantic air masses over the study area. In addition, the majority of these pure maritime cases were observed during the wet season from November to January. This result is in agreement with the study performed at the island of Lampedusa, in the central Mediterranean, showing that pure maritime situations are usually observed during Atlantic air advection (Pace et al., 2006). However, clean maritime conditions observed over Alborán Island during the analyzed period are more frequent than those observed over Lampedusa. Pace et al. (2006) showed that clean maritime conditions are rather rare over the central Mediterranean due to the large impact of natural and anthropogenic sources. The difference in the occurrences of clean maritime conditions at these two sites can be explained by their different locations. Alborán is closer to the Atlantic Ocean than Lampedusa is, and the Atlantic air masses reaching Lampedusa may be influenced more by anthropogenic aerosol during their passage over Mediterranean Sea and continents.

Threshold values for δ_a(500 nm) and α(440–870) have been widely used in remote sensing to identify marine aerosol type. For example, Smirnov et al. (2003) used δ_a(500 nm) $\leq$ 0.15 and α(440–870) $\leq$ 1 and Sayer et al. (2012a, b) proposed δ_a(500 nm) $\leq$ 0.2 and 0.2 $\leq$ α(440–870) $\leq$ 1 while Toledano et al. (2007) used δ_a(500 nm) $\leq$ 0.15 and α(440–870) $\leq$ 0.6 for identifying pure maritime situations. However, the proposed threshold values for δ_a(500 nm) and α(440–870) to identify maritime aerosol type are purely empirical. Therefore, not all observations that meet these thresholds will represent the pure maritime aerosol. In fact, in Alboran Island we found measurements that fulfill these

criteria but that are not associated with pure maritime conditions. In this sense, in Fig. 3 we show the δ_a(500 nm) and α(440–870) observed on 26 August, 2011. During this day, the δ_a(500 nm) values ranged from 0.06 to 0.13 with mean daily value of 0.09 ± 0.01 and α(440–870) was in the range 0.3–0.6, indicating clean atmospheric condition dominated by coarse particles. Thus, according to the above criteria this day is classified as pure maritime case. However, the back trajectory analysis and Meteosat Second Generation (MSG) satellite image (Thieuleux et al., 2005) for 26 August revealed the presence of dust over Alborán Island (Fig. 3). Therefore, care must be taken when using δ_a(500 nm) and α(440–870) thresholds for discriminating the pure maritime cases since dusty situations with low dust loads can be confused with pure maritime conditions. Additional information such as air mass back trajectory or satellite images is needed for better identifying the pure maritime cases.

As can be seen in Fig. 2, there were several days strongly influenced by aerosols, with δ_a(500 nm) values exceeding 0.3. High aerosol loads (δ_a(500 nm) > 0.3) over Alborán Island were observed on 30 of the 160 analyzed days. All these events were observed from July to October. In 27 of these cases, the mean daily α(440–870) values were lower than 0.8 and fine mode fraction (FMF) lower than 0.5; suggesting predominance of coarse particles as either sea salt or dust transported from desert areas. According to the analyses of back trajectories and MODIS satellite images (not shown), all these 27 cases were related to dust intrusions from North Africa. It is important to note that in these dust events, the δ_F(500 nm) values were also relatively high (for this remote site) and ranged from 0.07 to 0.20 with mean value of 0.12 ± 0.03. These results highlight a considerable contribution of fine mode particles (either dust or anthropogenic or both) to the aerosol population (FMF ranged from 20 to 52 %) during these dust events. Back trajectory analysis for dusty days with highest fine aerosol load revealed that the air masses reaching the study area at low levels (at 500 or 1500 m level) have originated over Europe and the Mediterranean Sea. However, during desert dust events with lowest fine aerosol loads, none of the air masses affecting the study area come from Europe or Mediterranean Sea, which points out significant contribution of anthropogenic particles to the fine mode fraction of δ_a(500 nm) during desert dust events associated with large loads of fine aerosol particles.

The remaining high aerosol load events were observed from 30 September to 4 October 2011 (Fig. 4). During these days, the high aerosol loads were associated with relatively high α(440–870) values that reached the highest α(440–870) value (about 1.6) during the entire study on 4 October. During these days, the δ_F(500 nm) values were also high (> 0.19) and reached the highest mean daily value of 0.33 on 4 October. This behavior suggests a predominance of fine particles transported from continental industrial/urban areas as there is no local anthropogenic activity in Alborán. The high $\delta_a(\lambda)$ values observed in this event were associated with persis-

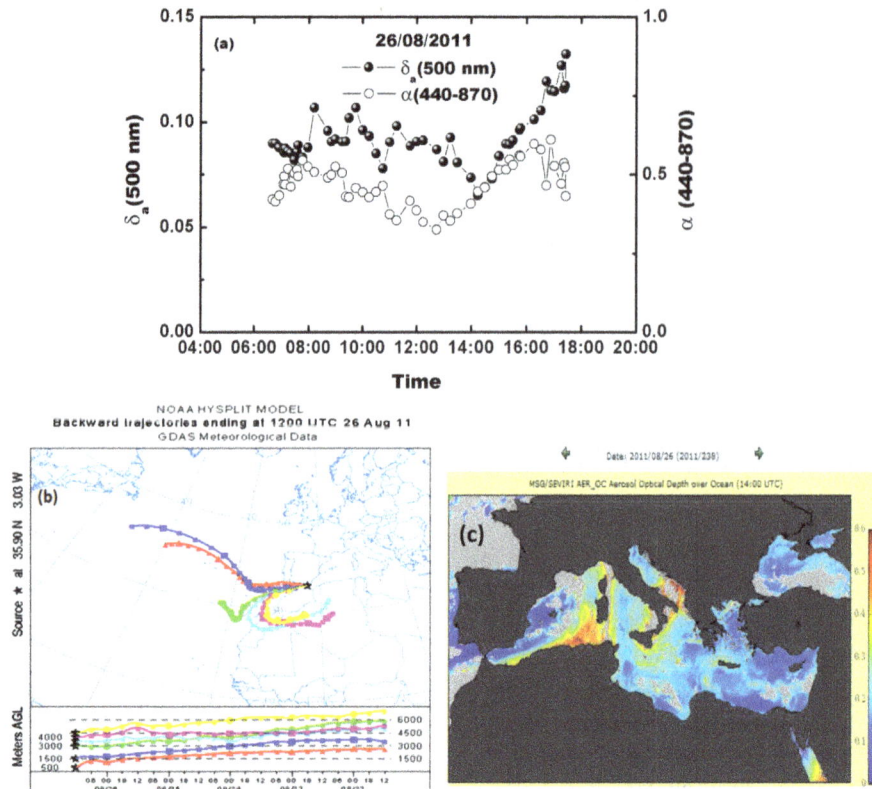

Figure 3. (a) Aerosol optical depth at 500 nm and Ångstrom exponent in the range 440–870 nm, **(b)** backward trajectories ending at 12 UTC over Alborán Island at height altitudes of 500, 1500, 3000 and 4000 m, **(c)** MSG satellite image for 26 August 2011 (http://www.icare. univ-lille1.fr).

tent intense high pressure systems centered over the Azores, which favor transport of anthropogenic particles emitted in Europe to Alborán Island. Indeed, on this day the air mass ending at 1500 m a.g.l. (Fig. 4c) came from central Europe and traveled at low altitude on the last 3 days before its arrival at Alborán Island, over an area with a great sulfate surface concentration according to Navy Aerosol Analysis and Prediction System (NAAPS) model (Fig. 4d, f). Therefore, these air masses might pick up fine anthropogenic particles in their way to Alborán Island, which may explain the high values of both $\delta_a(500\,\mathrm{nm})$ and $\alpha(440–870)$ parameters observed during this event. Thus, the desert dust transport appears to be a main cause of high aerosol loads while transport from central European urban areas is associated with occasional large aerosol loads over Alboran Island. These results are in accordance with those reported by Fotiadi et al. (2006) for Crete, who found the highest values of $\delta_a(\lambda)$ primarily during southeasterly winds, associated with coarse dust aerosols, and to a lesser extent to northwesterly winds associated with fine aerosols originated in urban industrial European areas.

3.2 Monthly variation of aerosol properties over Alborán Island

Figure 5 shows the monthly mean values of $\delta_a(500\,\mathrm{nm})$, $\delta_F(500\,\mathrm{nm})$ and $\delta_C(500\,\mathrm{nm})$ as well as $\alpha(440–870)$ and FMF with the corresponding standard deviations for the analyzed period. The monthly average data are calculated from daily averaged data. The largest values of $\delta_a(500\,\mathrm{nm})$, reflecting high aerosol load, were observed during July–October while the lowest values (0.06–0.08) were measured from November to January (Fig. 5a). On the other hand, the monthly mean values of $\alpha(440–870)$ and FMF were lower than 1.0 and 0.5 respectively, indicating a relatively high abundance of coarse particles in each month of the analyzed period, except in October (Fig. 5b). For October, the mean $\alpha(440–870)$ was 1.1 ± 0.4 and the FMF 0.63 ± 0.20, indicating an increase in fine particle contribution during this month (Fig. 5b). It is also worth noting that both $\delta_F(500\,\mathrm{nm})$ and $\delta_C(500\,\mathrm{nm})$ showed a pronounced increase during July–October, suggesting increased loads of both fine and coarse particles during these months (Fig. 5a). Moreover, $\delta_C(500\,\mathrm{nm})$ reached its maximum in August while $\delta_F(500\,\mathrm{nm})$ peaked in October (Fig. 5a).

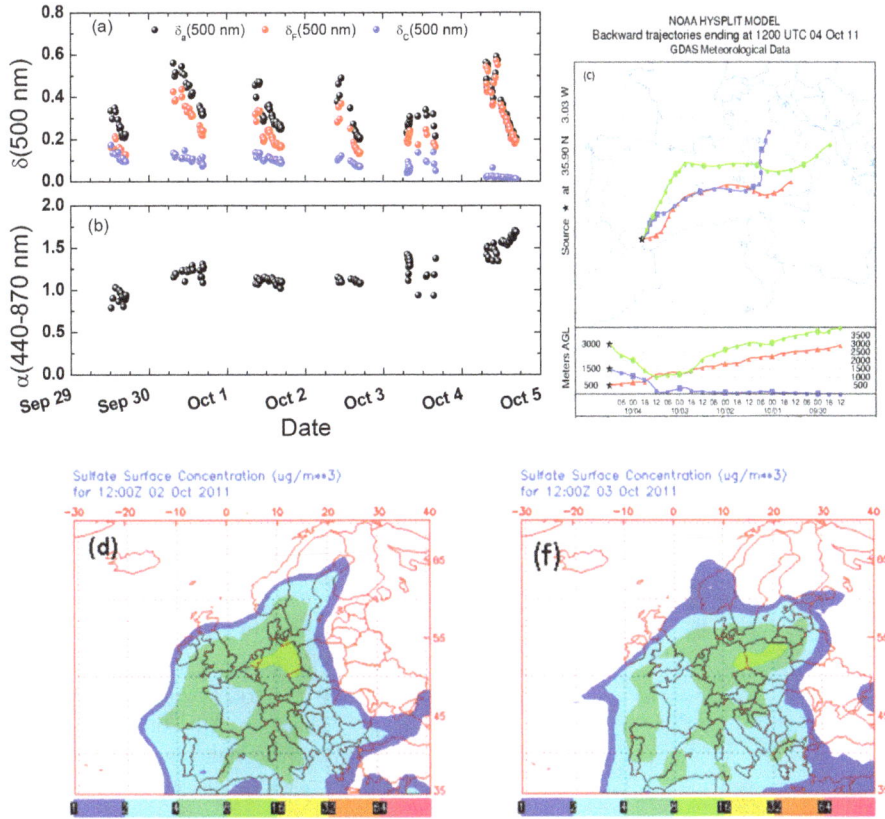

Figure 4. (a) Total, fine and coarse aerosol optical depths at 500 nm and **(b)** Ångstrom exponent in the range 440–870 nm obtained at Alborán Island during 29 September–5 October 2011. **(c)** Backward trajectories ending at 12:00 UTC on 4 October 2011 over Alborán Island at altitudes of 500, 1500 and 3000 m. **(d)** and **(f)** NAAPS maps for sulfate surface concentrations for 2 and 3 October 2011 at 12:00 UTC

This pronounced change in aerosol loads from summer to winter in 2011 is primarily due to the seasonal change in atmospheric circulation over the Mediterranean (Fig. 5c). The increased coarse aerosol load observed during July–October was associated with the high frequency of desert dust intrusions in summer in comparison to November–January (Fig. 5c). In fact, 40, 70, 41 and 14 % of measurement days in July, August, September and October were associated with Saharan dust intrusions, while in November–January there was no Saharan dust intrusion (Fig. 5c). Moreover, the air mass recirculation over the western Mediterranean especially in summer (Millan et al., 1997) along with the increased photochemical activity due to the high insolation during this season may favor the accumulation of fine aerosols that can explain the high fine particle loads during July–October in comparison with November–January. In addition, the presence of these fine aerosol particles may be favored by pollution transport from Europe and coastal urban industrial areas in northeast Africa. In this sense, the highest fine mode aerosol optical depth observed in October was associated with the increase in the frequency of air masses coming from European urban areas (see for example Fig. 4). The low aerosol loads registered in November–January can be explained by the

high frequency of clean Atlantic air advection (70–100 %) and the absence of Saharan dust intrusions (Fig. 5c) as well as efficient wet removal aerosol processes due to cloudy conditions and precipitation in this period. These results highlight the important role of the large scale circulation on monthly aerosol variation over Alborán Island.

3.3 Spatial variability of aerosol properties over western Mediterranean region

AERONET data of level 2 from Alborán and three AERONET stations surrounding the island (see Fig. 1) obtained from 1 July 2011 to 23 January 2012 are considered in this study to investigate the spatial variation of aerosol optical properties over the western Mediterranean region. For analyzing the spatial aerosol variability we compared the aerosol data obtained over Alborán during 1 July 2011–23 January 2012 with those observed over these nearby sites using only time coincident measurements.

Temporal evolutions of daily mean values of $\delta_a(500\,\text{nm})$ from 1 July 2011 to 23 January 2012 obtained over Alborán Island and Málaga stations are shown in Fig. 6a. Daily mean data were calculated only from time coincident measurements for direct comparison. Málaga is located approx-

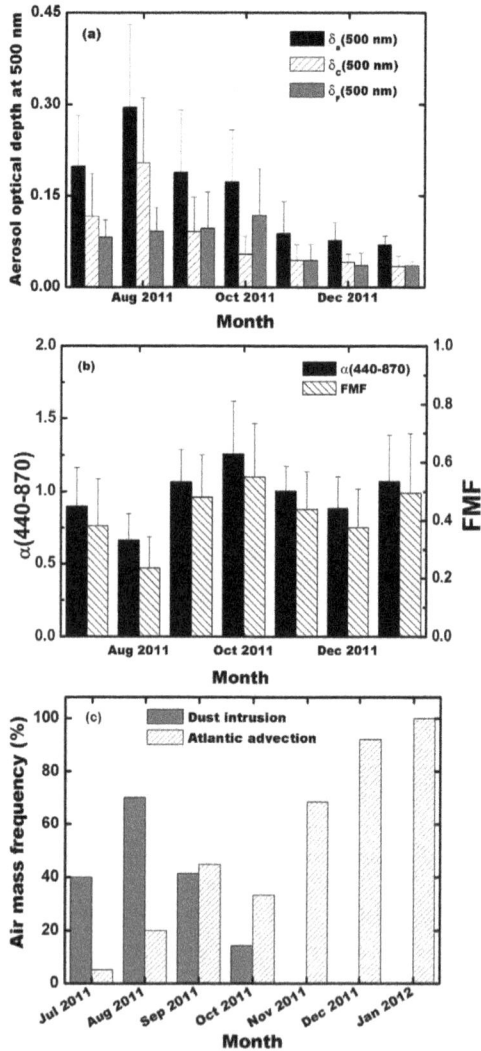

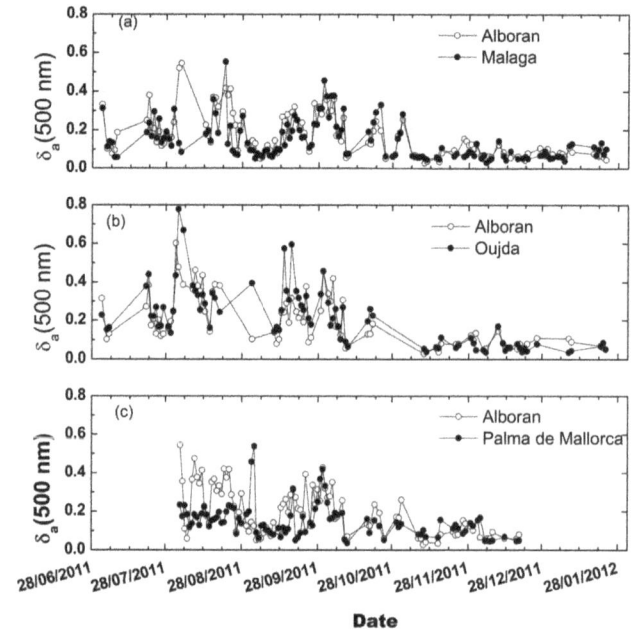

Figure 6. Temporal evolutions of daily mean values of $\delta_a(500\,\text{nm})$ from 1 July 2011 to 23 January 2012 obtained over **(a)** Alborán Island and Málaga, **(b)** Alborán Island and Oujda and **(c)** Alborán Island and Palma de Mallorca. Daily mean data were calculated only from time coincident measurements.

Figure 5. Monthly variations of **(a)** total, coarse and fine mode optical depths at 500 nm and **(b)** fine mode fraction and Ångstrom exponent in the range 440–870 nm obtained at Alborán Island from July 2011 to January 2012. The error bars are standard deviations. **(c)** Monthly relative frequency of Saharan dust intrusions and Atlantic air mass advections over Alborán Island from July 2011 to January 2012.

imately 150 km northwest of Alborán. The temporal variations of daily mean values of $\delta_a(500\,\text{nm})$ were similar for both sites on most days of the analyzed period, indicating similarities in the processes that control the aerosol load over both sites. In fact, high correlation in $\delta_a(500\,\text{nm})$ with correlation coefficient, R, of 0.75 between these two sites was found. Similar results were obtained when comparing $\delta_a(500\,\text{nm})$ over Alborán with those in Oujda ($R = 0.8$) and Palma de Mallorca ($R = 0.6$), Fig. 6b, c. However, large differences are also present on some days (e.g., on 8 August 2011 at Alboran Island we registered $\delta_a(500\,\text{nm})$ above 0.5 while at Málaga the values were below 0.1). These differ-

ences are due, in large part, to the differences in the times of occurrence and intensity of Saharan dust intrusions over these sites. In fact, the correlation in $\delta_F(500\,\text{nm})$ between Alborán Island and Málaga, $R = 0.86$, was higher than the correlation in $\delta_C(500\,\text{nm})$, $R = 0.65$. Similar results were obtained when comparing the aerosol properties over Alborán with those in Oujda ($R = 0.82$ for $\delta_F(500\,\text{nm})$ and $R = 0.70$ for $\delta_C(500\,\text{nm})$) and Palma de Mallorca ($R = 0.67$ for $\delta_F(500\,\text{nm})$ and $R = 0.32$ for $\delta_C(500\,\text{nm})$).

Table 2 shows average values of $\delta_a(\lambda)$, $\alpha(440\text{–}870)$, $\delta_F(500\,\text{nm})$ and FMF as well as the number of measurement days for each comparison (Alborán–Málaga, Alborán–Oujda and Alborán–Palma de Mallorca). Only days with coincident measurements obtained at Alborán and at each one of the additional AERONET stations from 1 July 2011 to 23 January 2012 were used for direct comparisons. For $\lambda > 500\,\text{nm}$, values of $\delta_a(\lambda)$ were slightly larger over Alborán than over Málaga (Table 2). Indeed, the mean $\delta_a(1020\,\text{nm})$ value obtained at Alborán was 35 % larger than that observed over Málaga. This indicates that the coarse particles levels were significantly larger over Alborán in comparison with Málaga during the analyzed period. In fact, the mean $\delta_C(500\,\text{nm})$ for the entire analyzed period was slightly higher (0.09 ± 0.08) at Alborán in comparison with Málaga (0.06 ± 0.05). The lower coarse particles load over Málaga as compared to Alborán is likely due to the higher frequency of Saharan dust outbreaks over Alborán as compared to Málaga and also to dust deposition in its way from Alborán to Málaga. On the

Table 2. Average values and standard deviations of $\delta_a(\lambda)$, $\alpha(440–870)$, $\delta_F(500\,nm)$ and FMF from 1 July 2011 to 23 January 2012 for Alborán Island, Málaga, Oujda and Palma de Mallorca. Only days with coincident measurements at Alborán and at each one of the additional AERONET stations are used for direct comparison.

	Alborán	Málaga	Alborán	Palma de Mallorca	Alborán	Oujda
$\delta_a(1020\,nm)$	0.09 ± 0.09	0.06 ± 0.05	0.13 ± 0.10	0.06 ± 0.04	0.13 ± 0.11	0.16 ± 0.17
$\delta_a(870\,nm)$	0.10 ± 0.09	0.08 ± 0.06	0.14 ± 0.11	0.08 ± 0.05	0.14 ± 0.11	0.18 ± 0.18
$\delta_a(670\,nm)$	0.12 ± 0.10	0.09 ± 0.07	0.16 ± 0.12	0.10 ± 0.06	0.16 ± 0.12	0.19 ± 0.18
$\delta_a(500\,nm)$	0.16 ± 0.11	0.14 ± 0.09	0.20 ± 0.13	0.14 ± 0.07	0.20 ± 0.13	0.23 ± 0.19
$\delta_a(440\,nm)$	0.18 ± 0.12	0.16 ± 0.10	0.23 ± 0.14	0.18 ± 0.09	0.22 ± 0.14	0.25 ± 0.19
$\delta_a(380\,nm)$	0.21 ± 0.13	0.20 ± 0.12	0.26 ± 0.15	0.21 ± 0.10	0.25 ± 0.15	0.29 ± 0.20
$\delta_a(340\,nm)$	0.23 ± 0.14	0.23 ± 0.13	0.29 ± 0.16	0.24 ± 0.11	0.28 ± 0.16	0.30 ± 0.20
$\alpha(440–870)$	0.9 ± 0.4	1.0 ± 0.3	0.8 ± 0.4	1.2 ± 0.4	0.8 ± 0.4	0.8 ± 0.4
$\delta_F(500\,nm)$	0.09 ± 0.06	0.09 ± 0.06	0.09 ± 0.07	0.09 ± 0.06	0.09 ± 0.07	0.09 ± 0.06
FMF	0.50 ± 0.15	0.53 ± 0.13	0.47 ± 0.18	0.60 ± 0.14	0.47 ± 0.19	0.47 ± 0.19
Number of coincident days	141	141	93	93	101	101

other hand, for $\lambda < 500\,nm$, the mean value of $\delta_a(\lambda)$ over Alborán was almost similar to that over Málaga (Table 2). It is interesting to note that the mean $\delta_F(500\,nm)$ value for the entire studied period observed over Alborán (0.09 ± 0.06) was similar to that obtained (0.09 ± 0.06) over the Málaga urban coastal site, suggesting similar concentrations of fine particles over both sites. This result is quite surprising because Málaga is a coastal city with significant local anthropogenic emissions in comparison to Alborán where there are no local anthropogenic activities. As we commented before, Alborán Island is located near an important shipping route and hence it is expected to be highly influenced by ship emissions. Thus, these results suggest that emissions from ships and/or from urban-industrial areas in Mediterranean countries could play in Alborán a similar role to that played by anthropogenic particles in Málaga. Further studies using chemical analysis of particles sampled in situ are needed to evaluate this hypothesis.

The comparison of the aerosol properties obtained at Oujda and Alborán Island is also shown in Table 2. In this case, the $\delta_a(\lambda)$ at all wavelengths were lower at Alborán than at Oujda, indicating lower aerosol concentrations over Alborán. However, $\delta_F(500\,nm)$ was similar over Oujda and Alborán (Table 2), indicating similar fine particle loading over both sites. This result is again surprising because Oujda is an urban site with significant local anthropogenic emissions in comparison to Alborán Island where there is no local anthropogenic activities. These results also point to the significant role that anthropogenic emissions from traffic ships and/or Mediterranean countries may play over Alborán. On the other hand, $\delta_C(500\,nm)$ obtained over Oujda (0.14 ± 0.15) was higher than that observed over Alborán (0.11 ± 0.10), indicating higher coarse particle concentrations over Oujda. The large coarse particle load over Oujda may result from its proximity to dust sources and local dust resuspension.

The mean $\delta_a(\lambda)$ values at all wavelengths over Alborán were higher than those observed over Mallorca, especially at the larger wavelengths which are more influenced by coarse particles (Table 2). However, as in the other cases, $\delta_F(500\,nm)$ was very similar over both sites (Table 2) in spite of the large distance (about 650 km) separating the sites and site characteristic differences. These results suggest homogeneous spatial distribution of fine particle loads over the four studied sites in spite of the large differences in local sources. On the other hand, the observed decrease in $\delta_a(500\,nm)$ from south (Alborán) to north (Mallorca) may be attributed to the proximity of Alborán Island to the dust sources in north Africa as compared to Mallorca. A gradient in dust load from south to north in western Mediterranean has also been reported by other authors (e.g., Moulin et al., 1998; Barnaba and Gobbi, 2004). Overall, based on the above comparisons it may be concluded that $\delta_C(\lambda)$ showed a south-to-north decrease in this region of western Mediterranean, while the fine mode aerosol optical depth was fairly similar over these sites.

3.4 Variability of aerosol properties during a MAN cruise

From 26 July to 13 November 2011 the Maritime Aerosol Network acquired measurements over the whole Mediterranean Sea, Black Sea and Atlantic Ocean from the ship *Nautilus 11*. Figure 7 shows $\delta_a(500\,nm)$, $\delta_F(500\,nm)$, $\delta_C(500\,nm)$ and FMF obtained during this cruise. The measurements made over the Mediterranean Sea were divided (on the basis of the differences in the aerosol sources and air masses affecting each area) into three regions: western, central and eastern Mediterranean. As can be seen from Fig. 7, all the analyzed aerosol properties showed large variability with no evident pattern during the cruise period. This large variability in aerosol properties during this cruise can be explained

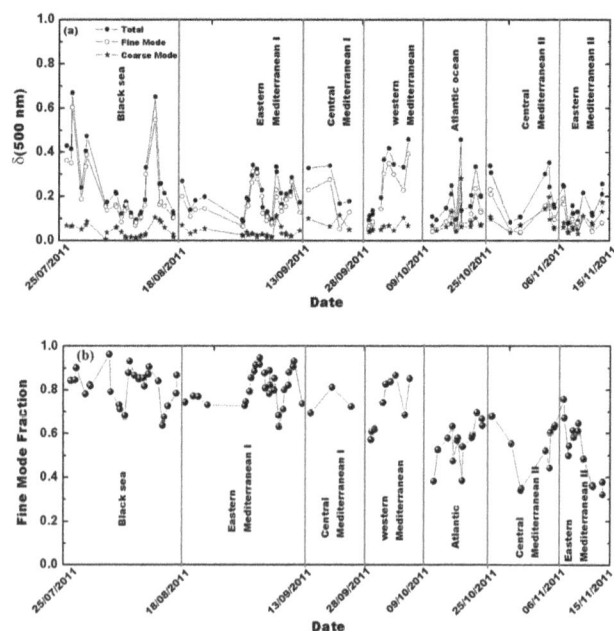

Figure 7. Temporal evolutions of $\delta_a(500\,\text{nm})$, $\delta_F(500\,\text{nm})$, $\delta_C(500\,\text{nm})$, and FMF obtained onboard the *Nautilus* ship. The data belong to the Maritime Aerosol Network (MAN) and were acquired between 26 July and 13 November 2011.

by the different aerosol sources and air masses that affected each region during the measurement period (see below). For the entire cruise period, the $\delta_a(500\,\text{nm})$ varied from 0.08 to 0.70 with a mean value of 0.22 ± 0.12. On the other hand, $\delta_F(500\,\text{nm})$ also showed large variability and ranged between 0.04 and 0.60 with a mean value of 0.16 ± 0.10 while $\delta_C(500\,\text{nm})$ fluctuated within the range 0.01–0.30 with mean value of 0.06 ± 0.04. For 85 % of the measurements, the fine mode fraction was in the range 0.52–0.96, indicating the predominance of situations dominated by fine mode particles during this cruise.

The highest $\delta_a(500\,\text{nm})$ values ranging from 0.20 to 0.46 with a mean value of 0.35 ± 0.09 were observed over the western Mediterranean Sea during the cruise period 28 September–08 October (Table 3). Also, $\delta_F(500\,\text{nm})$ values were highest (varying in the range of 0.14–0.40 with a mean value of 0.29 ± 0.09) over the western Mediterranean. These high aerosol loads were associated with high FMF values in the range 0.70–0.87, which show the predominance of fine anthropogenic particles over this area during this period (Fig. 7b). According to the back trajectory analyses, the air masses that affected the western Mediterranean region during this period come from European urban-industrial areas which explains the observed large values of $\delta_a(500\,\text{nm})$ and the predominance of fine mode particles (see for example Fig. 4c). The aerosol loads were also relatively high over the Black Sea ($\delta_a(500\,\text{nm})$, ranging from 0.08 to 0.68 with a mean value of 0.25 ± 0.16 during 26 July–15 August cruise period) and were strongly dominated by fine particles as

showed by FMF values ranging from 0.64 to 0.94. The large values of $\delta_a(500\,\text{nm})$ and those of $\delta_F(500\,\text{nm})$ ($\delta_F(500\,\text{nm})$ in the range 0.07–0.60) and the predominance of the fine mode over the Black Sea during this cruise period was associated, according to the HYSPLIT back trajectory analyses, with air masses coming from northeastern Europe (Figure not shown); this region has been identified as a strong source of pollutants and biomass burning particles during summer (e.g., Barnaba and Gobbi, 2004). In contrast, the lowest $\delta_a(500\,\text{nm})$ values (varying in the range 0.08–0.26 with mean value of 0.14 ± 0.06) were observed over the eastern Mediterranean at the end of the cruise (5–13 November). These low $\delta_a(500\,\text{nm})$ values were associated with FMF ranging between 0.30 and 0.64, showing a predominance of coarse aerosol over this area during this period. It is worth noting that the aerosol loads over the eastern Mediterranean during 5–13 November decreased drastically in comparison with the aerosol levels observed in the same region during the cruise period from 18 August to 13 September (Table 3). The decrease was more pronounced for the fine particle load; $\delta_F(500\,\text{nm})$ decreased from 0.16 ± 0.07 in the first measurements over the eastern Mediterranean to 0.07 ± 0.02 in the last ones. In contrast, $\delta_C(500\,\text{nm})$ showed an increase from 0.04 ± 0.02 during 18 August–12 September to 0.08 ± 0.04 during 5–13 November. This drastic change may be explained by the seasonal changes in the meteorological conditions. In this sense, the last measurements over the eastern Mediterranean Sea were obtained during the end of autumn when aerosol wet deposition is more effective and secondary aerosol formation is less important than in summer, which may explain the lower aerosol loads observed at the end of the expedition.

4 Conclusions

AERONET sun photometer measurements obtained over Alborán Island and three adjacent sites in the western Mediterranean were analyzed in order to investigate the temporal and spatial variations of columnar aerosol properties over this poorly explored region.

Within the analyzed period the daily average values of $\delta_a(500\,\text{nm})$ over Alborán Island ranged from 0.03 to 0.54 with a mean and standard deviation of 0.17 ± 0.12, indicating high aerosol load variation. The observed mean $\delta_a(500\,\text{nm})$ value over Alborán Island was significantly higher than reported for open oceanic areas not affected by long range aerosol transport (0.06–0.08). The $\alpha(440–870)$ values were lower than 1 for 70 % of the measurement days, suggesting that coarse particles dominated the aerosol population over the Alborán Island for the majority of the measurement days.

High aerosol loads over Alborán were mainly associated with desert dust transport from arid areas in North Africa and occasional advection of anthropogenic fine particles from central European urban-industrial areas. The aerosol opti-

Table 3. Mean values of $\delta_a(500\,\text{nm})$, $\delta_F(500\,\text{nm})$, $\delta_C(500\,\text{nm})$, $\alpha(440–870)$ and FMF obtained over the Black Sea; the western, central and eastern Mediterranean Sea; and the Atlantic Ocean during the *Nautilus* ship cruise from 26 July to 13 November 2011.

Region	$\delta_a(500\,\text{nm})$	$\delta_F(500\,\text{nm})$	$\delta_C(500\,\text{nm})$	$\alpha(440–870)$	FMF
Black Sea (26 July–15 August)	0.25 ± 0.16	0.21 ± 0.14	0.04 ± 0.03	1.76 ± 0.30	0.82 ± 0.08
Eastern Mediterranean I (18 August–12 September)	0.20 ± 0.08	0.16 ± 0.07	0.04 ± 0.02	1.74 ± 0.20	0.81 ± 0.08
Central Mediterranean I (13–28 September)	0.18 ± 010	0.12 ± 0.09	0.06 ± 0.03	1.27 ± 0.40	0.66 ± 0.08
Western Mediterranean (28 September–8 October)	0.35 ± 0.09	0.29 ± 0.09	0.07 ± 0.02	1.50 ± 0.13	0.80 ± 0.07
Atlantic Ocean (9–19 October)	0.19 ± 0.10	0.11 ± 0.05	0.09 ± 0.06	1.08 ± 0.25	0.56 ± 0.09
Central Mediterranean II (25 October–5 November)	0.22 ± 0.10	0.13 ± 0.07	0.09 ± 0.04	1.05 ± 0.30	0.57 ± 0.13
Eastern Mediterranean II (5–13 November)	0.14 ± 0.06	0.07 ± 0.02	0.08 ± 0.04	0.90 ± 0.35	0.49 ± 0.12

cal depth values of fine mode during dust events were also relatively high (for this remote site), suggesting that the fine mode particles also have considerable influence on optical properties during these dust events. Background maritime conditions over Alborán characterized by low aerosol load and Ångström exponent ($\delta_a(500\,\text{nm}) < 0.15$ and $\alpha(440–870) < 1$) were observed on about 40 % of the measurement days during the analyzed period; almost all of these days were characterized by advection of clean Atlantic air masses over the study area.

The mean value of $\delta_F(500\,\text{nm})$ over Alborán Island was comparable to the observations over the other three nearby AERONET stations, suggesting homogeneous spatial distribution of fine particle loads over the four studied sites in spite of the large differences in local sources. A northward decreases in $\delta_C(\lambda)$ was found which was probably associated with increased desert dust deposition from south to north or decreased dust frequency from south to north.

Aerosol properties acquired on board the ship *Nautilus 11* within Maritime Aerosol Network over the whole Mediterranean Sea, Black Sea and Atlantic Ocean from July to November 2011 showed large variability and no evident pattern was found. In 85 % of the measurements, the fine mode fraction was in the range 0.52–0.96, indicating the predominance of fine mode particles over the cruise areas during the monitoring period. The highest $\delta_a(500\,\text{nm})$ and $\delta_F(500\,\text{nm})$ mean values of 0.35 ± 0.09 and 0.29 ± 0.09 during the cruise period were observed over the western Mediterranean Sea, which were related to polluted air masses coming from European urban-industrial areas. In contrast, the lowest $\delta_a(500\,\text{nm})$ values (mean value of 0.14 ± 0.06) during this cruise were observed over the eastern Mediterranean Sea on the final days of the cruise in autumn, when aerosol wet deposition is more effective and secondary aerosol formation is less important than in summer.

Acknowledgements. This work was supported by the Andalusia Regional Government through projects P12-RNM-2409 and P10-RNM-6299, by the Spanish Ministry of Science and Technology through projects CGL2010-18782, and CGL2013-45410-R; and by the EU through ACTRIS project (EU INFRA-2010-1.1.16-262254). CIMEL Calibration was performed at the AERONET-EUROPE calibration center, supported by ACTRIS (European Union Seventh Framework Program (FP7/2007-2013) under grant agreement no. 262254. The authors gratefully acknowledge the outstanding support received from Royal Institute and Observatory of the Spanish Navy (ROA). The authors are grateful to the AERONET, MAN, and field campaign PIs for the production of the data used in this research effort. We would like to express our gratitude to the NASA Goddard Space Flight Center, NOAA Air Resources Laboratory and Naval Research Laboratory for the HYSPLIT model. We would like to acknowledge the constructive comments of A. Smirnov about the AERONET data. Finally, we also thank A. Kowalski for revising the manuscript.

Edited by: O. Dubovik

References

Alados-Arboledas, L., Lyamani, H., and Olmo, F. J.: Aerosol size properties at Armilla, Granada (Spain), Q. J. R. Meteorol. Soc., 129, 1395–1413, 2003.

Alados-Arboledas, L., Muller, D., Guerrero-Rascado, J. L., Navas-Guzman, F., Perez-Ramirez, D., and Olmo, F. J.: Optical and microphysical properties of fresh biomass burning aerosol retrieved by Raman lidar, and star-and sun-photometry, Geophys. Res. Lett., 38, L01807, doi:10.1029/2010gl045999, 2011.

Antón, M., Valenzuela, A., Cazorla, A., Gil, L. E., Fernandez-Galvez, J., Lyamani, H., Foyo-Moreno, I., Olmo, F. J., and Alados-Arboledas, L.: Global and diffuse shortwave irradiance during a strong desert dust episode at Granada (Spain), Atmos. Res., 118, 232–239, doi:10.1016/j.atmosres.2012.07.007, 2012.

Balis, D., Amiridis, V., Zerefos, C., Gerasopoulos, E., Andreae, M. O., Zanis, P., Kazantzidis, A., Kazadzis, S., and Papayannis, A.: Raman lidar and Sunphotometric measurements of aerosol optical properties over Thessaloniki, Greece during a biomass burning episode, Atmos. Environ., 37, 4529–4538, 2003.

Barnaba, F. and Gobbi, G. P.: Aerosol seasonal variability over the Mediterranean region and relative impact of maritime, continental and Saharan dust particles over the basin from MODIS data in the year 2001, Atmos. Chem. Phys., 4, 2367–2391, doi:10.5194/acp-4-2367-2004, 2004.

Becagli, S., Sferlazzo, D. M., Pace, G., di Sarra, A., Bommarito, C., Calzolai, G., Ghedini, C., Lucarelli, F., Meloni, D., Monteleone, F., Severi, M., Traversi, R., and Udisti, R.: Evidence for heavy fuel oil combustion aerosols from chemical analyses at the island of Lampedusa: a possible large role of ships emissions in the Mediterranean, Atmos. Chem. Phys., 12, 3479–3492, doi:10.5194/acp-12-3479-2012, 2012.

Boselli, A., Caggiano, R., Cornacchia, C., Madonna, F., Mona, L., Macchiato, M., Pappalardo, G., and Trippetta, S.: Multi year Sun photometer measurements for aerosol characterization in a Central Mediterranean site, Atmos. Res., 104–105, 98–110, doi:10.1016/j.atmosres.2011.08.002, 2012.

Di Biagio, C., di Sarra, A., and Meloni, D.: Large atmospheric shortwave radiative forcing by Mediterranean aerosols derived from simultaneous ground-based and spaceborne observations and dependence on the aerosol type and single scattering albedo, J. Geophys. Res., 115, D10209, doi:10.1029/2009JD012697, 2010.

Di Iorio, T., Sarra, A. D., Junkermann, W., Cacciani, M., Fiocco, G., and Fuà, D.: Tropospheric aerosols in the Mediterranean: 1. Microphysical and optical properties, J. Geophys. Res., 108, 4316, doi:10.1029/2002JD002815, 2003.

Di Sarra, A., Pace, G., Meloni, D., De Silvestri, L., Piacentino, S., and Monteleone, F.: Surface shortwave radiative forcing of different aerosol types in the central Mediterranean, Geophys. Res. Lett., 35, L02714. doi:10.1029/2007GL032395, 2008.

Draxler, R. R. and Rolph, G. D.: HYSPLIT (Hybrid Single-Particle Lagrangian Integrated Trajectory). Model access via NOAA ARL READY website http://ready.arl.noaa.gov/HYSPLIT.php (last access: May 2012), 2003.

Dubovik, O., Holben, B., Eck, T. F., Smirnov, A., Kaufman, Y. J., King, M. D., Tanre, D., and Slutsker, I.: Variability of absorption and optical properties of key aerosol types observed in worldwide locations, J. Atmos. Sci., 59, 590–608, 2002.

Estellés, V., Martinez-Lozano, J. A., Utrillas, M. P., and Campanelli, M.: Columnar aerosol properties in Valencia (Spain) by ground-based Sun photometry, J. Geophys. Res.-Atmos., 112, D11201, doi:10.1029/2006jd008167, 2007.

Estellés, V., Campanelli, M., Smyth, T. J., Utrillas, M. P., and Martínez-Lozano, J. A.: Evaluation of the new ESR network software for the retrieval of direct sun products from CIMEL CE318 and PREDE POM01 sun-sky radiometers, Atmos. Chem. Phys., 12, 11619–11630, doi:10.5194/acp-12-11619-2012, 2012.

Formenti, P., Andreae, M., Andreae, T., Galani, E., Vasaras, A., Zerefos, C., Amiridis, V., Orlovsky, L., Karnieli, A., Wendisch, M., Wex, H., Holben, B., Maenhaut, W., and Lelieveld, J.: Aerosol optical properties and large-scale transport of air masses: observations at a coastal and a semiarid site in the eastern Mediterranean during summer 1998, J. Geophys. Res., 106, 9807–9826, doi:10.1029/2000JD900609, 2001.

Forster, P., Ramaswamy, V., Artaxo, P., Berntsen, T., Betts, R., Fahey, D. W., Haywood, J., Lean, J., Lowe, D. C., Myhre, G., Nganga, J., R. Prinn, Raga, G., Schulz, M., and Dorland, R. V.: Changes in Atmospheric Constituents and in Radiative Forcing, Climate Change 2007: The Physical Science Basis, Contribution of Working Group I to the Fourth Assessment Report of the Intergovernmental Panel on Climate Change, edited by: Solomon, S., Qin, D., Manning, M., Chen, Z., Marquis, M., Averyt, K. B., Tignor, M., and Miller, H. L., Cambridge University Press, Cambridge, UK and New York, NY, USA, 2007.

Fotiadi, A., Hatzianastassiou, N., Drakakis, E., Matsoukas, C., Pavlakis, K. G., Hatzidimitriou, D., Gerasopoulos, E., Mihalopoulos, N., and Vardavas, I.: Aerosol physical and optical properties in the Eastern Mediterranean Basin, Crete, from Aerosol Robotic Network data, Atmos. Chem. Phys., 6, 5399–5413, doi:10.5194/acp-6-5399-2006, 2006.

Foyo-Moreno I., Alados, I., Antón, M., Fernández-Gálvez, J., Cazorla, A., and Alados-Arboledas, L.: Estimating aerosol characteristics from solar irradiance measurements at an urban location in southeastern Spain. J. Geophys. Res., 119, 1845–1859, doi:10.1002/2013JD020599, 2014.

Gerasopoulos, E., Andreae, M. O., Zerefos, C. S., Andreae, T. W., Balis, D., Formenti, P., Merlet, P., Amiridis, V., and Papastefanou, C.: Climatological aspects of aerosol optical properties in Northern Greece, Atmos. Chem. Phys., 3, 2025–2041, doi:10.5194/acp-3-2025-2003, 2003.

Haywood, J. and Boucher, O.: Estimates of the direct and indirect radiative forcing due to tropospheric aerosols: a review, Rev. Geophys., 38, 513–543, 2000.

Haywood, J. M. and Shine K. P.: Multi-spectral calculations of the direct radiative forcing of tropospheric sulphate and soot aerosols using a column model, Q. J. Roy. Meteor. Soc., 123, 1907–1930, 1997.

Holben, B. N., Eck, T. F., Slutsker, I., Tanre, D., Buis, J. P., Setzer, A., Vermote, E., Reagan, J. A., Kaufman, Y. J., Nakajima, T., Lavenu, F., Jankowiak, I., and Smirnov, A.: AERONET – A federated instrument network and data archive for aerosol characterization, Remote Sens. Environ., 66, 1–16, 1998.

Horvath, H., Alados Arboledas, L., Olmo, F.J., Jovanovic, O., Gangl, M., Sanchez, C., Sauerzopf, H., and Seidl, S.: Optical characteristics of the aerosol in Spain and Austria and its effect on radiative forcing, J. Geophys. Res., 107, 4386, doi:10.1029/2001JD001472, 2002.

Kaufman, Y. J., Wald, A. E., Remer, L. A., Gao, B. C., Li, R. R., and Flynn, L.: The MODIS 2.1-mu m channel – Correlation with visible reflectance for use in remote sensing of aerosol, IEEE T. Geosci. Remote. Sens., 35, 1286–1298, 1997.

Kaufman, Y. J., Koren, I., Remer, L. A., Rosenfeld, D., and Rudich, Y.: The effect of smoke, dust, and pollution aerosol on shallow cloud development over the Atlantic Ocean, P. Natl. Acad. Sci. USA, 102, 11207–11212, 2005.

Knobelspiesse, K. D., Pietras, C., Fargion, G. S., Wang, M. H., Frouin, R., Miller, M. A., Subramaniam, S., and Balch, W. M.: Maritime aerosol optical thickness measured by handheld sunphotometers, Remote Sens. Environ., 93, 87–106, 2004.

Kubilay, N., Cokacar, T., and Oguz T.: Optical properties of mineral dust outbreaks over the north-eastern Mediterranean, J. Geophys. Res., 108, 4666, doi:10.1029/2003JD003798, 2003.

Lelieveld, J., Berresheim, H., Borrmann, S., Crutzen, P. J., Dentener, F. J., Fischer, H., de Gouw, J., Feichter, J., Flatau, P., Heland, J., Holzinger, R., Korrmann, R., Lawrence, M., Levin, Z., Markowicz, K., Mihalopoulos, N., Minikin, A., Ramanathan, V., de Reus, M., Roelofs, G.-J., Scheeren, H. A., Sciare, J., Schlager, H., Schultz, M., Siegmund, P., Steil, B., Stephanou, E., Stier, P., Traub, M., Williams, J., and Ziereis, H.: Global air Pollution crossroads over the Mediterranean, Science, 298, 794–799, 2002.

Lyamani, H., Olmo, F. J., and Alados-Arboledas, L.: Saharan dust outbreak over southeastern Spain as detected by sun photometer, Atmos. Environ., 39, 7276–7284, doi:10.1016/j.atmosenv.2005.09.011, 2005.

Lyamani, H., Olmo, F. J., Alcantara, A., and Alados-Arboledas, L.: Atmospheric aerosols during the 2003 heat wave in southeastern-Spain I: Spectral optical depth, Atmos. Environ., 40, 6453–6464, doi:10.1016/j.atmosenv.2006.04.048, 2006a.

Lyamani, H., Olmo, F. J., Alcantara, A., and Alados-Arboledas, L.: Atmospheric aerosols during the 2003 heat wave in-southeastern Spain II: Microphysical columnar properties and radiative forcing, Atmos. Environ., 40, 6465–6476, doi:10.1016/j.atmosenv.2006.04.047, 2006b.

Mallet, M., Roger, J. C., Despiau, S., Dubovik, O., and Putaud, J. P.: Microphysical and optical properties of aerosol particles in urban zone during ESCOMPTE, Atmos. Res., 69, 73–97, 2003.

Mallet, M., Dubovik, O., Nabat, P., Dulac, F., Kahn, R., Sciare, J., Paronis, D., and Léon, J. F.: Absorption properties of Mediterranean aerosols obtained from multi-year ground-based remote sensing observations, Atmos. Chem. Phys., 13, 9195–9210, doi:10.5194/acp-13-9195-2013, 2013.

Markowicz, K. M., Flatau, P. J., Ramana, M. V., and Crutzen, P. J.: Absorbing mediterranean aerosols lead to a large reduction in the solar radiation at the surface, Geophys. Res. Lett., 29, 1968, doi:10.1029/2002GL015767, 2002.

Meloni, D., di Sarra, G., Biavati, G., DeLuisi, J. J., Monteleone, F., Pace, G., Piacentino, S., and Sferlazzo, D. M.: Seasonal behaviour of Saharan dust events at the Mediterranean island of Lampedusa in the period 1999–2005, Atmos. Environ., 41, 3041–3056, 2007.

Meloni, D., di Sarra, A., Monteleone, F., Pace, G., Piacentino, S., and Sferlazzo, D. M.: Seasonal transport patterns of intense Saharan dust events at the Mediterranean island of Lampedusa, Atmos. Res., 88, 134–148, 2008.

Millán M. M., Salvador, R., Mantilla, E., and Kallos, G.: Photooxidant dynamics in the Mediterranean basin in summer: results from European research projects., Geophys. Res. Lett., 102, 8811–8823, 1997.

Moulin, C., Lambert, C. E., Dayan, U., Masson, V., Ramonet, M., Bousquet, P., Legrand, M., Balkanski, Y. J., Guelle, W., Marticorena, B., Bergametti, G., and Dulac, F.: Satellite climatology of African dust transport in the Mediterranean atmosphere, J. Geophys. Res., 103, 13137–13144, 1998.

O'Neill, N. T., Eck, T. F., Smirnov, A., Holben, B. N., and Thulasiraman, S.: Spectral discrimination of coarse and fine mode optical depth, J. Geophys. Res.-Atmos., 108, 4559, doi:10.1029/2002jd002975, 2003.

Pace, G., Meloni, D., and di Sarra, A.: Forest fire aerosol over the Mediterranean basin during summer 2003, J. Geophys. Res.-Atmos., 110, D21202, doi:10.1029/2005jd005986, 2005.

Pace, G., di Sarra, A., Meloni, D., Piacentino, S., and Chamard, P.: Aerosol optical properties at Lampedusa (Central Mediterranean). 1. Influence of transport and identification of different aerosol types, Atmos. Chem. Phys., 6, 697–713, doi:10.5194/acp-6-697-2006, 2006.

Pandolfi, M., Gonzalez-Castanedo, Y., Alastuey, A., de la Rosa, J. D., Mantilla, E., Sánchez de la Campa, A., Querol, X., Pey, J., Amato, F., and Moreno, T.: Source apportionment of PM_{10} and $PM_{2.5}$ at multiple sites in the strait of Gibraltar by PMF: impact of shipping emissions, Environ. Sci. Pollut Res., 18, 260–269, doi:10.1007/s11356-010-0373-4, 2011.

Papadimas C. D., Hatzianastassiou, N., Mihalopoulos, N., Querol, X., and Vardavas, I.: Spatial and temporal variability in aerosol properties over the Mediterranean basin based on 6 yr (2000–2006) MODIS data, J. Geophys. Res., 113, D11205, doi:10.1029/2007JD009189, 2008.

Papadimas, C. D., Hatzianastassiou, N., Matsoukas, C., Kanakidou, M., Mihalopoulos, N., and Vardavas, I.: The direct effect of aerosols on solar radiation over the broader Mediterranean basin, Atmos. Chem. Phys., 12, 7165–7185, doi:10.5194/acp-12-7165-2012, 2012.

Pérez-Ramírez, D., Lyamani, H., Olmo, F. J., Whiteman, D. N., and Alados-Arboledas, L.: Columnar aerosol properties from sun-and-star photometry: statistical comparisons and day-to-night dynamic, Atmos. Chem. Phys., 12, 9719–9738, doi:10.5194/acp-12-9719-2012, 2012.

Saha, A., Mallet, M., Roger, J. C., Dubuisson, P., Piazzola, J., and Despiau, S.: One year measurements of aerosol optical properties over an urban coastal site: Effect on local direct radiative forcing, Atmos. Res., 90, 195–202, 2008.

Sayer, A. M., Smirnov, A., Hsu, N. C., and Holben, B. N.: A pure marine aerosol model, for use in remote sensing applications, J. Geophys. Res., 117, D05213, doi:10.1029/2011JD016689, 2012a.

Sayer, A. M., Smirnov, A., Hsu, N. C., Munchak, L. A., and Holben, B. N.: Estimating marine aerosol particle volume and number from Maritime Aerosol Network data, Atmos. Chem. Phys., 12, 8889–8909, doi:10.5194/acp-12-8889-2012, 2012b.

Smirnov, A., Holben, B. N., Kaufman, Y. J., Dubovik, O., Eck, T. F., Slutsker, I., Pietras, C., and Halthore, R.: Optical properties of atmospheric aerosol in maritime environments, J. Atmos. Sci., 59, 501–523, 2002.

Smirnov, A., Holben, B. N., Dubovik, O., Frouin, R., Eck, T. F., and Slutsker I.: Maritime component in aerosol optical models derived from Aerosol Robotic Network data, J. Geophys. Res., 108, 4033, doi:10.1029/2002JD002701, 2003.

Smirnov, A., Holben, B. N., Slutsker, I., Giles, D. M., Mc-Clain, C. R., Eck, T. F., Sakerin, S. M., Macke, A., Croot, P., Zibordi, G., Quinn, P. K., Sciare, J., Kinne, S., Harvey, M., Smyth, T. J., Piketh, S., Zielinski, T., Proshuninsky, A., Goes, J. I., Nelson, N. B., Larouche, P., Radionov, V. F., Goloub, P., Moorthy, K. K., Matarresse, R., Robertson, E. J., and Jourdin, F.: Maritime Aerosol Network as a component of Aerosol Robotic Network, J. Geophys. Res., 112, D06204, doi:10.1029/2008JD011257, 2009.

Smirnov, A., Holben, B. N., Giles, D. M., Slutsker, I., O'Neill, N. T., Eck, T. F., Macke, A., Croot, P., Courcoux, Y., Sakerin, S. M.,

Smyth, T. J., Zielinski, T., Zibordi, G., Goes, J. I., Harvey, M. J., Quinn, P. K., Nelson, N. B., Radionov, V. F., Duarte, C. M., Losno, R., Sciare, J., Voss, K. J., Kinne, S., Nalli, N. R., Joseph, E., Krishna Moorthy, K., Covert, D. S., Gulev, S. K., Milinevsky, G., Larouche, P., Belanger, S., Horne, E., Chin, M., Remer, L. A., Kahn, R. A., Reid, J. S., Schulz, M., Heald, C. L., Zhang, J., Lapina, K., Kleidman, R. G., Griesfeller, J., Gaitley, B. J., Tan, Q., and Diehl, T. L.: Maritime aerosol network as a component of AERONET – first results and comparison with global aerosol models and satellite retrievals, Atmos. Meas. Tech., 4, 583–597, doi:10.5194/amt-4-583-2011, 2011.

Sumner, G., Homar, V., and Ramis, C.: Precipitation seasonality in Eastern and Southern coastal Spain, Int. J. Climatol., 21, 219–247, 2001.

Valenzuela, A., Olmo, F. J., Lyamani, H., Antón, M., Quirantes, A., and Alados-Arboledas, L.: Classification of aerosol radiative properties during African desert dust intrusions over southeastern Spain by sector origins and cluster analysis, J. Geophys. Res.-Atmos., 117, D06214, doi:10.1029/2011JD016885, 2012a.

Valenzuela, A., Olmo, F. J., Lyamani, H., Antón, M., Quirantes, A., and Alados-Arboledas, L.: Aerosol radiative forcing during African desert dust events (2005–2010) over Southeastern Spain, Atmos. Chem. Phys., 12, 10331–10351, doi:10.5194/acp-12-10331-2012, 2012b.

Viana, M., Amato, F., Alastuey, A., Querol, X., Moreno, T., Dos Santos, S. G., Herce, M. D., and Fernández-Patier, R.: Chemical tracers of particulate emissions from commercial shipping, Environ. Sci. Technol., 43, 7472–7477, 2009.

Thieuleux, F., Moulin, C., Breon, F. M., Maignan, F., Poitou, J., and Tanre D.: Remote sensing of aerosols over the oceans using MSG/SEVIRI imagery, Ann. Geophys., European Geosciences Union (EGU), 23, 3561–3568, 2005.

Toledano, C., Cachorro, V. E., Berjon, A., de Frutos, A. M., Sorribas, M., de la Morena, B. A., and Goloub, P.: Aerosol optical depth and Ångstrom exponent climatology at El Arenosillo AERONET site (Huelva, Spain), Q. J. Roy. Meteorol. Soc., 133, 795–807, doi:10.1002/qj.54, 2007.

Uplifting of carbon monoxide from biomass burning and anthropogenic sources to the free troposphere in East Asia

K. Ding[1,6], J. Liu[1,2,6], A. Ding[1,6], Q. Liu[1,6], T. L. Zhao[3], J. Shi[4], Y. Han[1], H. Wang[5], and F. Jiang[5]

[1]School of Atmospheric Sciences, Nanjing University, Nanjing, Jiangsu 210093, China
[2]University of Toronto, Toronto, Ontario, M5S 3G3, Canada
[3]Nanjing University of Information Science and Technology, Nanjing, Jiangsu 210044, China
[4]Institute of Remote Sensing Applications, Chinese Academy of Sciences, Beijing 100101, China
[5]International Institute for Earth System Sciences, Nanjing University, Nanjing, Jiangsu 210093, China
[6]Collaborative Innovation Center of Climate Change, Jiangsu 210093, China

Correspondence to: J. Liu (jliu@nju.edu.cn)

Abstract. East Asia has experienced rapid development with increasing carbon monoxide (CO) emission in the past decades. Therefore, uplifting CO from the boundary layer to the free troposphere in East Asia can have great implications on regional air quality around the world. It can also influence global climate due to the longer lifetime of CO at higher altitudes. In this study, three cases of high CO episodes in the East China Sea and the Sea of Japan from 2003 to 2005 are examined with spaceborne Measurements of Pollution in the Troposphere (MOPITT) data, in combination with aircraft measurements from the Measurement of Ozone and Water Vapor by Airbus In-Service Aircraft (MOZAIC) program. High CO abundances of 300–550 ppbv are observed in MOZAIC data in the free troposphere during these episodes. These are among the highest CO abundances documented at these altitudes. On average, such episodes with CO over 400 ppbv (in the 2003 and 2004 cases) and between 200 and 300 ppbv (in the 2005 case) may occur 2–5 and 10–20 % in time, respectively, in the respective altitudes over the region. Correspondingly, elevated CO is shown in MOPITT daytime data in the middle to upper troposphere in the 2003 case, in the lower to middle troposphere in the 2004 case, and in the upper troposphere in the 2005 case. Through analyses of the simulations from a chemical transport model GEOS-Chem and a trajectory dispersion model FLEXPART, we found different CO signatures in the elevated CO and distinct transport pathways and mechanisms for these cases. In the 2003 case, emissions from large forest fires near Lake Baikal dominated the elevated CO, which had been rapidly transported upward by a frontal system from the fire plumes. In the 2004 case, anthropogenic CO from the North China Plain experienced frontal lifting and mostly reached ~ 700 hPa near the East China Sea, while CO from biomass burning over Indochina experienced orographic lifting, lee-side-trough-induced convection, and frontal lifting through two separate transport pathways, leading to two distinct CO enhancements around 700 and 300 hPa. In the 2005 case, the observed CO of ~ 300 ppbv around 300 hPa originated from anthropogenic sources over the Sichuan Basin and the North China Plain and from forest fires over Indochina. The high CO was transported to such altitudes through strong frontal lifting, interacting with convection and orographic lifting. These cases show that topography affects vertical transport of CO in East Asia via different ways, including orographic uplifting over the Hengduan Mountains, assisting frontal lifting in the North China Plain, and facilitating convection in the Sichuan Basin. In particular, topography-induced lee-side troughs over Indochina led to strong convection that assisted CO uplifting to the upper troposphere. This study shows that the new daytime MOPITT near-infrared (NIR) and thermal-infrared (TIR) data (version 5 or above) have enhanced vertical sensitivity in the free troposphere and may help qualitative diagnosis of vertical transport processes in East Asia.

1   Introduction

Carbon monoxide (CO) plays several important roles in the atmosphere. The oxidizing capability, an ability of the atmosphere to cleanse itself, is strongly influenced by the CO level in the troposphere. CO near the surface is a major pollutant. Under high NO_x conditions, CO is a precursor of ozone, while in low NO_x air masses, CO helps ozone destruction (Jacob, 1999; Holloway et al., 2000). As carbon dioxide (CO_2) is produced in both ozone production and destruction processes (Holloway et al., 2000), CO is linked to the global carbon cycle (Suntharalingam et al., 2004; Yurganov et al., 2008; Nassar et al., 2010) affecting climate change. With a lifetime of weeks to months, CO is a good tracer tracking transport of pollution. In the purview of these roles, it is important to understand processes influencing the CO distribution and variability in the atmosphere.

Although the main sources of atmospheric CO and its mean status are generally understood (Novelli et al., 1998; Jacob, 1999; Holloway et al., 2000), many processes influencing CO variations at different timescales are not well known. Uplifting CO from the boundary layer to the free troposphere (FT) is such a process, which usually occurs on the synoptic scale that spans hundreds to thousands of kilometers in space and lasts hours to days in time (Daley, 1991). Uplifted CO usually has a longer lifetime and can be transported fast by the upper layer winds over long distances through continents and between hemispheres in the troposphere (Stohl, 2001; Stohl et al., 2002; Damoah et al., 2004). Uplifting air mass from the surface to FT generally takes place by three processes: (1) frontal lifting, (2) orographic lifting, and (3) deep convection (Brown et al., 1984; Banic et al., 1986; Dickerson et al., 1987; Bethan et al., 1998; Pickering et al., 1998; Chung et al., 1999; Donnell et al., 2001; Kowol-Santen et al., 2001; Cooper et al., 2002; Liu et al., 2003; Miyazaki et al., 2003; Chan et al.; 2004; Mari et al., 2004; Li et al., 2005; Liu et al., 2006; Kar et al., 2008; Zhao et al., 2008; Ding et al., 2009; Randel et al., 2010; Chen et al., 2012).

East Asia has experienced rapid development with increasing CO emission in the past decades (Duncan et al., 2007). In addition to impacts on local air quality (Wang et al., 2010), continuing increase in CO emissions will lead to great impacts on regional air quality and climate of the world (Jaffe et al., 1999; Berntsen et al., 1999; Bertschi et al., 2004) because of an expected upward trend in pollution outflow from the region. East Asia is characterized by its unique and complex meteorology, topography, and land covers. Vertical transport of CO can be modulated by one or more of these conditions or by their interactions. For example, the likelihood of when and where extratropical cyclones are active is closely linked to the locations and frequency of frontal uplifting. Wet and dry convections prevail in different seasons in northern China because of the distinct climatological pattern in precipitation there (Dickerson et al., 2007). The topography there also

plays an important role in uplifting of CO alone and/or interplaying with frontal systems, aiding convection in mountainous regions (Liu et al., 2003; Ding et al., 2009). Recently, Lin et al. (2009) proposed a new mechanism that emphasizes the role of topography-induced lee-side troughs over Indochina in promoting strong convection. A variety of land cover types in East Asia diversifies CO sources there. In highly populated urban areas, such as those in the North China Plain, anthropogenic emissions are high. Large biomass burning, occurring in areas with abundant vegetation, can generate great amounts of CO for vertical transport when meteorological conditions become favorable. Two such areas are Southeast Asia and the boreal forested area in Russia (Wotawa et al., 2001; Schultz, 2002; Duncan et al., 2003). So far, our understanding of the impacts of these processes and their interactions on CO uplifting is still rather limited (Dickerson et al., 2007). The objectives of studying vertical transport of CO in East Asia are to better understand the vertical distribution of CO in the region, to advance the assessment of impacts of long-range transport of Asian CO on regions downwind, and to help improve simulating this process in atmospheric models on the synoptic scale, eventually leading to more realistic chemical weather forecast in the future (Lawrence et al., 2003).

Due to lack of continuous measurements, most studies on CO in East Asia are based on observations from periodic field campaigns (Jacob et al., 2003; Tsutsumi et al., 2003; Li et al., 2007; Ding et al., 2009) or simulations by chemical transport models (Berntsen et al., 1999; Bey et al., 2001) or both (Liu et al., 2003). CO measurements from satellites provide unprecedented data revealing CO variations over East Asia. One of the instruments is the Measurements of Pollution in the Troposphere (MOPITT) (Drummond, 1992; Drummond and Mand, 1996). MOPITT provides data of CO total column and CO vertical profiles at several altitude levels, which are retrieved using a nonlinear optimal estimation method theoretically based on the observed radiances and their weighting functions, the a priori information, and the retrieval averaging kernels (Rogers, 2000; Deeter et al., 2003). As a result, the MOPITT retrieval at one level can be influenced by CO at other levels and thus MOPITT vertical resolution is coarse, generally having only 2–3 pieces of independent information vertically in the troposphere. Therefore, MOPITT's vertical sensitivity was an issue with earlier versions of MOPITT data (Jacob et al., 2003). Nevertheless, a few studies (Deeter et al., 2004; Kar et al., 2004, 2006, 2008; Liu et al., 2006) demonstrated MOPITT's vertical sensitivity to some extent. Kar et al. (2004) found Asian summer monsoon plumes in MOPITT CO data as a strong enhancement of CO in the upper troposphere over India and southern China. Deeter et al. (2004) illustrated similar distributions of the rain rate and the ratio of MOPITT CO at 350 to at 850 hPa in the tropical eastern Pacific Ocean. Liu et al. (2006) observed large differences (20–40 ppbv) in MOPITT CO at 250 hPa between two cases of vertical transport of CO and attributed the differences to the respective weather systems. Furthermore, the MOPITT data in new versions that use both thermal-infrared (TIR) and near-infrared (NIR) radiances have offered enhanced vertical sensitivity (Worden et al., 2010; Deeter et al., 2012, 2013). Therefore, a detailed examination of MOPITT's vertical sensitivity in East Asia, especially for its ability in detecting vertical transport of high CO episodes, is desirable.

In this study, three cases of high CO episodes in East Asia from 2003 to 2005 are examined with MOPITT satellite data, in combination with aircraft measurements from the Measurement of Ozone and Water Vapor by Airbus In-Service Aircraft (MOZAIC) program (Marenco et al., 1998) (see Sects. 2 and 4). The vertical transport mechanisms are analyzed with simulations from a trajectory dispersion model FLEXPART (Stohl et al., 2005) and a chemical transport model GEOS-Chem (Bey et al., 2001), along with other meteorology data and satellite fire data (see Sects. 2 and 4). MOPITT data are analyzed in two ways. First, the vertical sensitivity of MOPITT is evaluated with the coincident MOZAIC data (see Sect. 3) and further illustrated with the three high CO episodes in comparison with the MOZAIC data (see Sect. 4). Second, the vertical variation in CO captured by MOPITT is used to diagnose vertical transport of CO (see Sect. 4). Discussion on the three cases is synthesized in Sect. 5 and the major conclusions are provided in Sect. 6.

2 Model and data

2.1 Satellite MOPITT CO data

MOPITT is the first space instrument that targets continuous measurements of tropospheric CO. MOPITT has been onboard of the Terra satellite since 1999, making scientific measurements since March 2000. Terra is flying in a sun synchronous polar orbit with an altitude of 705 km, crossing the equator at ∼ 10:45 and 22:45 LT and making 14–15 daytime and nighttime overpasses each day. MOPITT uses a cross-track scanning method with a swath of 29 pixels (4 pixels in a row), each pixel being 22 km × 22 km. Therefore, with a swath of ∼ 600 km, about one-third of the global area is covered in a day. Additionally, clouds can cause even more gaps in MOPITT daily data. This makes it challenging to use MOPITT data for synoptic studies. It takes 3 days to achieve a near-complete global coverage (Edwards et al., 1999) assuming no blockage from clouds.

MOPITT measures upwelling radiation in two narrow infrared spectral regions for CO retrieval: (1) a TIR band near 4.7 μm that has strong carbon monoxide absorption and (2) a NIR band near 2.3 μm that has weak CO absorption. MOPITT Version 5 retrieval products are significantly different from earlier products and offer three distinct products depending on application requirements. One of them is a TIR/NIR multispectral product, which has enhanced sensi-

tivity to CO in the lower-most troposphere (Worden et al., 2010; Deeter et al., 2012, 2013). Validations and evaluations of MOPITT data in various versions are documented in Emmons et al. (2004), Worden et al. (2010), and Deeter et al. (2012, 2013).

In this study, the MOPITT CO profiles (Level 2 data) were first compared with the coincident MOZAIC profiles. Advances of Version 5 (V5, a TIR/NIR multispectral product) from Version 4 (V4, a TIR-only product) data were assessed. Then, the V5 data were used in the case studies, in which MOPITT Level 2 data were gridded horizontally into 0.25° latitude × 0.25° longitude bins and vertically at the MOPITT resolution of 100 from the surface to 100 hPa.

2.2 Aircraft MOZAIC CO data

The MOZAIC program was initiated in 1993 by European scientists, aircraft manufacturers, and airlines to collect experimental data (Marenco et al., 1998). MOZAIC consists of automatic and regular measurements of ozone, CO, and water vapor by several long-range passenger airliners flying all over the world. The aim is to build a large database of measurements to allow studies of chemical and physical processes in the atmosphere.

In comparing MOPITT with MOZAIC CO data, coincident MOPITT and MOZAIC data from 2003 to 2005 were screened within a radius of $1.5°$ and within a 4 h period. The radius of $1.5°$ was applied to selected MOZAIC profiles at 500 hPa and the MOZAIC slant path was included in the radius. MOZAIC profile was smoothed by applying the MOPITT averaging kernels and the a priori profile for the co-located retrieved MOPITT profile to account for the bias introduced by the averaging kernels and the a priori. Therefore, the smoothed MOZAIC CO profile $\hat{x}^{\text{MOZAIC}}$ is derived by (Rogers, 2000)

$$\hat{x}^{\text{MOZAIC}} = x^{\text{MOPITT}} + \mathbf{A}(x^{\text{MOZAIC}} - x_a^{\text{MOPITT}}), \qquad (1)$$

where $\mathbf{A} = \delta\hat{x}/\delta x$ is the MOPITT averaging kernel matrix which describes the sensitivity of the MOPITT CO estimate to the true profile of CO, x^{MOZAIC} is the MOZAIC CO profile, which has been mapped to the MOPITT pressure grid. The quantity x_a^{MOPITT} is the MOPITT a priori, which is based on CO simulations from the MOZART model (Emmons et al., 2004).

The MOZAIC measurements usually extend from the surface to ∼ 250 hPa. When validating MOPITT data using Eq. (1), CO mixing ratios above 300 hPa was supplemented with CO from the GEOS-Chem chemical transport model (see Sect. 2.6) on the same location and day, similar to the treatments by Worden et al. (2010), who used the MOZART climatology simulations. Because CO above 250 hPa is lower than that in the middle and lower troposphere, the bias due to this treatment is expected to be low.

2.3 MODIS fire count data

The Moderate-resolution Imaging Spectroradiometer (MODIS) is the type of instrument which has been onboard of the Terra (EOS AM) satellite since 1999 and on the Aqua (EOS PM) satellite since 2002. The MODIS fire products include a validated daily global active fire product (MOD14 Terra and MYD14 Aqua) (Justice et al., 2002), generated using a global active fire detection algorithm that uses a multispectral contextual approach to exploit the strong emission of mid-infrared radiation from fires allowing subpixel fire detection (Giglio et al., 2003). The horizontal resolution is 1 km. The fire data are acquired from the Fire Information for Resource Management System (FIRMS) (Davies et al., 2009).

2.4 NCEP FNL meteorological data

The National Centers for Environmental Prediction (NCEP) final (FNL) global tropospheric analyses are on $1° \times 1°$ grids every 6 h (http://rda.ucar.edu/datasets/ds083.2/). Parameters in FNL include surface pressure, sea level pressure, geopotential height, temperature, sea surface temperature, potential temperature, relative humidity (RH), precipitable water, u and v winds, and vertical motion, available on the surface, at 26 levels from 1000 to 10 hPa, the tropopause, the boundary layer, and a few others. In addition to driving FLEXPART (see Sect. 2.5), the FNL data are used to analyze the meteorological conditions including the surface pressure, wind fields, and development of a cyclone. The data are generated from the global data assimilation system (GDAS).

2.5 The FLEXPART trajectory model

To diagnose the transport processes and trace CO sources, we used the FLEXPART model (Stohl et al., 2005), which is a Lagrangian Particle Dispersion Model developed at the Norwegian Institute for Air Research in the Department of Atmospheric and Climate Research. FLEXPART can be driven by meteorological input data generated from a variety of global and regional models. In this study, the simulations were driven by the NCEP FNL data. This model has been extensively validated (Stohl et al., 1998; Cristofanelli et al., 2003) and widely used in studies of the influence of various meteorological processes on pollution transport (Cooper et al., 2004, 2005, 2006; Hocking et al., 2007; Ding et al., 2009; Barret et al., 2011; He et al., 2011; Chen et al., 2012). In running FLEXPART, a large number of particles are released from defined locations (latitude, longitude, and altitude) at a time. Backward or forward trajectories of the particles are recorded in latitude (°), longitude (°), and altitude (km) every hour.

2.6 The GEOS-Chem chemical transport model

GEOS-Chem is a global three-dimensional chemical transport model (http://geos-chem.org). The model contains detailed description of tropospheric O_3-NO_x-hydrocarbon chemistry, including the radiative and heterogeneous effects of aerosols. It is driven by assimilated meteorological observations from the National Aeronautics and Space Administration (NASA) Goddard Earth Observing System (GEOS) from the Global Modeling and Assimilation Office (GMAO). In this study, GEOS-Chem version v9-1-3 was employed and executed in the full chemistry mode, which is driven by GEOS meteorology with temporal resolution of 6 h (3 h for surface meteorological variables), with a horizontal resolution of 2° latitude by 2.5° longitude and 47 vertical levels, including ~ 35 levels in the troposphere from 1000 to 100 hPa.

GEOS-Chem uses anthropogenic emissions from the Emissions Database for Global Atmospheric Research (EDGAR) global inventory (Olivier and Berdowski, 2001), which are updated with regional inventories, including the emission inventory in Asia (Streets et al., 2006; Zhang et al., 2009). The biomass burning emissions are from the Global Fire Emissions Data (GFEDv3) monthly inventories (van der Werf et al., 2010) and biogenic volatile organic compound (VOC) emissions are taken from the Model of Emissions of Gases and Aerosols from Nature (MEGAN) global inventory. Emissions from other natural sources (e.g., lightning, volcanoes) are also included.

The model has been extensively evaluated and used in studies of atmospheric chemistry and pollution transport (Bey et al., 2001; Heald et al., 2003; Liu et al., 2003, 2006; Zhang et al., 2006; Jones et al., 2009; Nassar et al., 2009; Kopacz et al., 2010; Jiang et al., 2011). GEOS-Chem can generally describe CO variability in the troposphere but somewhat underestimate the observations in the northern mid-latitudes possibly due to biases in the CO inventory or numerical diffusion in the model or both (Heald et al., 2003; Duncan et al., 2007; Nassar et al., 2009; Kopacz et al., 2010).

3 Comparison between MOPITT and MOZAIC CO profiles

MOPITT's vertical sensitivity can be described in terms of the averaging kernels (see Eq. 1) and the degree of freedom for signal (DFS). The averaging kernel matrix indicates the sensitivity of the MOPITT CO estimate to the true CO profile, with $\mathbf{I}$ (identity matrix) being the best, when true CO profiles are retrieved, and 0 being the worst, when MOPITT retrievals just take the a priori. In reality, the average kernel matrix is less than $\mathbf{I}$, implying some contribution of CO from other levels to the retrieved level so that the CO vertical structure cannot be fully resolved. DFS gives the number of independent pieces of information available vertically in the measurements and it is the sum of the diagonal elements of the averaging kernel matrix (Rogers, 2000). Figure 1 shows a yearly mean of DFS for daytime and nighttime, respectively, in East Asia for the V5 TIR/NIR data, indicating substantial increases in DFS compared to earlier MOPITT versions (Worden et al., 2010; Deeter et al., 2012). The daytime annual mean DFS in East Asia (Fig. 1a) ranges from 0.5 to 2.7, usually decreasing with latitude, similar to its distribution in other regions and on the global scale (Deeter et al., 2004; Worden et al., 2010). In the same latitudinal zones, the DFS is higher over land than over ocean. The daytime annual mean DFS is high in the Sichuan Basin, the eastern part of mainland China, the Indochina peninsula, and the Indian subcontinent. Over the mountain or valley regions, DFS is low, such as above the Tibetan Plateau. The stars indicate the cities where MOZAIC vertical measurements are available for validation of MOPITT data. The annual mean DFS is 1.65, 1.51, 1.60, and 1.64, respectively, in an area of $1° \times 1°$ around Beijing, Narita, Shanghai, and Hong Kong, with a maximum of 1.98, 1.64, 1.81, and 1.74 for the cities, respectively. The nighttime DFS values (Fig. 1b) are lower (from 0.5 to 1.5) than the daytime values, similar to that in Deeter et al. (2004) for an earlier MOPITT version. Spatially, nighttime DFS is high over regions where the daytime DFS is also high.

The general patterns of MOPITT averaging kernels have been documented (Pan et al., 1998; Emmons et al., 2004; Deeter et al., 2003, 2004, 2012; Kar et al., 2008; Worden et al., 2010). For V5 MOPITT data, the averaging kernels at the four cities are similar to these in Worden et al. (2010, in their Fig. 7). The difference in the averaging kernels between V4 and V5 can be as large as 0.14 in the surface and lower troposphere and as 0.10 in the upper troposphere (not shown).

Figure 2 shows the relative bias between MOPITT and the smoothed MOZAIC ($\hat{x}^{MOZAIC}$) profiles (see Eq. 1), which is also referred to as the "MOPITT estimate of in situ" in Worden et al. (2010) and the "transferred profile" in Emmons et al. (2004). For V5 data (in red), the mean bias is within ±20 % for all the cities. In all the altitude levels, the bias is the smallest (close to zero) around 500–400 hPa and increases upward and downward. The bias is mostly positive above 500–400 hPa, while below 500–400 hPa, it is positive at Beijing and Narita but negative at Shanghai and Hong Kong. Whether the sign change is related to the change in the geographic location (Shanghai and Hong Kong are both coastal cities) can be a subject for further study. The V4 data (in green) also show the smallest bias in the middle troposphere. In the lower troposphere, the bias in V5 is reduced by 5–10 % at Beijing and Narita. At Shanghai, the bias changes from positive in V4 to negative in V5, with a smaller magnitude, while at Hong Kong, the negative bias in V4 becomes larger in magnitude in V5. In the upper troposphere above 500–400 hPa, the bias in V5 at Beijing, Narita, and Shanghai changes to positive, with a magnitude similar to or larger than that in V4. At Hong Kong, the bias in V5 remains positive but the magnitude is enlarged. Deeter et al. (2013) com-

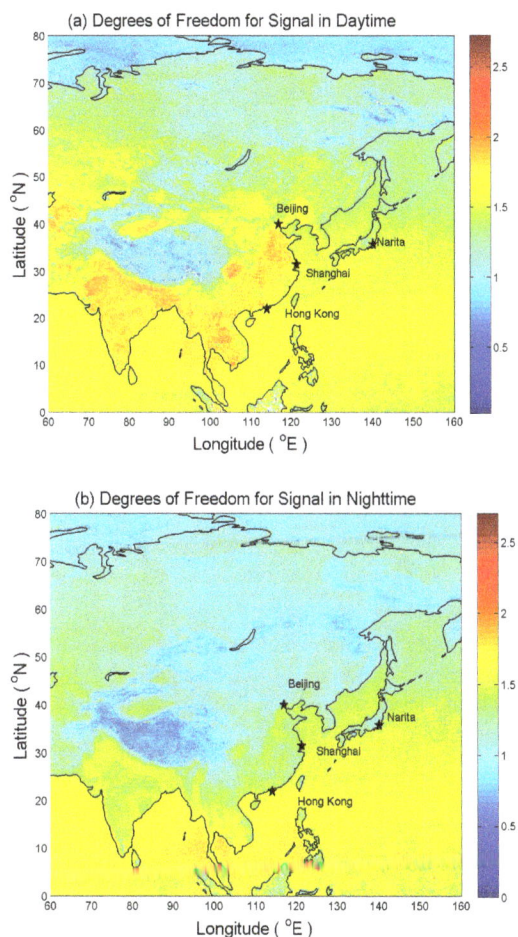

Figure 1. The degree of freedom for signal (DFS) of the MOPITT V5 TIR/NIR data over East Asia, averaged for 2005 during (**a**) daytime and (**b**) nighttime. Locations of four cities with MOZAIC CO measurements are indicated as stars. Note that the MOZAIC CO data from Narita also include a small portion of measurements from its surrounding cities at Osaka and Nagoya.

pared MOPITT data with the National Oceanic and Atmospheric Administration (NOAA) aircraft measurements over North America and data from the (High-performance Instrumented Airborne Platform for Environmental Research) HIAPER Pole-to-Pole Observations (HIPPO) field campaign data (Wofsy et al., 2011). They found a positive bias in MOPITT V5 TIR/NIR data at 400 (4 %) and 200 hPa (14 %). They also showed a latitude-dependent positive bias in the northern hemispherical upper troposphere in MOPITT V3 and V4 data. This study suggests an overall positive bias, agreeing with Deeter et al. (2013) in magnitude and sign, in MOPITT V5 data for the upper troposphere. As a comparison, we also validated MOPITT data in other cities around the globe and found that the mean bias in Europe or the United States is lower than that in East Asia, especially in the surface layer (not shown).

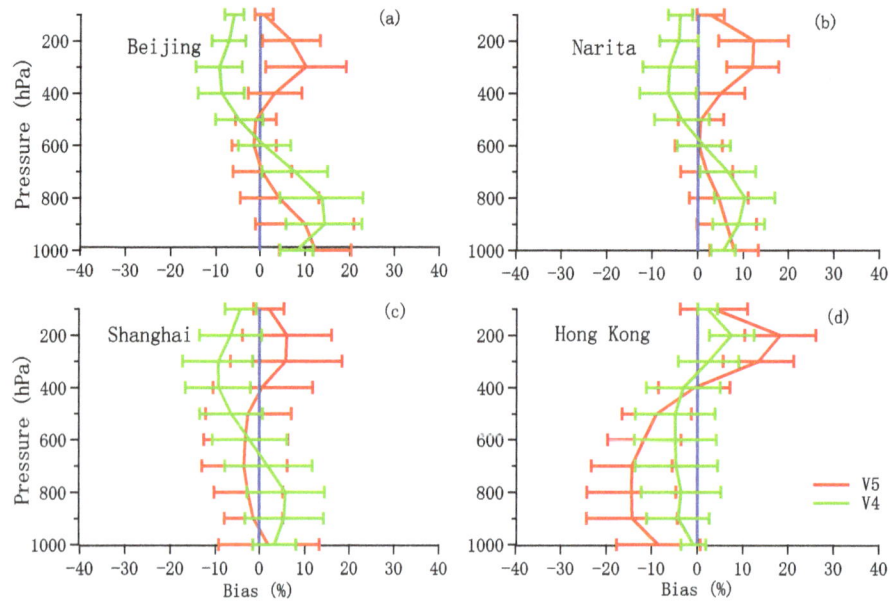

Figure 2. Relative bias of CO profiles (in %) between MOPITT and MOZAIC data (smoothed with the MOPITT averaging kernels; see Eq. 1) from 2003 to 2005 at Beijing, Narita, Shanghai, and Hong Kong for MOPITT V4 and V5 data. The number of profiles for the comparison is 18, 23, 11, and 15, respectively, at Beijing, Narita, Shanghai, and Hong Kong. The error bars indicate the interquartile range of the mean.

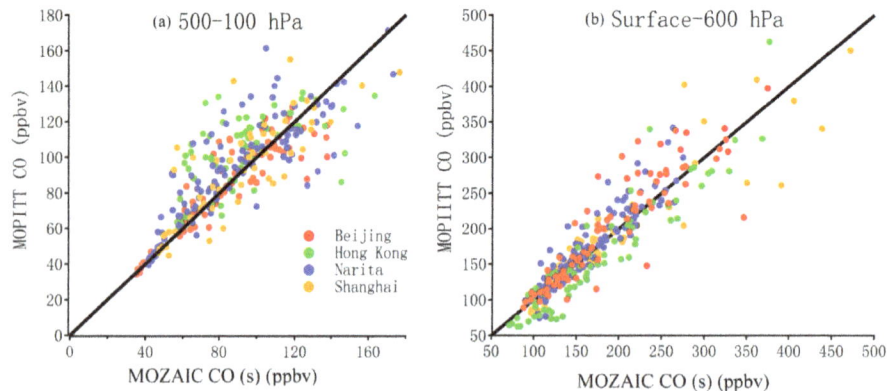

Figure 3. Correlation between MOPITT and MOZAIC data (smoothed with the MOPITT averaging kernels, see Eq. 1) from 2003 to 2005 at Beijing, Narita, Shanghai, and Hong Kong (**a**) from the middle to upper troposphere and (**b**) from the surface to the middle troposphere.

The correlation between MOPITT and smoothed MOZAIC data is shown in Fig. 3. From 500 to 100 hPa, the correlation coefficient between the two data sets is 0.92, 0.86, 0.83, 0.68 at Beijing, Narita, Shanghai, and Hong Kong, respectively (Fig. 3a), while from the surface to 600 hPa, the correlation becomes stronger, being 0.90, 0.92, 0.92, 0.94 at Beijing, Narita, Shanghai, and Hong Kong, respectively (Fig. 3b). The correlation coefficient between the two data is best in the middle troposphere (500–400 hPa, not shown).

4   Uplifting of CO to the free troposphere

Daily MOPITT and MOZAIC data from 2003 to 2005 were screened to find cases of high CO episodes observed by both MOPITT and MOZAIC at the same location and time. We found three cases of high CO in MOPITT data with close-by MOZAIC measurements, while it was hard to find such high CO episodes with exact coincident MOPITT and MOZAIC observations because of large gaps in MOPITT data and limited aircraft sampling coverages. In the three cases, high CO concentrations up to 300–500 ppbv were observed by MOZAIC in the free troposphere from 750 to 250 hPa.

In the following, we provide detailed analyses of each case, ordered by year of occurrence (Table 1). The cases oc-

Table 1. Characterization of the three cases.

Case	2003	2004	2005
Date	6 June 2003	18 March 2004	10 April 2005
Maximum CO (ppbv) in MOZAIC profiles	~ 550	~ 500	~ 300
CO peak height (hPa) in MOZAIC profiles	500–350	750–550	350–250
Maximum CO (ppbv) in MOPITT images	300–400	200–250	150–250
CO peak height (hPa) in MOPITT images	650–300	750–500	400–250
Peak CO area in MOPITT images	35–55° N, 125–145° E	20–32° N, 125–135° E	32–37° N, 130–140° E
Major CO sources	Large fires in Russia near Lake Baikal	Fires in the Indochina peninsula, anthropogenic emissions in the North China Plain	Fires in the Indochina peninsula, anthropogenic emissions in the North China Plain and the Sichuan Basin
Vertical transport mechanism	Frontal lifting	Convection, frontal lifting, and orographic lifting	Convection, frontal lifting, and orographic lifting
Outflow	West coast of Canada	West coast of the United States	West coast of Canada

curred over the East China Sea or the Sea of Japan or both. High CO was shown in MOPITT daytime data in the middle to upper troposphere in case 2003, in the lower to middle troposphere in case 2004, and in the upper troposphere in case 2005. The MOPITT and MOZAIC observations for the three cases are shown in Figs. 4–6, followed by analyses for each case with FLEXPART and GEOS-Chem simulations, in combination with MODIS fire data and NCEP FNL meteorological data. The cases occurred in spring and summer when cyclone activities are strong in East Asia (Chen et al., 1991; Yue and Wang, 2008). The main CO sources are identified as biomass burning or a combination of biomass burning and anthropogenic origins. The outflow of the high CO episodes finally reached the boundary layer at the west coast of the United States and Canada.

4.1   Case study I: 6 June 2003

On 6 June 2003, a large area ($\sim 400\,\text{km} \times 1500\,\text{km}$) of high CO up to 350 ppbv appeared in the MOPITT image over the Sea of Japan and the nearby continent in the middle to high troposphere (Fig. 4a). In Fig. 5a, the MOPITT CO profile averaged over the boxed area in Fig. 4a shows a broad enhancement from the monthly profile between 650 and 300 hPa, with peak CO abundances of ~ 300 ppbv around 550 hPa. The location and shape of the box was selected to ensure enough MOPITT samplings (> 30) at the closest upwind direction of MOZAIC measurements (the same for Fig. 4b and c). The large difference between the MOPITT a priori and the measurements over these altitudes in-

dicates MOPITT's capability of detecting pollution episodes with some degree of vertical sensitivity. The vertical sensitivity is demonstrated through (1) the strongest CO source among the three cases was shown as the largest magnitude (200–250 ppbv) of elevated CO from the a priori, (2) the altitude with the maximum CO enhancement was detected around the middle troposphere, in contrast to the other two cases which show the maximum in the lower-middle and upper troposphere, respectively, and (3) the elevated CO was over a broad range of altitudes as the vertical resolution of MOPITT is rather coarse, i.e., the annual mean DFS maximizes about 2.5 (Fig. 1). This CO peak was not shown in the MOPITT monthly mean profile, reflecting the episodic nature of this event. The high CO episode was also detected by a near-by MOZAIC measurement (Fig. 5b). A layer of elevated CO is apparent between 500 and 350 hPa, with a CO peak up to ~ 550 ppbv around 400 hPa. In addition, the MOZAIC relative humidity and ozone profiles are shown in Fig. 5b. Around the altitudes of CO buildup, elevated humidity followed the CO profile, while ozone also showed some enhancement.

A latitude–altitude cross section from MOPITT is shown in Fig. 6a. It is the average between two blue dashed lines in Fig. 4a. The arrows represent the winds in the meridional and vertical directions and the contour represents the zonal wind speed. Consistent with Fig. 4a, high CO up to 350 ppbv appeared in the middle to upper troposphere between 35 and 50° N.

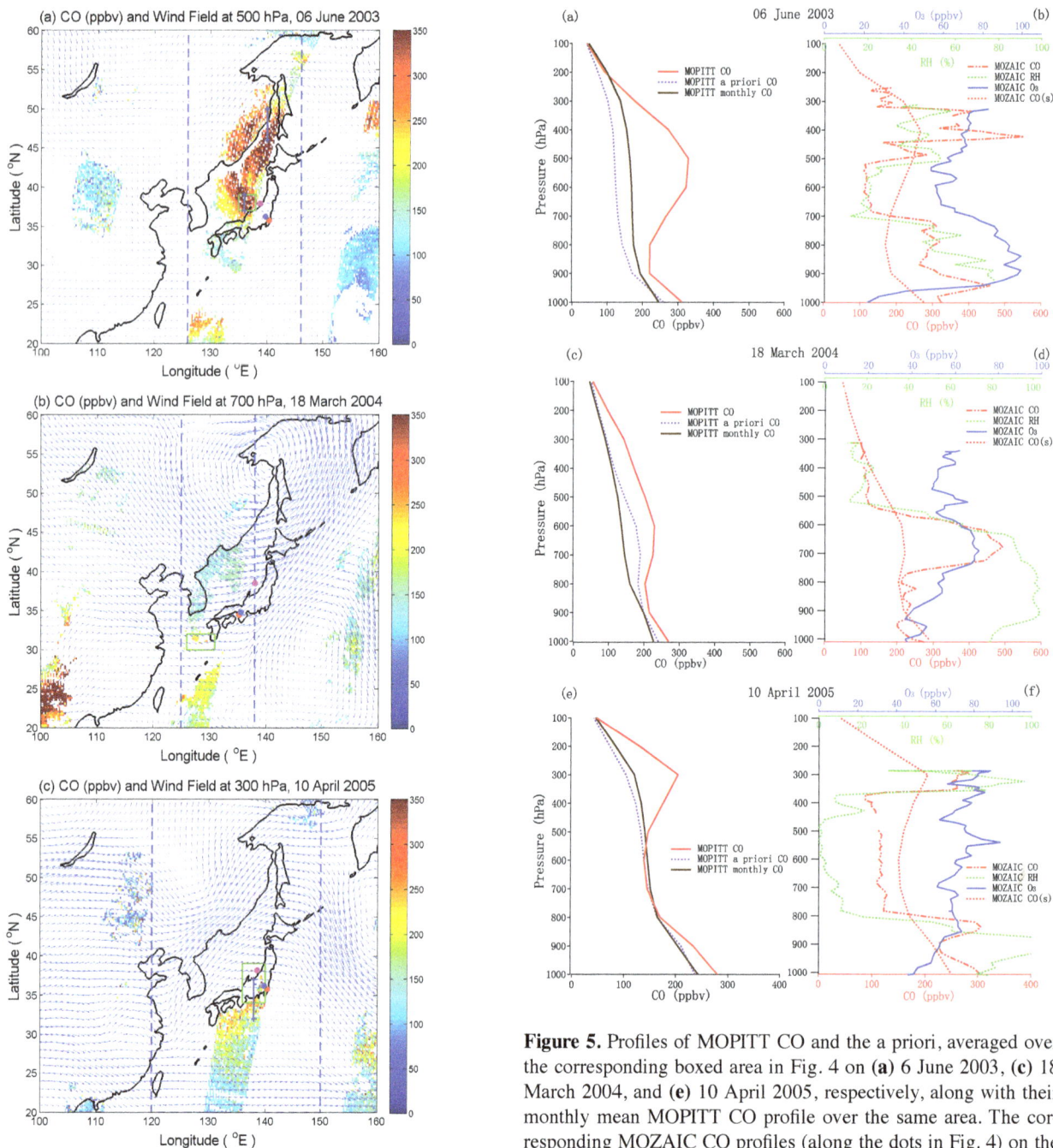

Figure 4. MOPITT CO mixing ratio (ppbv, in color) **(a)** on 6 June 2003 at 500 hPa, **(b)** on 18 March 2004 at 700 hPa, and **(c)** on 10 April, 2005 at 300 hPa. All are overlaid with horizontal winds (in arrows) at the same altitude. In each subfigure, the locations of MOZAIC data at 900, 600, and 300 hPa are indicated as red, blue and pink dots, respectively. The box indicates an area over which mean MOPITT CO profile is taken and displayed in Fig. 5. The box is selected to ensure enough MOPITT samplings at the closest upwind direction of MOZAIC measurements. The two blue dashed lines define the longitudinal zone, over which the CO abundances were averaged and shown in Fig. 6. The solid blue bars in Fig. 4a and c indicate the locations where particles were released and backward trajectories were simulated using FLEXPART (see text for detail).

Figure 5. Profiles of MOPITT CO and the a priori, averaged over the corresponding boxed area in Fig. 4 on **(a)** 6 June 2003, **(c)** 18 March 2004, and **(e)** 10 April 2005, respectively, along with their monthly mean MOPITT CO profile over the same area. The corresponding MOZAIC CO profiles (along the dots in Fig. 4) on the same day are shown in **(b)**, **(d)**, and **(f)**, respectively. The corresponding MOZAIC ozone and relative humidity profiles are also shown in **(b)**, **(d)**, and **(f)**. Note that the smoothed MOZAIC CO profiles (MOZAIC CO(s)) were calculated using the averaging kernels and the a priori in the boxed area in each case (see Sect. 5 for discussion).

To trace down the CO source, backward trajectories of the air particles were simulated using FLEXPART after releasing 30 000 and 7000 particles, respectively, from the locations of the large and small boxed areas in Fig. 6a (the same as the blue bars in Fig. 4a) on 6 June 2003 when CO

Figure 7. (a) Particle distribution between 6.25 and 10.25 km (∼ 550–250 hPa) during 1–6 June 2003. The particles were released from two locations (in pink lines) around 400 hPa (also see Figs. 4a and 6a) on 6 June 2003 and backward trajectories were calculated. The contour lines are the geopotential heights at 850 hPa on 3 June 2003. A cold front and a warm front are indicated by green and red lines, respectively. **(b)** the same as **(a)**, but between 0 and 3.5 km. The contour lines are the geopotential heights at 850 hPa on 2 June 2003. The circles, diamonds, and stars denote daily mean fire counts of 20–100, 100–300, and 300–500 per 2.5° × 2.5° grid area, respectively, from 31 May to 6 June.

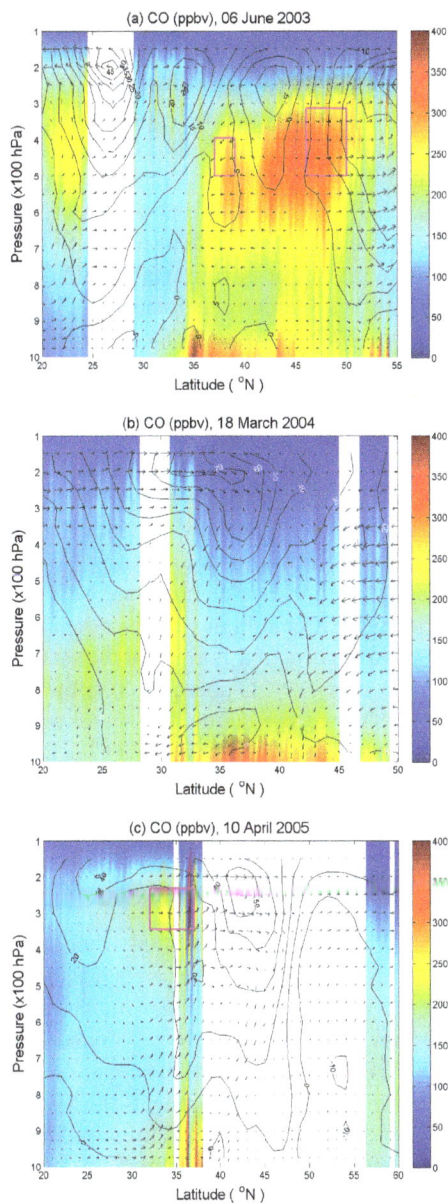

Figure 6. A latitude–altitude cross section of MOPITT CO averaged between the two blue dashed lines in Fig. 4 on **(a)** 6 June 2003, **(b)** 18 March 2004, and **(c)** 10 April 2005. The contour lines indicate U wind speed (m s^{-1}). Vectors are for wind directions in V and W. For a better illustration, W is enlarged by a factor of 100. The pink box(es) in **(a)** and **(c)** indicate the locations where particles were released and backward trajectories were simulated using FLEXPART (see text for detail).

was high in the MOPITT data. Because CO has a relatively long lifetime (weeks to months), it is assumed that CO is not removed in the backward trajectories. Figure 7 shows the distribution of particle concentration between 6.25 and 10.25 (∼ 500–250 hPa, Fig. 7a) and between 0 and 3.25 km (∼ 1000–650 hPa, Fig. 7b). The contour lines indicate the geopotential height at 850 hPa at 12:00 UTC on 3 June 2003

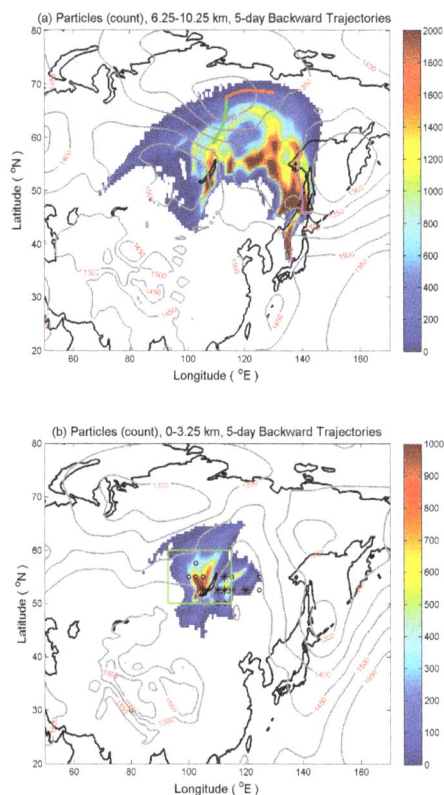

(Fig. 7a) and at 00:00 UTC on 2 June 2003 (Fig. 7b), respectively. The locations of large forest fires near Lake Baikal from MODIS fire data are indicated in Fig. 7b by the stars, diamonds, and circles, with fire counts of 20–100, 100–300, and 300–500 per 2.5° × 2.5° grid area, respectively, averaged daily from 31 May to 6 June. The high particle counts between 0 and 3 km in the vicinity of Lake Baikal match well with the location of fire counts (Fig. 7b). On 3 June 2013, there was a cyclone with a cold front (Fig. 7a) that rapidly lifted the CO originated from the fires along the warm conveyer belt (WCB) to the upper level. The particle distribution in the upper troposphere shows the transport pathway of the particles to the Sea of Japan. To further illustrate this, particles were released from the fire region near Lake Baikal (93–115° E, 50–60° N, 0–3 km, following Lavoué et al. (2000), who found an average injection height of Siberian fires of ∼ 3 km). Forward trajectories were simulated and the resul-

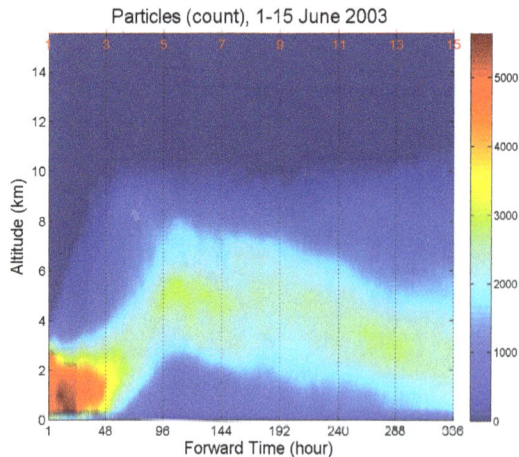

Figure 8. Vertical distribution of particles, varying with time from 1 June 2003 at 00:00 UTC to 15 June 2003 at 00:00 UTC. The particles were released from fire regions in Fig. 7b from the surface to 3 km on 1 June 2003 and forward trajectories were calculated (14 days). The forward time (in h) and date (in June) are indicated in the x axis on the bottom and the top, respectively.

tant vertical distribution of the particles, varying with time during 1–15 June 2003, is shown in Fig. 8. The released particles from the fires traveled along the isobars to northeast of Lake Baikal from 1 June to 3 June 2003 and then the particles were lifted to the upper layers (2–5 km) starting on 3 June at 12:00 LT (in 60–70 h) (Fig. 8). Then, the particles were transported further upward and eastward. On 6 June (in 120–140 h), a large amount of particles appeared in a layer of 3–8 km (Figs. 8 and 4a). The altitudes with high particle concentrations agree well the MOPITT data between 650 and 350 hPa (Figs. 4a and 6a).

It is the cyclone with a front northeast of Lake Baikal that transported the CO up along the WCB (Figs. 7a and 8). Figure 5b shows that the relative humidity reached about 65 % in the MOZAIC measurement, suggesting the air mass indeed came from a WCB (Cooper et al., 2002). The MOZAIC ozone profile also shows elevated ozone at the same altitudes but the shape does not exactly follow the ones of CO and humidity, implying complexity of chemical processes involved. The polluted air reached as high as 9 km although most particles remained at heights of about 3–8 km (Fig. 8). After being lifted to higher altitudes, the polluted air was transported by strong westerlies over long distances. Figure 8 shows that the particles were further transported to the east and sink slowly after 7 June. Around 14 June 2003, the particles reached the east coast of Canada (0–5 km). The satellite MODIS data show a large number of hot spots near Lake Baikal in May and June 2003. Earlier studies have shown that forest fires in Asia can impact air quality in North America (Jaffe et al., 2004; Liang et al., 2004; Oltmans et al., 2010). This case illustrates again the role that WCBs played in the intercontinental transport of pollution for such high CO. Notice that the

FLEXPART simulation was made by using the FNL meteorological data, which may not have considered the buoyancy force due to fires. Such buoyancy force can lift CO plumes even faster and higher.

Our analyses are consistent with Nédélec et al. (2005), who examined 320 MOZAIC flight routes from Europe to Asia in 2003 and reported the observations of high CO up to 800 ppbv above 8 km ($\sim$ 350 hPa) on 3 and 4 June 2003 around 57° N (northeast of Lake Baikal). With different data sets, i.e., along-track scanning radiometer (ATSR) fire data, the Total Ozone Mapping Spectrometer (TOMS) aerosols data, and the MODIS cloud data, Nédélec et al. (2005) also attributed the high CO at these altitudes to front lifting of CO from large forest fires near Lake Baikal. The time and location of frontal lifting of CO in our FLEXPART simulations match well with the observations of high CO by Nédélec et al. (2005). Furthermore, this study provides a more explicit description on the CO transport pathways (Figs. 7 and 8). We also found this rare case demonstrate MOPITT's capability of detecting extreme high CO episodes through relative variations in vertical and horizontal dimensions. Corresponding to the strongest CO source among the three cases, MOPITT data showed the largest horizontal area with CO plumes (Fig. 4), the deepest vertical CO buildup with the highest abundances (Fig. 6), and the biggest enhancement of 200–250 ppbv from the a priori (Fig. 5).

4.2 Case study II: 18 March 2004

This case occurred on 18 March 2004 when high CO appeared in the MOPITT data in the lower and middle troposphere over the East China Sea (Fig. 4b). The elevated CO of 200–250 ppbv is observed between 750 and 550 hPa vertically in MOPITT data (Fig. 5c). The departure of the MOPITT CO profile from its a priori reflects the MOPITT's vertical sensitivity (Fig. 5c). The MOPITT monthly mean, like for the other two cases, follows a typical CO profile pattern with CO concentrations being the highest near the surface and decreasing gradually with altitude. The CO on 18 March 2004 was 50 ppbv higher than the monthly mean above 800 hPa. A layer of elevated CO appeared in the MOZAIC profile between 750 and 550 hPa with a peak of 500 ppbv around 650 hPa (Fig. 5d). The high RH ($\sim$ 90– 100 %) below 600 hPa in the MOZAIC data suggests that the air mass experienced some uplifting process that enhanced its humidity, likely from a WCB. The MOZAIC ozone peaked ($\sim$ 70 ppbv) around the same altitudes as CO, implying that ozone may be produced in the air mass carrying high CO during the transport process. Figure 6b shows a latitude–altitude cross section averaged between the two blue dashed lines in Fig. 4b. Around 30° N, elevated CO levels ($\sim$ 200 ppbv) are evident around 700 hPa.

This case was simulated with GEOS-Chem to identify the sources of CO and to explore the transport mechanisms. The MODIS fire data suggest biomass burning over north-

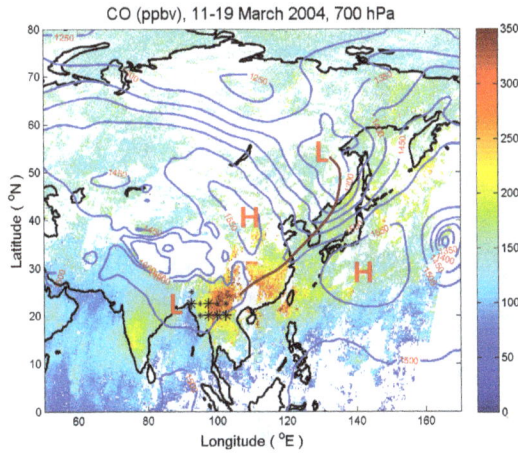

Figure 9. MOPITT CO mixing ratio at 700 hPa from 11 to 19 March 2004, overlaid with the geopotential height at 850 hPa on 17 March 2004 in blue contour and with a front shown by a brown solid line. The large and small stars denote daily mean fire counts of 100–200 and over 200 per $2.5° \times 2.5°$ grid area during the period, respectively. L and H indicate a low and high pressure system, respectively.

ern Indochina peninsula to be a source for the observed high CO (Fig. 9). The time series of fire counts over area of 20–25° N and 92–105° E peaked on 12 March 2004. Correspondingly, high CO of ~ 300 ppbv appeared in the MOPITT composite of 11–18 March 2004 at 700 hPa over northern Indochina peninsula (Fig. 9). This source was also recognized in the GEOS-Chem simulation (Fig. 10b). In addition, the anthropogenic source concentrated over the North China Plain (approximately 30–40° N, 110–125° E) was identified as another source (Fig. 10c). The fire-induced CO spread larger areas from south to north than the anthropogenic CO. Figure 11 shows the latitude–altitude cross sections of the GEOS-Chem simulations of CO, fire-induced CO, and anthropogenic CO, respectively, along 130° E on 18 March 2004. CO abundances from both sources were high around 700 hPa (Fig. 11b and c) between 25 and 35° N across 130° E where MOPITT also observed high CO (Fig. 6b).

The different CO distributions for the two sources in three dimensions (Figs. 10 and 11) reflect rather different transport pathways and uplifting mechanisms. We found that the transport of the fire-induced CO can be divided into four processes. First, the CO was orographically lifted along the Hengduan Mountains from the surface to ~ 750 hPa. The lifted CO is shown in Fig. 12 around 100° E on a longitude–altitude cross section along 22° N. Then, the uplifted CO experienced two separate transports. In the second process, part of the lifted CO was further transported upward to 400–300 hPa, shown as a bulb in Fig. 12 around 105° E. This is due to strong convection, possibly caused by a frontal system developed on 17 March 2004 (Fig. 9), and interplay with the lee-side troughs east of the Hengduan Mountains.

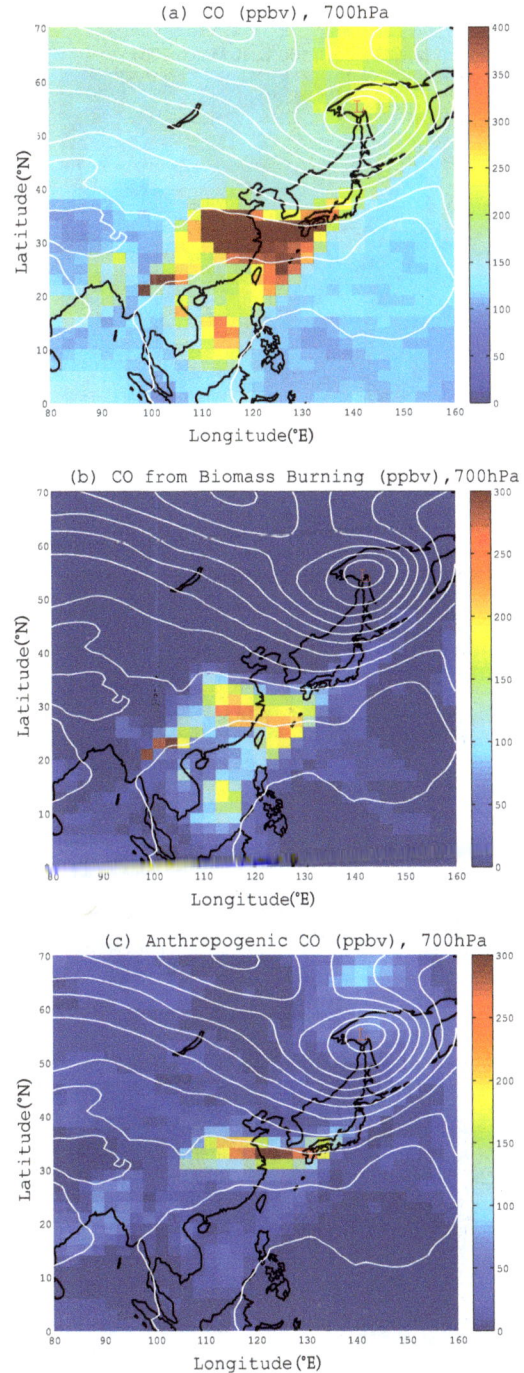

Figure 10. (a) CO, **(b)** CO from biomass burning, and **(c)** CO from the anthropogenic source on 17 March 2004 at 00:00 UTC, simulated by GEOS-Chem. The geopotential height at 700 hPa is indicated with white contours. L indicates a low pressure system.

The vertical velocity reached $0.2\,\mathrm{m\,s^{-1}}$ in FNL data around this level (not shown). The ECMWF (European Centre for Medium-Range Weather Forecasts) data also show northeastward airflow from Indochina peninsula with high potential energy (warm and wet) available for strong convection.

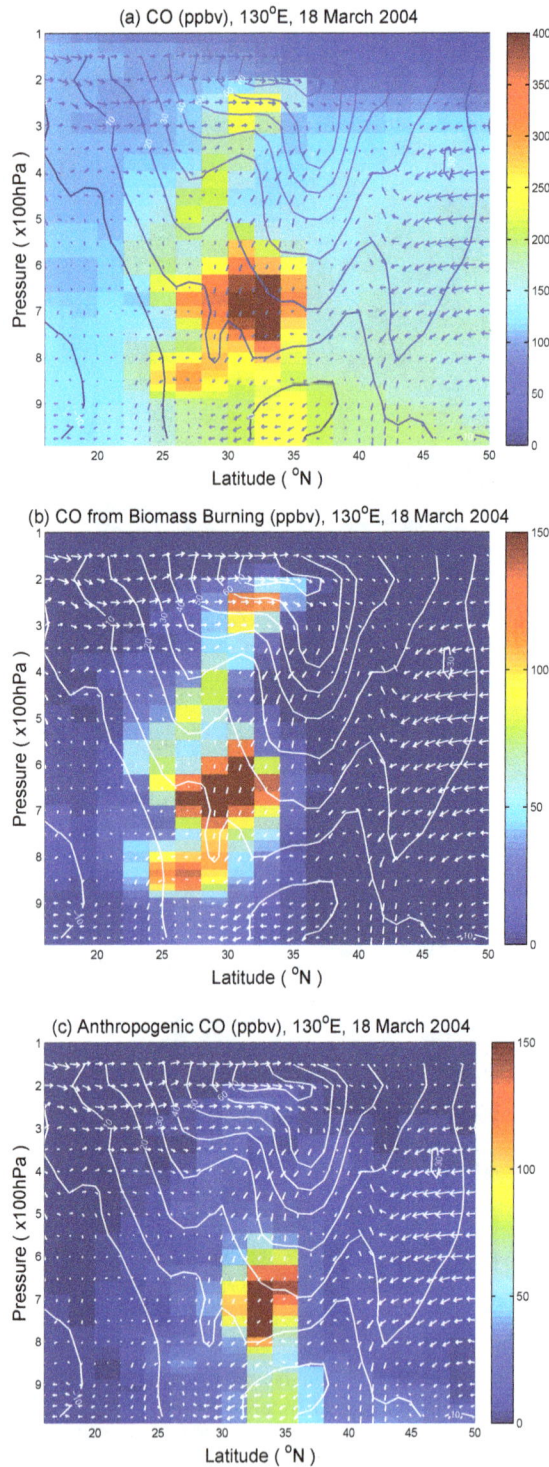

Figure 11. Latitude–altitude cross sections along 130° E of (**a**) CO, (**b**) CO from biomass burning, and (**c**) anthropogenic CO on 18 March 2004 at 00:00 UTC, simulated by GEOS-Chem. The contour lines indicate U wind speed (m s^{-1}). Vectors are for wind directions in V and W. For a better illustration, W is enlarged by a factor of 100.

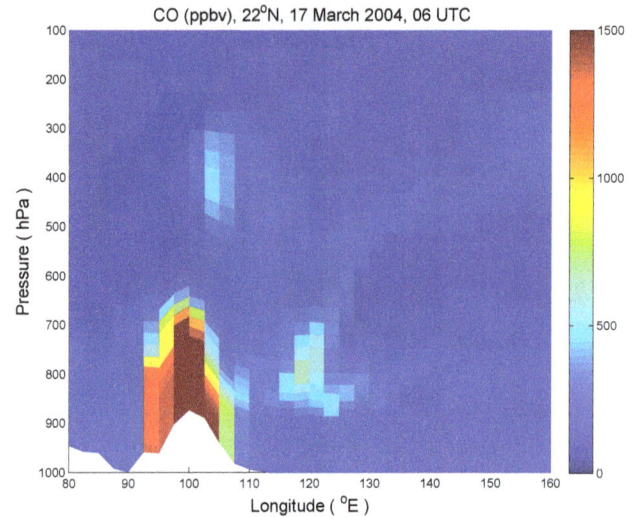

Figure 12. A longitude–altitude cross section of CO along 22° N on 17 March 2004 at 06:00 UTC, simulated by GEOS-Chem. The topography of the Hengduan Mountains is indicated in white.

All of these suggest that the strong convection over the lee-side troughs rapidly lifted CO up to ~ 350 hPa. In fact, the orographic lifting and topography-induced convection are quite common in this region; therefore, high CO often appears at these two altitudinal levels in March as simulated by GEOS-Chem (not shown). On 17 March, the lifted CO was with even higher concentrations (~ 500 ppbv) around 400 hPa than the monthly mean because of its fire origin and the presence of the lee-side troughs. In the third process, the uplifted CO around 400–300 hPa (near 105° E in Fig. 12) was transported northeastward by strong winds along the front in the upper troposphere, reaching the East China Sea (near 30° N, 130° E) on 18 March. This transport enables high CO from forest fires in southern Asia in low latitudes to rapidly reach the upper troposphere in the mid-latitudes. In the fourth process, paralleling to the second and third, part of the orographically uplifted CO was afloat around ~ 700 hPa because of lee-side-trough-induced convection. This CO was transported eastward along the isobars of the low pressure system around 700 hPa (Figs. 10 and 12). This process occurred at lower altitudes than processes two and three. The transport was slower and it took a longer time (from 15 to 18 March) for the CO to reach the East China Sea. Processes two and three brought CO to the upper troposphere (200–300 hPa in Fig. 11b), while process four increased CO in the lower to middle troposphere (700–500 hPa in Fig. 11b). For the anthropogenic CO from the North China Plain, the vertical transport was mainly carried out by frontal lifting on 17 March 2004 (Fig. 9) and then the uplifted CO was transported eastward along 30° N (Fig. 10c). Consequently, the total CO shows a buildup centered near 700 hPa around 30° N and 130° E, mostly coming from the two CO sources (Fig. 11a–c).

The Hengduan Mountains run mainly north to south, with elevations ranging from 1300 to 6000 m. This topography provides a favorable condition for the formation of the lee-side troughs if meteorology is satisfied. Such troughs promote vertical transport of CO on the east side of the mountains (in the second and fourth processes), while the orographic lifting occurred on the west side of the mountains (in the first process). The lee-side troughs occur most and least frequently in spring and summer, respectively. Interannual variation of the lee-side troughs is also observed.

Comparison of the vertical CO distributions between MOPITT and GEOS-Chem (Fig. 6b vs. Fig. 11a) suggests that MOPITT can generally capture vertical transport of CO from forest fires and anthropogenic sources, although the magnitude of CO in MOPITT data was lower and there were also substantial gaps in the MOPITT images due to convective clouds. In the MOPITT data, high CO of ~ 200 ppbv reached up to 200 hPa. In the lower to middle troposphere, elevated CO (~ 200 ppbv) was centered around 650–700 hPa. These features are similar to the GEOS-Chem simulations. Note that the CO buildup around 300 hPa in the GEOS-Chem simulation (Fig. 11a and b) was reflected in the MOPITT data (Fig. 6b), but not as obvious as in the simulation since the MOPITT retrievals are smoothed with the averaging kernels. This CO is also shown as a little bump around 300 hPa in MOPITT vertical profile in Fig. 5c. This buildup is missing in the MOZAIC profile (Fig. 5d) because the aircraft flew towards the north and outside the region with high CO (Fig. 4b).

As the backward trajectories, starting from the boxed area at 700 hPa in Fig. 4b, indicated the most particles came from the large fire in the Indochina peninsula starting from 11 March 2004, we released air particles in FLEXPART over the fire regions from the surface to 1 km on 11 March 2004, and forward trajectories were simulated to track down the air parcels until 18 May 2004 at 02:00 LT. Taking the same zonal means as for Fig. 6b, it is found that the vertical distribution of particle concentrations is similar to that in Fig. 6b with the highest particle concentrations between 4 and 5 km (not show). As simulated by FLEXPART, the outflow of the high CO finally reached the west coast of the United States with particles mainly distributed around 5 km in altitude. High CO observed in East Asia in this case appeared the mostly southerly among the three (Fig. 6), leading to a most southerly outflow.

The strong lee-side-trough-induced convection described in the fourth process was first proposed by Lin et al. (2009) who found that the lee-side troughs above the Indochina peninsula play a significant role in uplifting ozone there. In this study, we found these lee-side troughs can promote lifting of CO even up to the upper troposphere (in the second process, Fig. 12). It is the interplay of the lee-side troughs and the cyclone in the northeast of China which formed a front system that transported CO from the Indochina peninsula upward.

4.3 Case study III: 10 April 2005

In this case, MOPITT observed high CO of ~ 250 ppbv at 300 hPa near the east coast of Japan on 10 April 2005 (Fig. 4c). Like for the other cases, the mean MOPITT profile was taken over a boxed area (in Fig. 4c) upwind of the MOZAIC measurement for comparison. The MOPITT vertical profile clearly shows a CO peak around 300 hPa, where it departs from the MOPITT monthly mean (Fig. 5e). Compared with the other cases, MOPITT CO peaked at higher altitudes, illustrating some MOPITT vertical sensitivity even at these altitudes. In Fig. 5f, a sharp peak of 300 ppbv in MOZAIC CO is shown around 350 hPa. This peak can also be reproduced in the GEOS-Chem simulation with a lower CO abundance of ~ 200 ppbv (not shown). The profile of relative humidity follows closely that of CO, with values up to 90–100 % around 350 hPa, implying that the elevated CO was lifted to this level from the lower troposphere by a cyclone system along its WCB. However, the MOZAIC ozone profile varies differently from the CO profile. We found this was connected to a strong stratospheric intrusion introduced by the cyclone. HYSPLIT (HYbrid Single-Particle Lagrangian Integrated Trajectory) simulations suggest that a large amount of air mass plunged around 4 April from 9 to 3–4 km over northwest of China, bringing high ozone to the lower troposphere (not shown). Another piece of evidence for a stratospheric intrusion is suggested by the low humidity between 700 and 400 hPa. Such a downwelling of stratospheric air on the back side of cyclones was also reported by Miyazaki et al. (2003). Figure 6c shows an altitude–latitude cross section averaged between 120 and 150° E (between two dashed lines in Fig. 4c). High CO of 200–250 ppbv appeared between 300 and 200 hPa around 35° N. This is a rare case in which MOPITT reports such high CO (200–250 ppbv) at these high altitudes (around 300 hPa). Documented CO abundances observed by MOPITT at these altitudes were ~ 130 ppbv over the Indian summer monsoon seasons (Kar et al., 2004), 110–150 ppbv in North America from the forest fires, chemical, and anthropogenic sources (Liu et al., 2005, 2006), and ~ 150 ppbv in spring at Hong Kong (Zhou et al., 2013).

The MODIS fire data show that there were indeed large fires over Indochina peninsula in 3–10 April 2005, shown as stars in Fig. 13. Using GEOS-Chem, CO from fire and anthropogenic sources was simulated to identify their respective contributions and transport pathways.

The entire process of vertical and horizontal transport of CO was well reproduced by GEOS-Chem (Fig. 14). Figure 14a provides the CO distribution in the lower troposphere on 8 April 2005, while Fig. 14b and c show the CO distribution on the next day and the day after in the middle and upper troposphere, respectively. The geopotential heights at 750, 450, and 250 hPa are overlaid with the CO images for each layer accordingly. On 8 April 2005, there was a cyclone developing in the east of Lake Baikal located around 110–

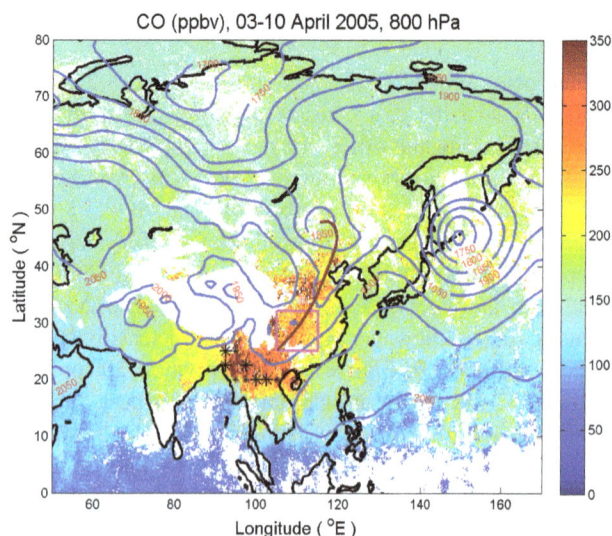

Figure 13. MOPITT CO mixing ratio at 800 hPa from 3 to 10 April 2005, overlaid with the geopotential height at 850 hPa on 9 April 2005 at 00:00 UTC in blue contour and with a front as a brown solid line. The large and small stars denote daily mean fire counts of 100–200 and over 200 per $2.5° \times 2.5°$ grid area during the period, respectively. The boxed area was identified as a major CO source region from the FLEXPART simulation (see text for detail).

120° E, 45–55° N. The surface CO was transported upward and northeastward along the WCB (Fig. 14a). On 9 April the cyclone moved to the east (Figs. 13 and 14b). The high CO shows a comma shape along WCB at the mid-troposphere; this shape is typical for a mature cyclone system with a WCB (Cooper et al., 2002). On 10 April, the cyclone further moved eastward and reached the Sea of Japan (Fig. 14c). The GEOS-Chem simulation shows accumulation of high CO over the ridge of high pressure and along the front at the upper troposphere. The GEOS-Chem simulations suggest that the outflow of the high CO reached Canada on 16 April.

The combined effects of cyclone activities, topography, and CO from different sources and locations are reflected in distinct CO signatures along the WCB. Figure 15 shows the CO from the fires (Fig. 15a and c) and from the anthropogenic source (Fig. 15b and d) in the middle and upper troposphere, respectively, overlaid with the geopotential height at 450 (Fig. 15a and b) and 250 hPa (Fig. 15c and d), respectively. In the middle troposphere (500–400 hPa), a large amount of CO from the fires in the Indochina peninsula was uplifted to this level through orographic lifting and strong convection on the west and east side of the Hengduan Mountains, respectively. This CO distributed along the middle part of the WCB on 9 April 2005 and was transported eastward on 10 April 2005 (Fig. 15a). One source of the anthropogenic CO was concentrated around the North China Plain (Ding et al., 2009) where high CO was evident in MOPITT data (Fig. 13, 35–40° N, 100–120° E). On 8 April, this CO was uplifted along the WCB and further transported to

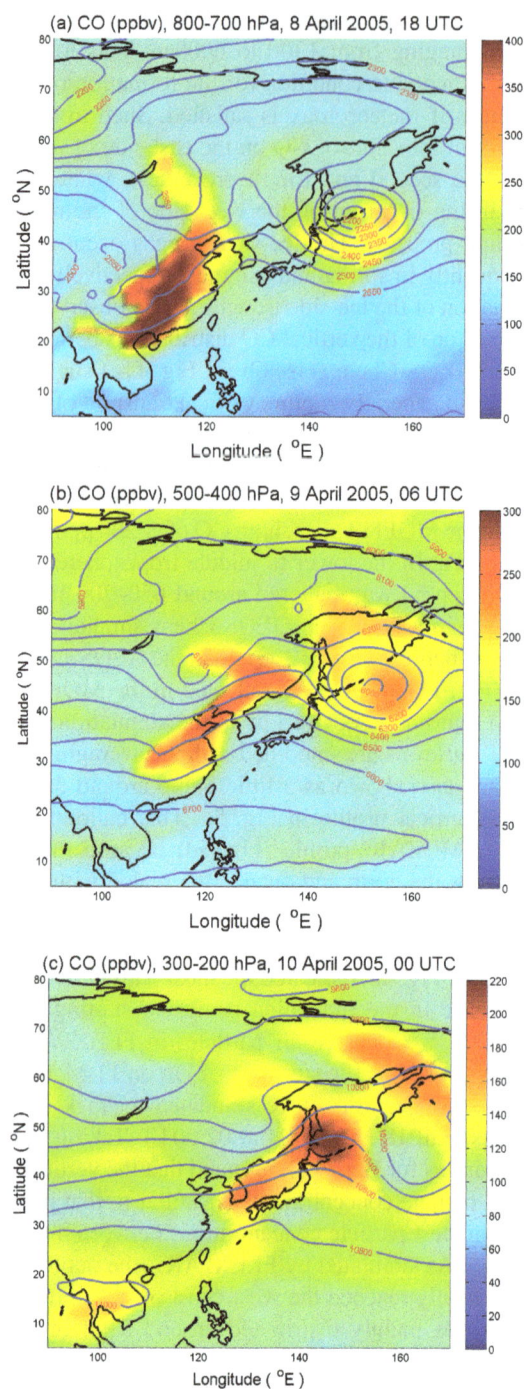

Figure 14. The GEOS-Chem simulated CO (**a**) on 8 April 2005 in the lower troposphere (800–700 hPa), (**b**) on 9 April in the middle troposphere (500–400 hPa), and (**c**) on 10 April in the upper troposphere (300–200 hPa). The contours are the geopotential height at 850, 450, and 250 hPa, respectively.

the middle troposphere, coming across sudden elevated terrains on the way and forming the head of the comma in the cyclone system (Figs. 14b and 15b). The topography's role was noticed by Liu et al. (2003), who found a ring of conver-

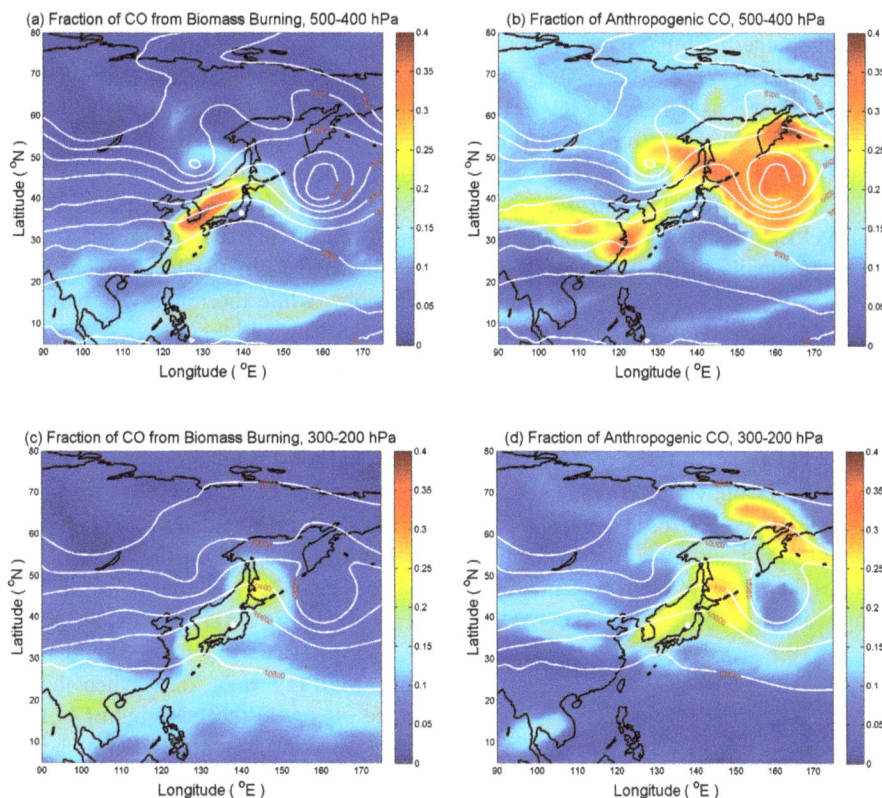

Figure 15. The GEOS-Chem simulated fractional CO (**a**) from biomass burning and (**b**) from the anthropogenic source on 10 April 2005 at 00:00 UTC in the middle troposphere (500–400 hPa). (**c**) and (**d**) are the same as for (**a**) and (**b**), respectively, but in the upper troposphere (300–200 hPa). The geopotential height at 450 and 250 hPa is overlaid with the CO images in the middle and upper troposphere, respectively. White dots indicate the location of MOZAIC measurements.

gence around the North China Plain associated with elevated terrain, and by Ding et al. (2009), who speculated possible topography lifting in the North China Plain. In the southern end of the WCB (near 30° N, 120° E in Fig. 15b), the CO came from the anthropogenic source in the vicinity of the Sichuan Basin (∼ 26–34° N, 102–110° E). This CO was transported vertically to 500 hPa on 8 April at 18:00 UTC to 9 April at 00:00 UTC. Air pollution often accumulates in the Sichuan Basin because of its special topography. The development of small-scale cyclones there is well known as the southwest vortex or Sichuan low (Tao and Ding, 1981). Accumulated pollutants there usually are transported to the free troposphere by such convection. The strong convection can last more than 6 h and peak at midnight (Yu et al., 2007). As this anthropogenic source is quite stable, its contribution should not be understated.

Interestingly, Lin et al. (2009) reported an observed ozone enhancement from ozonesonde data at 4 km in Taiwan on 11 April 2005. They proposed a new transport mechanism from their study as discussed in Sect. 4.2, in which they attributed the elevated ozone to the biomass burning in Indochina. Similarly, CO from biomass burning was also apparent over Taiwan at the middle troposphere in the GEOS-Chem simula-

tion (Fig. 15a), although the maximum CO enhancement was north of Taiwan at this altitude.

The white dot in Fig. 15 indicates the location where MOZAIC passed over. It is clear that MOZAIC measurement at 200–300 hPa was within the WCB, while it was at a distance from the WCB at 500–400 hPa. This is consistent with the MOZAIC CO, ozone, RH profiles shown in Fig. 5f, suggesting that MOZAIC in fact measured air from the stratosphere at these altitudes. As the wind in the upper troposphere was stronger than in the lower troposphere, the WCB-transported CO reached further east in the upper levels (Fig. 15). The simulations suggest that over the boxed area in the MOPITT image in Fig. 4c at 300 hPa, the fire and anthropogenic sources contributed approximately 15 and 20 % CO, respectively. It is noteworthy that there were large gaps in MOPITT data north of 33° N (Fig. 4c) where CO abundances may be even higher than the MOPITT CO south of 33° N as suggested by the GEOS-Chem simulation (Fig. 14c). These gaps were caused by clouds associated with the cyclone system. The complication due to clouds is a problem with an optical instrument like MOPITT. This is why this case is rare in which high CO was observed by both MOPITT and MOZAIC under a frontal system.

In this case, the strong part of the front (close to the centre of the cyclones) swept southern China, where CO was high (Fig. 13). Along the front (30–40° N, 100–120° E), the temperature gradient at 925 hPa was as high as 4.9 °C per degree. Strong ascents occurred ahead of the front, with vertical velocity being $\sim 0.05 \, \mathrm{m \, s^{-1}}$ at 900 and $\sim 0.20 \, \mathrm{m \, s^{-1}}$ at 750 hPa, increasing with altitude until 300–250 hPa where the maximum vertical velocity was $0.26 \, \mathrm{m \, s^{-1}}$. Consequently, the high CO can be rapidly lifted to the upper troposphere in this case.

FLEXPART was also used to trace down high CO in the MOPITT image by releasing air particles in the boxed area in Fig. 6c (indicated by a bar in Fig. 4c). We found that the most CO came from the southwest part of China (boxed area in Fig. 13) where MOPITT CO composite of 3–10 April 2005 shows high CO of 250–300 ppbv at 800 hPa. This CO was lifted along the WCB described above. This agrees with the GEOS-Chem simulation which attributed the major CO source in the upper troposphere to the anthropogenic CO, likely from the Sichuan Basin (Fig. 15d).

5 Discussion

New insights gained from this study and suggestions for future work are discussed as follows.

5.1 Observations of high CO episodes

In the three CO episodes, high CO abundances 300–550 ppbv were observed by MOZAIC in the free troposphere (Fig. 5). The CO abundances are among the highest documented at these altitudes in East Asia. Ding et al. (2009) observed high CO episode of ~ 1185 ppbv at 2.6 km (850–700 hPa) over the North China Plain in summer 2007. Nédélec et al. (2005) found CO up to 800 ppbv above 8 km (~ 400 hPa) near the fire region of Lake Baikal on 3 and 4 June 2003. Highest CO concentrations during TRACE-P were between 250 and 300 ppbv from 2 to 12 km (Heald et al., 2003; Liu et al., 2003; Miyazaki et al., 2003). Occurrences of such high CO episodes are not by chance. They reflect the uniqueness and complexity of meteorology, orography, vegetation covers, and CO sources in East Asia. For example, in all the cases, biomass burning occurred from regions with dense vegetation covers and with most active forest fires in East Asia (Schultz, 2002; Duncan et al., 2003). These fires are usually most active in summer in boreal forest in Russia (like in case 2003) and in spring in the southern East Asia (like cases 2004 and 2005) thus enhancing chances of high CO episodes in these seasons.

The frequency of occurrences of such high CO is illustrated in Table 2. As the three cases occurred near Japan, MOZAIC data around the vicinity of Narita from 2001 to 2006 are summarized, showing occurrences of various CO abundance ranges in the boundary layer (the surface–850),

the lower (850–600), middle (600–400), and upper (400–200 hPa) troposphere. Among all the data in the upper troposphere, CO abundances occurred 93 times (17 %) between 200 and 300, 19 times (4 %) between 300 and 400, and 6 times (1 %) over 400 ppbv. In the middle troposphere, the fraction of occurrences of CO within 200–300, 300–400, and over 400 ppbv was 14, 3, and 2 %, respectively. In the boundary layer, the highest occurrences of CO abundances (38 % of all the data in the layer) were within a range of 200–300 ppbv, while the range was within 100–200 ppbv in the lower (47 %), middle (74 %), and upper troposphere (66 %). Seasonally, there were more high CO episodes in the higher altitudes in spring and summer than in fall and winter.

The frequency of such high CO episodes is also examined in the GEOS-Chem simulations and MOPITT observations in the vicinity of Narita (126–140° E, 30–40° N) in 2005 (Table 3). A count is added to a CO range if the daily maximum CO in the area (126–140° E, 30–40° N) falls into that CO range. Thus, the total counts for all the CO ranges at a given layer are 365 in 2005 for GEOS-Chem, while the counts are 281 for MOPITT due to missing data. To minimize noise in daily MOPITT data, only when there are at least 10 data in the area with the maximum CO falling into a given CO range, a count is added. GEOS-Chem can simulate CO up to 400 ppbv in the upper troposphere, while the maximum CO in MOPITT is lower so that different CO ranges are used in Table 3. Overall, MOZAIC, MOPITT, and GEOS-Chem all show a high frequency of high CO (larger than 200 ppbv) at the surface, progressively shifting to a high frequency of low CO (less than 200 ppbv) at the upper troposphere. Between 400 and 200 hPa, CO with 200–300 ppbv occurred 1.2 times every 10 days in GEOS-Chem, which was slightly lower than in MOPITT (1.8 times) and MOZAIC (1.7 times). Overall, MOZAIC observed 2–5 % more vertical transport of high CO (> 300 ppbv) to the upper troposphere than GEOS-Chem, while the latter simulated 10–20 % more frequently the transport to the middle and lower troposphere with similar or lower CO abundances.

It is likely that on average, the extremely high CO episodes (~ 500 ppbv), such as the 2003 and 2004 cases (Fig. 5), occurred 2–5 times per 100 days in their respective altitudes over the East China Sea and the Sea of Japan (Tables 2 and 3). With a lower CO abundances of 200–300 ppbv (case 2005), the frequency for the air mass to be transported to 400–200 hPa is 1–2 times per 10 days (Tables 2 and 3). The frequency can be even higher in spring and summer, approximately once a week (Table 2). Significant impacts of such vertical transport can be expected on the air quality downwind and on the global climate. The transport mechanisms and CO source contributions revealed in this study can also be applicable for CO episodes with lower CO abundances or at lower altitudes.

Table 2. Occurrences of various CO ranges at different altitudes in the MOZAIC measurements in the vicinity of Narita from 2001 to 2006.

Season	Pressure (hPa)	Occurrence						Fractional occurrence (%)					
		CO range (ppbv)						CO range (ppbv)					
		0–100	100–200	200–300	300–400	>400	All	0–100	100–200	200–300	300–400	>400	All
All	400–200	67	354	93	19	6	539	12	66	17	4	1	100
	600–400	36	359	69	15	8	487	7	74	14	3	2	100
	850–600	17	180	150	31	6	384	4	47	39	8	2	100
	Surface–850	4	60	142	83	88	377	1	16	38	22	23	100
Spring	400–200	11	96	28	12	2	149	7	64	19	8	1	100
	600–400	1	101	29	9	3	143	1	71	20	6	2	100
	850–600	0	38	55	20	5	118	0	32	47	17	4	100
	Surface–850	0	14	44	27	29	114	0	12	39	24	25	100
Summer	400–200	14	132	41	4	4	195	7	68	21	2	2	100
	600–400	13	138	22	1	2	176	7	78	13	1	1	100
	850–600	14	80	50	6	0	150	9	53	33	4	0	100
	Surface–850	4	32	48	30	36	150	3	21	32	20	24	100
Fall	400–200	30	61	15	2	0	108	28	56	14	2	0	100
	600–400	20	50	11	2	1	84	24	60	13	2	1	100
	850–600	3	30	17	3	0	53	6	57	32	6	0	100
	Surface–850	0	10	20	12	11	53	0	19	38	23	21	100
Winter	400–200	12	65	9	1	0	87	14	75	10	1	0	100
	600–400	2	70	7	3	2	84	2	83	8	4	2	100
	850–600	0	32	28	2	1	63	0	51	44	3	2	100
	Surface–850	0	4	30	14	12	60	0	7	50	23	20	100

5.2 The role of topography

East Asia's topography varies significantly across its vast width, increasing from east to west, with a variety of terrains. This study found that topography there affected the three cases in different ways. In addition to its general function in orographic lifting (in cases 2004 and 2005), topography also interplay with frontal systems and enhance the uplifting substantially in the North China Plain (in cases 2004 and 2005). It is notable that CO transports from south to north along elevated terrain over China (in case 2005, Fig. 14a). Under the influence of the Tibetan Plateau, the southwest vortex (or the Sichuan low) is formed (Tao and Ding, 1981) and can facilitate strong convection in the Sichuan Basin (in case 2005).

In particular, topography-induced convection due to the lee-side troughs east of the Hengduan Mountains, proposed by Lin et al. (2009), offers a new mechanism for vertical transport of pollution from the region (in cases 2004 and 2005). Lin et al. (2009) mainly aimed at pollution transport to the lower and middle troposphere. Extending from Lin et al. (2009), this study found such a mechanism to be plausible in explaining pollution transport to the upper troposphere. We found that the impacts of the topography-induced convection on vertical CO transport vary substantially from year to year. A study on such interannual variation is underway.

5.3 The implications of WCB trends on uplifting of CO

Extratropical cyclones and associated frontal activities are important in lifting CO from the boundary layer to the free troposphere. This also applies to other air pollutants. Zhao et al. (2008) found that the influence of Asian dust storms on North American ambient particulate matter levels is highly related to the height to which the frontal cyclones in East Asia can lift dust. Although many functions and characteristics of WCBs have been recognized by earlier studies, we found some details new or unique for the three cases. In case 2004, it is the interplay of the lee-side troughs and the cyclone in the northeast of China that transported CO from the Indochina peninsula upward. The high CO in this case appeared the most southerly among the three, leading to a most southerly outflow. In case 2005, the downwelling of stratospheric air on the back side of cyclones was recognized. The CO along various parts of the WCB was identified to be of fire origin from Indochina and anthropogenic origin from Sichuan and the North China Plain. The source allocation was sensitive to the location of the front. Comparing cases 2004 and 2005, we found that uplifting of CO to the upper troposphere became more possible when large CO sources coincided with the strongest part of a WCB.

In East Asia, cyclones occur most frequently in two regions in spring and summer: one over the lee sides of the Altai-Sayan and the other in the East China Sea and the Sea of Japan (Chen et al., 1991; Yue and Wang, 2008). These are the locations and seasons where and when we can expect

Table 3. Occurrences of various CO ranges in GEOS-Chem simulations and MOPITT observations in the vicinity of Narita (126–140° E 30–40° N) in 2005.

Pressure (hPa)	GEOS-Chem: fractional occurrence (%)						Pressure (hPa)	MOPITT: fractional occurrence (%)				
	CO range (ppbv)							CO range (ppbv)				
	0–100	100–200	200–300	300–400	>400	All		0–100	100–200	200–250	>250	All
200–100	12	87	1	0	0	100	200–100	38	45	11	6	100
400–200	0	87	12	1	0	100	400–200	4	68	18	10	100
600–400	0	53	35	11	1	100	600–400	6	86	6	1	100
850–600	0	20	46	25	8	100	800–600	7	67	20	6	100
1000–850	0	7	46	35	11	100	1000–800	2	16	18	64	100

similar events to happen in the future. Chen et al. (1991) suggested a decline in cyclonic events in East Asia from 1957 to 1977 and no such decline from 1977 to 1987. Recently, an analysis for a longer term from 1951 to 2010 based on ensembles of twentieth century reanalysis (20CR) showed a decreasing trend in the northern part of the Sea of Japan and an increasing trend over the southern part of the Sea of Japan and the lee side of the Altai-Sayan in summer (Wang et al., 2013). The implications of these trends on uplifting of CO deserve further investigation. It would be helpful to conduct statistical analysis of the CO source distribution along WCBs in East Asia in the future.

5.4 Model simulations of pollution transport

Pollution transport can be tracked computationally with Eulerian and Lagrangian approaches, as represented by GEOS-Chem and FLEXPART models, respectively. GEOS-Chem can not only track transport of CO (a physical process) but also consider chemical reactions during the transport while FLEXPART can visualize transport pathways and pin down source regions effectively, without considering chemical functions in the meantime. GEOS-Chem can also fill the gaps in MOPITT satellite data (Figs. 10, 11, and 14). We found that GEOS-Chem simulates the observed aircraft and satellite CO well in cases 2004 and 2005 but cannot fully reproduce the elevated CO in MOZAIC data in case 2003. The simulated CO plume is with lower mixing ratios and at lower altitudes than in the MOZAIC data. This is possibly due to an underestimated fire inventory or conservative parameterizations in simulating large forest fires or both in GEOS-Chem. Nassar et al. (2009) reported underestimated CO over the 2006 Indonesia fire region by GEOS-Chem, in comparison with the Tropospheric Emission Spectrometer (TES) observations. FLEXPART can generally simulate the three cases, strikingly well sometimes in agreement with observed details in space and time, although discrepancies between FLEXPART and satellite and aircraft observations can be found in various places on small scales. FLEXPART simulates strong sources well but sometimes omits weak sources.

5.5 Applications of MOPITT data

We analyzed MOPITT data from two aspects: vertical sensitivity on the synoptic scale. Both are challenging and have not been studied adequately. Large gaps due to clouds and the limited MOPITT swath make application of MOPITT on the synoptic-scale difficult. Thus, application of MOPITT data over East Asia were mostly focused on monthly or seasonal scales (Tanimoto et al., 2008; Zhao et al., 2010; Hao et al., 2011; Liu et al., 2011; Zhou et al., 2013; Su et al., 2012). This study shows that even with large gaps, daily MOPITT data can capture vertical disturbances of CO on the synoptic scale, which are usually diluted on longer timescales. This study also suggests the importance of filling the gaps with other satellite data or in designing new satellite instruments, for the purpose of detecting such variation over large areas on regional and global scales.

Typically for satellite remote-sensing products, the MOPITT retrieval at a specific pressure level is influenced by CO from other levels and thus its retrieval at that pressure level can be biased. However, MOPITT can more accurately measure the average CO mixing ratio over a thick layer, resulting in a coarse vertical resolution. It was suggested that the vertical variation in CO cannot be fully resolved in earlier applications of MOPITT data (Jacob et al., 2003). This study addressed the MOPITT vertical sensitivity with new MOPITT V5 data and found enhanced vertical sensitivity in V5 data in the free troposphere, even in the upper troposphere, in addition to in the boundary layer emphasized by Worden et al. (2010) and Deeter et al. (2012). The enhanced DFSs and the averaging kernels in V5 illustrated by Worden et al. (2010) and Deeter et al. (2012) are supported (Figs. 1 and 5 and Sect. 3).

In Fig. 5, the smoothed MOZAIC profiles were calculated using the averaging kernels and the a priori in an area upwind of the MOZAIC measurement within 0–10° distance for each case as there were no MOPITT data available at the exact locations of the MOZAIC measurements. Although this may introduce some bias, the averaging kernel smoothed MOZAIC profiles in V5 show more vertical structure in CO than an earlier version of MOPITT data in Jacob et al. (2003, in their Fig. 4). Overall, this study found (1) MOPITT can differentiate the magnitude of CO plumes originated from strong or weak sources (Figs. 4, 5, and 6); (2) MOPITT can distinguish elevated CO in the lower, middle, and upper troposphere (Figs. 4, 5, and 6); (3) the shape of CO plumes in vertical direction matches with simulations of GEOS-Chem and FLEXPART, sometimes remarkably well (Figs. 6, 8, and 11); and (4) there is more vertical structure in CO in new V5 than in earlier versions of MOPITT data (Fig. 5).

It is the relative variations in MOPITT CO data that help diagnose CO transport vertically or horizontally. This study suggests using MOPITT data quantitatively with caution, especially at altitudes with high CO plumes because, as illustrated in Figs. 5 and 6, the magnitude of elevated CO in the MOPITT data could be lower than that in the MOZAIC data at the altitudes where CO peaked. Therefore, the vertical variation of CO, even enhanced in V5, is still much smoothed in MOPITT data. MOPITT can distinguish elevated CO in different layers of the free troposphere, yet sometimes cannot specify the exact altitude of elevated CO shown in the MOZAIC measurements (Figs. 5 and 6). One limitation for MOPITT's application of vertical transport is the complication of clouds, which often accompany frontal systems. As shown in cases 2004 and 2005, CO is usually high in cloudy areas. Therefore, the magnitude of CO abundances can be underestimated by MOPITT in these areas.

6 Conclusions

East Asia is characterized by its unique and complex meteorology, topography, vegetation covers, and CO sources. The characteristics are reflected in uplifting of CO illustrated in three high CO episodes during 2003–2005 in this study. Through integrated analyses of observations from the airborne MOZAIC and spaceborne MOPITT instruments and simulations from a trajectory dispersion model FLEXPART (Stohl et al., 2005) and a chemical transport model GEOS-Chem (Bey et al., 2001), this study draws the following conclusions.

1. In the three CO episodes, high CO abundances of 300–550 ppbv were observed by MOZAIC in the free troposphere over the East China Sea and the Sea of Japan. These are among the highest CO abundances ever documented at these altitudes. The three cases occurred when and where meteorology was favorable and CO sources were strong. It is likely that on average, the extremely high CO episodes ($\sim$ 500 ppbv) like cases 2003 and 2004 occurred 2–5 times every 100 days in their respective altitudes over the region, while in case 2005, episodes with a lower CO abundances (200–300 ppbv) occurred 1–2 times per 10 days between 400 and 200 hPa. CO episodes in even lower altitudes and with even lower abundance occurred more frequently in the region.

2. GEOS-Chem and FLEXPART simulations reveal different CO signatures from biomass burning and anthropogenic sources in the CO enhancement in the three cases, reflecting different transport pathways and mechanisms and locations of both sources. In case 2003, CO from large forest fires near Lake Baikal dominated the elevated CO. In case 2004, anthropogenic CO came from the North China Plain and mostly reached $\sim$ 700 hPa near the East China Sea, while CO from biomass burning in Indochina was transported through two separate pathways, leading to two distinct CO enhancements around 700 and 300 hPa. In case 2005, along a WCB over the East China Sea and the Sea of Japan, anthropogenic CO from the North China Plain and from the Sichuan Basin prevailed in the northern and southern part of the WCB, while CO from biomass burning in Indochina was mostly distributed in the middle part of the WCB.

3. Topography in East Asia influences vertical transport of CO in different ways. In particular, topography-induced lee-side troughs east of the Hengduan Mountains over Indochina lead to strong convection. This new mechanism proposed by Lin et al. (2009) is supported by this study in explaining CO transport to the middle troposphere and further extended for CO transport to the upper troposphere. Strong convection from the Sichuan Basin also plays an important role in vertically transporting anthropogenic CO. The topography interacting with frontal activities can enhance the vertical transport of CO substantially in the North China Plain.

4. Extratropical cyclones and associated frontal activities are important mechanism in lifting CO from the boundary layer to the free troposphere, as illustrated by the three cases and earlier studies. East Asia is one of two regions between 25 and 45° N with most frequent WCB events (Eckhardt et al., 2004). Inside East Asia, there are two regions where cyclones occur most frequently: one over the lee sides of the Altai-Sayan and the other in the East China Sea and the Sea of Japan, occurring mostly in spring and summer over both regions (Chen et al., 1991). The seasons and locations of the three high CO episodes just match well with these two areas and active cyclone seasons, which may not happen by chance.

5. Biomass burning is identified as an important source for all three episodes, suggesting that CO from sporadic fire activities can provide additional CO to less varying anthropogenic emission and enhance chances of high CO episodes. The fire regions shown in this study are the places with dense vegetation covers and with the most active forest fires in East Asia.

6. The MOPITT's vertical sensitivity is found to be enhanced in its new V5 NIR/TIR data in the free troposphere, even in the upper troposphere. The daytime

V5 data can detect synoptic disturbances of weather systems on horizontal variation of CO. The data also show more vertical structure than earlier versions and can distinguish CO enhancements at different layers of the troposphere, although the detected high CO is over a broad range in altitudes and lacks detailed vertical structure in comparison with the aircraft observations. Because the CO retrieval at a certain pressure level is often smoothed by the MOPITT averaging kernels, the MOPITT retrievals usually underestimate elevated CO at altitudes with peak CO plumes. The complication of clouds within frontal systems can generate large gaps in MOPITT data and cause underestimation of CO statistically in these regions. Nevertheless, MOPITT data can be used to qualitatively help diagnose vertical transport processes, with caution on their absolute CO values. On average, MOPITT slightly overestimates the background CO in the upper troposphere.

Acknowledgements. The authors gratefully acknowledge the following data and modeling tools. The satellite CO data are provided by the MOPITT team and acquired from the NASA Langley Research Center Atmospheric Science Data Center. The MOZAIC CO data are from the European Commission, Airbus, and the Airlines (Lufthansa, Austrian, Air France) who have carried free of charge the MOZAIC equipment and perform the maintenance since 1994. The final Analysis Data (FNL) were obtained from NOAA CDC. The GEOE-Chem model is developed and managed by the Atmospheric Chemistry Modeling Group at Harvard University with support from the NASA Atmospheric Chemistry Modeling and Analysis Program (ACMAP). The FLEXPART model development team consists of Andreas Stohl, Sabine Eckhardt, Harald Sodemann, and John Burkhart at the Norwegian Institute for Air Research (NILU). Insights and critiques from two anonymous reviewers are highly appreciated. Financial support is provided by an open fund from the Institute of Remote Sensing and Digital Earth, Chinese Academy of Sciences (OFSLRSS201107), the Key Basic Research Program (2010CB950704, 2014CB441203), and the Natural Science Foundation of China (41375140).

Edited by: T. Röckmann

References

Banic, C. M., Isaac, G. A., Cho, H. R., and Iribane, J. V.: The distribution of pollutants near a frontal surface: a comparison between field experiment and modeling, Water Air Soil Poll., 30, 171–177, 1986.

Barret, B., Le Flochmoen, E., Sauvage, B., Pavelin, E., Matricardi, M., and Cammas, J. P.: The detection of post-monsoon tropospheric ozone variability over south Asia using IASI data, Atmos. Chem. Phys., 11, 9533–9548, doi:10.5194/acp-11-9533-2011, 2011.

Berntsen, T. K., Karlsdóttir, S., and Jaffe, D. A.: Influence of Asian emissions on the composition of air reaching the north

western United States, Geophys. Res. Lett., 26, 2171–2174, doi:10.1029/1999GL900477, 1999.

Bertschi, I. B., Jaffe, D. A., Jaeglé, L., Price, H. U., and Dennison, J. B.: PHOBEA/ITCT 2002 airborne observations of trans-Pacific transport of ozone, CO, VOCs and aerosols to the northeast Pacific: impacts of Asian anthropogenic and Siberian Boreal fire emissions, J. Geophys. Res., 109, D23S12, doi:10.1029/2003JD004328, 2004.

Bethan, S., Vaughan, G., Gerbig, C., Volz-Thoms, A., Richer, H., and Tiddeman, D. A.: Chemical air mass differences near fronts, J. Geophys. Res., 103, 13413–13434, 1998.

Bey, I., Jacob, D. J., Yantosca, R. M., Logan, J. A., Field, B. D., Fiore, A. M., Li, Q., Liu, H. Y., Mickley, L. J., and Schultz, M. G.: Global modeling of tropospheric chemistry with assimilated meteorology: model description and evaluation, J. Geophys. Res., 106, 23073–23095, 2001.

Brown, R. M., Daum, P. H., Schwartz, S. E., and Hjelmfelt, M. R.: Variations in the chemical composition of clouds during frontal passage, in: The Meteorology of Acid Deposition, edited by: Samson, P. J., Air Pollut. Control Assoc., Pittsburgh, Pa., 202–212, 1984.

Chan, D., Yuen, C. W., Higuchi, K., Shashkov, A., Liu, J., Chen, J., and Worthy, D.: On the CO_2 exchange between the atmosphere and the biosphere: the role of synoptic and mesoscale processes, Tellus B, 56, 194–212, 2004.

Chen, B., Xu, X. D., Yang, S., and Zhao, T. L.: Climatological perspectives of air transport from atmospheric boundary layer to tropopause layer over Asian monsoon regions during boreal summer inferred from Lagrangian approach, Atmos. Chem. Phys., 12, 5827–5839, doi:10.5194/acp-12-5827-2012, 2012.

Chen, S., Kuo, Y., Zhong, P., and Bai, Q.: Synoptic climatology of cyclogenesis over East Asia, 1958–1987, Mon. Weather Rev., 119, 1407–1418, 1991.

Chung, K. K., Chan, J. C. L., Ng, C. N., Lam, K. S., and Wang, T.: Synoptic conditions associated with high carbon monoxide episodes at coastal station in Hong Kong, Atmos. Environ., 33, 3099–3095, 1999.

Cooper, O. R., Moody, J. L., Parrish, D. D., Trainer, M., Ryerson, T. B., Holloway, J. S., Hübler, G., Fehsenfeld, F. C., and Evans, M. J.: Trace gas composition of midlatitude cyclones over the western North Atlantic Ocean: a conceptual model, J. Geophys. Res., 107, 4056, doi:10.1029/2001JD000901, 2002.

Cooper, O. R., Forster, C., Parrish, D., Dunlea, E., Habler, G., Fehsenfeld, F., Holloway, J., Oltmans, S., Johnson, B., Wimmers, A., and Horowitz, L.: On the life-cycle of a stratospheric intrusion and its dispersion into polluted warm conveyor belts, J. Geophys. Res., 109, D23S09, doi:10.1029/2003JD004006, 2004.

Cooper, O. R., Stohl, A., Hubler, G., Hsie, E. Y., Parrish, D. D., Tuck, A. F., Kiladis, G. N., Oltmans, S. J., Johnson, B. J., Shapiro, M., Moody, J. L., and Lefohn, A. S.: Direct transport of midlatitude stratospheric ozone into the lower troposphere and marine boundary layer of the tropical Pacific Ocean, J. Geophys. Res., 110, D23310, doi:10.1029/2005JD005783, 2005.

Cooper, O. R., Stohl, A., Trainer, M., Thompson, A., Witte, J. C., Oltmans, S. J., Johnson, B. J., Merrill, J., Moody, J. L., Tarasick, D., Nédélec, P., Forbes, G., Newchurch, M. J., Schmidlin, F. J., Johnson, B. J., Turquety, S., Baughcum, S. L., Ren, X., Fehsenfeld, F. C., Meagher, J. F., Spichtinger, N., Brown, C. C., McKeen, S. A., McDermid, I. S., and Leblanc, T.: Large upper tropospheric ozone enhancements above mid-latitude North America during summer: in situ evidence from the IONS and MOZAIC ozone monitoring network, J. Geophys. Res., 111, D24S05, doi:10.1029/2006JD007306, 2006.

Cristofanelli, P., Bonasoni, P., Collins, W., Feichter, J., Forster, C., James, P., Kentarchos, A., Kubik, P. W., Land, C., Meloen, J., Roelofs, G. J., Siegmund, P., Sprenger, M., Schnabel, C., Stohl, A., Tobler, L., Tositti, L., Trickl, T., and Zanis, P.: Stratosphere-to-troposphere transport: a model and method evaluation, STACCATO special section of J. Geophys. Res., 108, 8525, doi:10.1029/2002JD002600, 2003.

Daley, R.: Atmospheric Data Analysis, Cambridge University Press, Cambridge, 454 pp., 1991.

Damoah, R., Spichtinger, N., Forster, C., James, P., Mattis, I., Wandinger, U., Beirle, S., Wagner, T., and Stohl, A.: Around the world in 17 days – hemispheric-scale transport of forest fire smoke from Russia in May 2003, Atmos. Chem. Phys., 4, 1311–1321, doi:10.5194/acp-4-1311-2004, 2004.

Davies, D. K., Ilavajhala, S., Wong, M. M., and Justice, C. O.: Fire information for resource management system: archiving and distributing MODIS active fire data, IEEE T. Geosci. Remote, 47, 72–79, 2009.

Deeter, M. N., Emmons, L. K., Francis, G. L., Edwards, D. P., Gille, J. C., Warner, J. X., Khattatov, B., Ziskin, D., Lamarque, J.-F., Ho, S.-P., Yudin, V., Attié, J.-L., Packman, D., Chen, J., Mao, D., and Drummond, J. R.: Operational carbon monoxide retrieval algorithm and selected results for the MOPITT instrument, J. Geophys. Res., 108, 4399, doi:10.1029/2002JD003186, 2003.

Deeter, M. N., Emmons, L. K., Edwards, D. P., Gille, J. C., and Drummond, J. R.: Vertical resolution and information content of CO profiles retrieved by MOPITT, Geophys. Res. Lett., 31, L15112, doi:10.1029/2004GL020235, 2004.

Deeter, M. N., Worden, H. M., Edwards, D. P., Gille, J. C., and Andrews, A. E.: Evaluation of MOPITT retrievals of lower-tropospheric carbon monoxide over the United States, J. Geophys. Res., 117, D13306, doi:10.1029/2012JD017553, 2012.

Deeter, M. N., Martínez-Alonso, S., Edwards, D. P., Emmons, L. K., Gille, J. C., Worden, H. M., Pittman, J. V., Daube, B. C., and Wofsy, S. C.: Validation of MOPITT Version 5 thermal-infrared, near-infrared, and multispectral carbon monoxide profile retrievals for 2000–2011, J. Geophys. Res., 118, 6710–6725, doi:10.1002/jgrd.50272, 2013.

Dickerson, R. R., Huffman, G. J., Luke, W. T., Nunnermacker, L. J., Pickering, K. E., Leslie, A. C. D., Lindsey, C. G., Slinn, W. G. N., Kelly, T. J., Daum, P. H., Delany, A. C., Greenberg, J. P., Zimmerman, P. R., Boatman, J. F., Ray, J. D., and Stedman, D. H.: Thunderstorms – an important mechanism in the transport of air pollutants, Science, 235, 4787, 460–464, 1987.

Dickerson, R. R., Li, C., Li, Z., Marufu, L., T., Stehr, J. W., McClure, B., Krotkov, N., Chen, H., Wang, P., Xia, X., Ban, X., Gong, F., Yuan, J., and Yang, J.: Aircraft observations of dust and pollutants over northeast China: insight into the meteorological mechanisms of transport, J. Geophys. Res., 112, D24S90, doi:10.1029/2007JD008999, 2007.

Ding, A., Wang, T., Xue, L., Gao, J., Stohl, A., Lei, H., Jin, D., Ren, Y., Wang, X., Wei, X., Qi, Y., Liu, J., and Zhang, X.: Transport of north China air pollution by midlatitude cyclones: case study of aircraft measurements in summer 2007, J. Geophys. Res., 114, D08304, doi:10.1029/2008JD011023, 2009.

Donnell, E. A., Fish, D. J., Dicks, E. M., and Thorpe, A. J.: Mechanisms for pollutant transport between the boundary layer and the free troposphere, J. Geophys. Res., 106, 7847–7856, 2001.

Drummond, J. R. and Mand, G. S.: The measurements of pollution in the troposphere (MOPITT) instrument: overall performance and calibration requirements, J. Atmos. Ocean. Tech., 13, 314–320, 1996.

Drummond, J. R.: Measurements of pollution in the troposphere (MOPITT), in: The Use of EOS for Studies of Atmospheric Physics, edited by: Gille, J. C. and Visconti, G., the Netherlands, New York, 77–101, 1992.

Duncan, B. N., Martin, R. V., Staudt, A. C., Yevich, R., and Logan, J. A.: Interannual and seasonal variability of biomass burning emissions constrained by satellite observations, J. Geophys. Res., 108, 4040, doi:10.1029/2002JD002378, 2003.

Duncan, B. N., Logan, J. A., Bey, I., Megretskaia, I. A., Yantosca, R. M., Novelli, P. C., Jones, N. B., and Rinsland, C. P.: Global budget of CO, 1988-1997: source estimates and validation with a global model, J. Geophys. Res., 112, D22301, doi:10.1029/2007JD008459, 2007.

Eckhardt, S., Stohl, A., Wernli, H., James, P., Forster, C., and Spichtinger, N.: A 15-Year climatology of warm conveyor belts, J. Climate, 17, 218–237, 2004.

Edwards, D. P., Halvorson, C. M., and Gille, J. C.: Radiative transfer modeling for the EOS Terra satellite measurements of pollution in the troposphere (MOPITT instrument), J. Geophys. Res., 104, 16755–16775, 1999.

Emmons, L. K., Deeter, M. N., Gille, J. C., Edwards, D. P., Attie, J.-L., Warner, J., Ziskin, D., Khattatov, B., Yudin, V., Lamarque, J. F., Ho, S.-P., Mao, D., Chen, J. S., Drummond, J., Novelli, P., Sachse, G., Coffey, M. T., Hannigan, J. W., Gerbig, C., Kawakami, S., Kondo, Y., Takegawa, N., Baehr, J., and Ziereis, H.: Validation of MOPITT CO retrievals with aircraft in situ profiles, J. Geophys. Res., 109, D03309, doi:10.1029/2003JD004101, 2004.

Giglio, L., Descloitres, J., Justice, C. O., and Kaufman, Y. J.: An enhanced contextual fire detection algorithm for MODIS, Remote Sens. Environ., 87, 273–282, 2003.

Hao, H., Valks, P., Loyola, D., Chen, Y. F., and Zimmer, W.: Space-based measurements of air quality during the World Expo 2010 in Shanghai, Environ. Res. Lett., 6, 044004, doi:10.1088/1748-9326/6/4/044004, 2011.

He, H., Tarasick, D. W., Hocking, W. K., Carey-Smith, T. K., Rochon, Y., Zhang, J., Makar, P. A., Osman, M., Brook, J., Moran, M. D., Jones, D. B. A., Mihele, C., Wei, J. C., Osterman, G., Argall, P. S., McConnell, J., and Bourqui, M. S.: Transport analysis of ozone enhancement in Southern Ontario during BAQS-Met, Atmos. Chem. Phys., 11, 2569–2583, doi:10.5194/acp-11-2569-2011, 2011.

Heald, C. L., Jacob, D. J., Fiore, A. M., Emmons, L. K., Gille, J. C., Deeter, M. N., Warner, J., Edwards, D. P., Crawford, J. H., Hamlin, A. J., Sachse, G. W., Browell, E. V., Avery, M. A., Vay, S. A., Westberg, D. J., Blake, D. R., Singh, H. B., Sandholm, S. T., Talbot, R. W., and Fuelberg, H. E.: Asian outflow and transpacific transport of carbon monoxide and ozone pollution: an integrated satellite, aircraft and model perspective, J. Geophys. Res., 108, 4804, doi:10.1029/2003JD003507, 2003.

Hocking, W. K., Carey-Smith, T. K., Tarasick, D. W., Argall, P. S., Strong, K., Rochon, Y., Zawadzki, I., and Taylor, P. A.: Detection of stratospheric ozone intrusion by wind profiler radars, Nature, 450, 281–284, doi:10.1038/nature06312, 2007.

Holloway, T., Levy II, H., and Kasibhatla, P.: Global distribution of carbon monoxide, J. Geophys. Res., 105, 12123–12147, doi:10.1029/1999JD901173, 2000.

Jacob, D. J.: Introduction to Atmospheric Chemistry, Princeton University Press, Princeton, New Jersey, 1999.

Jacob, D. J., Crawford, J. H., Kleb, M. M., Connors, V. S., Bendura, R. J., Raper, J. L., Sachse, G. W., Gille, J. C., Emmons L., and Heald, C. L.: Transport and Chemical Evolution over the Pacific (TRACE-P) aircraft mission: design, execution, and first results, J. Geophys. Res., 108, 9000, doi:10.1029/2002JD003276, 2003.

Jaffe, D., Anderson, T., Covert, D., Kotchenruther, R., Trost, B., Danielson, J., Simpson, W., Berntsen, T., Karlsdottir, S., Blake, D., Harris, J., Carmichael, G., and Uno, I.: Transport of Asian air pollution to North America, Geophys. Res. Lett., 26, 711–714, 1999.

Jaffe, D., Bertschi, I., Jaegle, L., Novelli, P., Reid, J. S., Tanimoto, H., Vingarzan, R., and Westphal, D. L.: Long-range transport of Siberian biomass burning emissions and impact on surface ozone in western North America, Geophys. Res. Lett., 31, L16106, doi:10.1029/2004GL020093, 2004.

Jiang, Z., Jones, D. B. A., Kopacz, M., Liu, J., Henze, D. K., and Heald, C.: Quantifying the impact of model errors on top-down estimates of carbon monoxide emissions using satellite observations, J. Geophys. Res., 116, D15306, doi:10.1029/2010JD015282, 2011.

Jones, D. B. A., Bowman, K. W., Logan, J. A., Heald, C. L., Liu, J., Luo, M., Worden, J., and Drummond, J.: The zonal structure of tropical O_3 and CO as observed by the Tropospheric Emission Spectrometer in November 2004 – Part 1: Inverse modeling of CO emissions, Atmos. Chem. Phys., 9, 3547–3562, doi:10.5194/acp-9-3547-2009, 2009.

Justice, C. O., Giglio, L., Korontzi, S., Owens, J., Morisette, J. T., Roy, D., Descloitres, J., Alleaume, S., Petitcolin, F., and Kaufman, Y.: The MODIS fire products, Remote Sens. Environ., 83, 244–262, 2002.

Kar, J., Bremer, H., Drummond, J. R., Rochon, Y. J., Jones, D. B. A., Nichitiu, F., Zou, J., Liu, J., Gille, J. C., Edwards, D. P., Deeter, M. N., Francis, G., Ziskin, D., and Warner, J.: Evidence of vertical transport of carbon monoxide from measurements of pollution in the troposphere (MOPITT). Geophys. Res. Lett., 31, L23105, doi:10.1029/2004GL021128, 2004.

Kar, J., Drummond, J. R., Jones, D. B. A., Liu, J., Nichitiu, F., Zou, J., Gille, J. C. Edwards, D. P., and Deeter, M. N.: Carbon monoxide (CO) maximum over the Zagros mountains in the Middle East: signature of mountain venting?, Geophys. Res. Lett., 33, L15819, doi:10.1029/2006GL026231, 2006.

Kar, J., Jones, D. B. A., Drummond, J. R., Attie, J. L., Liu, J., Zou, J., Nichitiu, F., Seymour, M. D., Edwards, D. P., Deeter, M. N., Gille, J. C., and Richter, A.: Measurement of low-altitude CO over the Indian subcontinent by MOPITT, J. Geophys. Res., 113, D16307, doi:10.1029/2007JD009362, 2008.

Kopacz, M., Jacob, D. J., Fisher, J. A., Logan, J. A., Zhang, L., Megretskaia, I. A., Yantosca, R. M., Singh, K., Henze, D. K., Burrows, J. P., Buchwitz, M., Khlystova, I., McMillan, W. W., Gille, J. C., Edwards, D. P., Eldering, A., Thouret, V., and Nedelec, P.: Global estimates of CO sources with high resolution by adjoint inversion of multiple satellite datasets (MOPITT,

AIRS, SCIAMACHY, TES), Atmos. Chem. Phys., 10, 855–876, doi:10.5194/acp-10-855-2010, 2010.

Kowol-Santen, J., Beekmann, M., Schmitgen, S., and Dewey, K.: Tracer analysis of transport from the boundary layer to the free atmosphere, Geophys. Res. Lett., 28, 2907–2910, 2001.

Lavoué, D., Liousse, C., Cachier, H., Stocks, B. J., and Goldammer, J. G.: Modeling of carbonaceous particles emitted by boreal and temperate wildfires at northern latitudes, J. Geophys. Res., 105, 26871–26890, doi:10.1029/2000JD900180, 2000.

Lawrence, M. G., Rasch, P. J., von Kuhlmann, R., Williams, J., Fischer, H., de Reus, M., Lelieveld, J., Crutzen, P. J., Schultz, M., Stier, P., Huntrieser, H., Heland, J., Stohl, A., Forster, C., Elbern, H., Jakobs, H., and Dickerson, R. R.: Global chemical weather forecasts for field campaign planning: predictions and observations of large-scale features during MINOS, CONTRACE, and INDOEX, Atmos. Chem. Phys., 3, 267–289, doi:10.5194/acp-3-267-2003, 2003.

Li, Q. B., Jacob, D. J., Park, R. J., Wang, Y. X., Heald, C. L., Hudman, R., Yantosca, R. M., Martin, R. V., and Evans, M. J.: North American pollution outflow and the trapping of convectively lifted pollution by upper-level anticyclone, J. Geophys. Res., 110, D10301, doi:10.1029/2004JD005039, 2005.

Li, Z., Chen, H., Cribb, M., Dickerson, R., Holben, B., Li, C., Lu, D., Luo, Y., Maring, H., Shi, G., Tsay, S.-C., Wang, P., Wang, Y., Xia, X., Zheng, Y., Yuan, T., and Zhao, F.: Preface to special section on East Asian Studies of Tropospheric Aerosols: an International Regional Experiment (EAST-AIRE), J. Geophys. Res., 112, D22S00, doi:10.1029/2007JD008853, 2007.

Liang, Q., Jaegle, L., Jaffe, D. A., Weiss-Penzias, P., Heckman, A., and Snow, J. A.: Long-range transport of Asian pollution to the northeast Pacific: seasonal variations and transport pathways of carbon monoxide, J. Geophys. Res., 109, D23S07, doi:10.1029/2003JD004402, 2004.

Lin, C.-Y., Hsu, H.-M., Lee, Y. H., Kuo, C. H., Sheng, Y.-F., and Chu, D. A.: A new transport mechanism of biomass burning from Indochina as identified by modeling studies, Atmos. Chem. Phys., 9, 7901–7911, doi:10.5194/acp-9-7901-2009, 2009.

Liu, C., Beirle, S., Butler, T., Liu, J., Hoor, P., Jöckel, P., Penning de Vries, M., Pozzer, A., Frankenberg, C., Lawrence, M. G., Lelieveld, J., Platt, U., and Wagner, T.: Application of SCIAMACHY and MOPITT CO total column measurements to evaluate model results over biomass burning regions and Eastern China, Atmos. Chem. Phys., 11, 6083–6114, doi:10.5194/acp-11-6083-2011, 2011.

Liu, H. Y., Jacob, D. J., Bey, I., Yantosca, R. M., Duncan, B. N., and Sachse, G. W.: Transport pathways for Asian combustion outflow over the Pacific: interannual and seasonal variations, J. Geophys. Res., 108, 8786, doi:10.1029/2002JD003102, 2003.

Liu, J., Drummond, J. R., Li, Q., Gille, J. C., and Ziskin, D. C.: Satellite mapping of CO emission from forest fires in northwest America using MOPITT measurements, Remote Sens. Environ., 95, 502–516, 2005.

Liu, J., Drummond, J. R., Jones, D. B. A., Cao, Z., Bremer, H., Kar, J., Zou, J., Nichitiu, F., and Gille, J. C.: Large horizontal gradients in atmospheric CO at the synoptic scale as seen by spaceborne measurements of pollution in the troposphere, J. Geophys. Res., 111, D02306, doi:10.1029/2005JD006076, 2006.

Mari, C., Evans, M. J., Palmer, P. I., Jacob, D. J., and Sachse, G. W.: Export of Asian pollution during two cold front episodes of the TRACE-P experiment, J. Geophys. Res., 109, D15S17, doi:10.1029/2003JD004307, 2004.

Miyazaki, Y., Kondo, Y., Koike, M., Fuelberg, H. E., Kiley, C. M., Kita, K., Takegawa, N., Sachse, G. W., Flocke, F., Weinheimer, A. J., Singh, H. B., Eisele, F. L., Zondlo, M., Talbot, R. W., Sandholm, S. T., Avery, M. A., and Blake, D. R.: Synoptic-scale transport of reactive nitrogen over the western Pacific in spring, J. Geophys. Res., 108, 8788, doi:10.1029/2002JD003248, 2003.

Marenco, A., Thouret, V., Nédélec, P., Smit, H., Helten, M., Kley, D., Karcher, F., Simon, P., Law, K., Pyle, J., Poschmann, G., Wrede, R. V., Hume, C., and Cook, T.: Measurement of ozone and water vapor by Airbus in-service aircraft: the MOZAIC airborne program, An overview, J. Geophys. Res., 103, 25631–25642, 1998.

Nassar, R., Logan, J. A., Megretskaia, I. A., Murray, L. T., Zhang, L., and Jones, D. B. A.: Analysis of tropical tropospheric ozone, carbon monoxide, and water vapor during the 2006 El Niño using TES observations and the GEOS-Chem model, J. Geophys. Res., 114, D17304, doi:10.1029/2009JD011760, 2009.

Nassar, R., Jones, D. B. A., Suntharalingam, P., Chen, J. M., Andres, R. J., Wecht, K. J., Yantosca, R. M., Kulawik, S. S., Bowman, K. W., Worden, J. R., Machida, T., and Matsueda, H.: Modeling global atmospheric CO_2 with improved emission inventories and CO_2 production from the oxidation of other carbon species, Geosci. Model Dev., 3, 689–716, doi:10.5194/gmd-3-689-2010, 2010.

Nédélec, P., Thpuret, V., Brioude, J., Sauvage, B., Cammas, J., Stohl, A.: Extreme CO concentrations in the upper troposphere over northeast Asia in June 2003 from the in situ MOZAIC aircraft data, Geophys. Res. Lett., 32, L14807, doi:10.1029/2005GL023141, 2005.

Novelli, P., Masarie, K. A., and Lang, P. M.: Distributions and recent changes of carbon monoxide in the lower troposphere, J. Geophys. Res., 103, 19015–19033, 1998.

Olivier, J. G. J. and Berdowski, J. J. M.: Global emission sources and sinks, in: The Climate System, edited by: Berdowski, J., Guicherit, R., and Heij, B. J., Swets & Zeitlinger, Lisse, the Netherlands, 33–77, 2001.

Oltmans, S. J., Lefohn, A. S., Harris, J. M., Tarasick, D. W., Thompson, A. M., Wernli, H., Johnson, B. J., Novelli, P. C., Montzka, S. A., Ray, J. D., Patrick, L. C., Sweeney, C., Jefferson, A., Dann, T., Davies, J., Shapiro, M., and Holben, B. N.: Enhanced ozone over western North America from biomass burning in Eurasia during April 2008 as seen in surface and profile observations, Atmos. Environ., 44, 4497–4509, 2010.

Pan, L., Gille, J. C., Edwards, D. P., Bailey, P. L., and Rodgers, C. D.: Retrieval of tropospheric carbon monoxide for the MOPITT experiment, J. Geophys. Res., 103, 32277–32290, 1998.

Pickering, K. E., Dickerson, R. R., Huffman, G. J., Boatman, J. F., and Schanot, A.: Trace gas transport in the vicinity of frontal convective clouds, J. Geophys. Res., 93, 759–773, doi:10.1029/JD093iD01p00759, 1998.

Randel, W. J., Park, M., Emmons, L., Kinnison, D., Bernath, P., Walker, K. A., Boone, C., and Pumphrey, H.: Asian monsoon transport of pollution to the stratosphere, Science, 328, 611–613, doi:10.1126/science.1182274, 2010.

Rogers, C. D.: Inverse Methods for Atmospheric Sounding, Theory and Practice, World Sci., 234 pp., River Edge, NJ, 2000.

Schultz, M. G.: On the use of ATSR fire count data to estimate the seasonal and interannual variability of vegetation fire emissions, Atmos. Chem. Phys., 2, 387–395, doi:10.5194/acp-2-387-2002, 2002.

Stohl, A.: A 1-year Lagrangian "climatology" of airstreams in the North Hemisphere troposphere and lowermost stratosphere, J. Geophys. Res., 106, 7263–7279, 2001.

Stohl, A., Hittenberger, M., and Wotawa, G.: Validation of the Lagrangian particle dispersion model FLEXPART against large scale tracer experiment data, Atmos. Environ., 24, 4245–4264, 1998.

Stohl, A., Eckhardt, S., Forster, C., James, P., and Spichtinger, N.: On the pathways and timescales of intercontinental air pollution transport, J. Geophys. Res., 107, 4684, doi:10.1029/2001JD001396, 2002.

Stohl, A., Forster, C., Frank, A., Seibert, P., and Wotawa, G.: Technical note: The Lagrangian particle dispersion model FLEXPART version 6.2, Atmos. Chem. Phys., 5, 2461–2474, doi:10.5194/acp-5-2461-2005, 2005.

Streets, D. G., Zhang, Q., Wang, L., He, K., Hao, J., Wu, Y., Tang, Y., and Carmichael, G. R.: Revisiting China's CO emissions after the Transport and Chemical Evolution over the Pacific (TRACE-P) mission: synthesis of inventories, atmospheric modeling, and observations, J. Geophys. Res., 111, D14306, doi:10.1029/2006JD007118, 2006.

Su, M., Lin, Y., Fan, X., Peng, L., Zhao, C.: Impacts of global emissions of CO, NO_x, and CH_4 on China tropospheric hydroxyl free radicals, Adv. Atmos. Sci., 29, 838–854, 2012.

Suntharalingam, P., Jacob, D. J., Palmer, P. I., Logan, J. A., Yantosca, R. M., Xiao, Y., Evans, M. J., Streets, D., Vay, S. A., and Sachse, G.: Improved quantification of Chinese carbon fluxes using CO_2/CO correlations in Asian outflow, J. Geophys. Res., 109, D18S18, doi:10.1029/2003JD004362, 2004.

Tao, S. and Ding, Y.: Observational evidence of the influence of the Qinghai-Xizang (Tibet) Plateau on the occurrence of heavy rain and severe convective storms in China, B. Am. Meteorol. Soc., 62, 2–30, 1981.

Tanimoto, H., Sawa, Y., Yonemura, S., Yumimoto, K., Matsueda, H., Uno, I., Hayasaka, T., Mukai, H., Tohjima, Y., Tsuboi, K., and Zhang, L.: Diagnosing recent CO emissions and ozone evolution in East Asia using coordinated surface observations, adjoint inverse modeling, and MOPITT satellite data, Atmos. Chem. Phys., 8, 3867–3880, doi:10.5194/acp-8-3867-2008, 2008.

Tsutsumi, Y., Makino, Y., and Jensen, J. B.: Vertical and latitudinal distributions of tropospheric ozone over the western Pacific: case studies from the PACE aircraft missions, J. Geophys. Res., 108, 4251, doi:10.1029/2001JD001374, 2003.

van der Werf, G. R., Randerson, J. T., Giglio, L., Collatz, G. J., Mu, M., Kasibhatla, P. S., Morton, D. C., DeFries, R. S., Jin, Y., and van Leeuwen, T. T.: Global fire emissions and the contribution of deforestation, savanna, forest, agricultural, and peat fires (1997–2009), Atmos. Chem. Phys., 10, 11707–11735, doi:10.5194/acp-10-11707-2010, 2010.

Wang, T., Nie, W., Gao, J., Xue, L. K., Gao, X. M., Wang, X. F., Qiu, J., Poon, C. N., Meinardi, S., Blake, D., Wang, S. L., Ding, A. J., Chai, F. H., Zhang, Q. Z., and Wang, W. X.: Air quality during the 2008 Beijing Olympics: secondary pollutants and regional impact, Atmos. Chem. Phys., 10, 7603–7615, doi:10.5194/acp-10-7603-2010, 2010.

Wang, X. L., Feng, Y., Compo, G. P., Swail, V. R., Zwiers, F. W., Allan, R. J., and Sardeshmukh, P. D.: Trends and low frequency variability of extra-tropical cyclone activity in the ensemble of twentieth century reanalysis, Clim. Dynam., 40, 2775–2800, doi:10.1007/s00382-012-1450-9, 2013.

Wofsy, S. C. and the HIPPO Science Team and Cooperating Modellers and Satellite Teams: HIAPER pole-to-pole observations (HIPPO): Fine-grained, global-scale measurements of climatically important atmospheric gases and aerosols, Phil. Trans. R. Soc. A, 369, 2073–2086, doi:10.1098/rsta.2010.0313, 2011.

Worden, H. M., Deeter, M. N., Edwards, D. P., Gille, J. C., Drummond, J. R., and Nédélec, P.: Observations of near-surface carbon monoxide from space using MOPITT multispectral retrievals, J. Geophys. Res., 115, D18314, doi:10.1029/2010JD014242, 2010.

Wotawa, G., Novelli, P. C., Trainer, M., and Granier, C.: Interannual variability of summertime CO concentrations in the Northern Hemisphere explained by boreal forest fires in North America and Russia, Geophys. Res. Lett., 28, 4575–4578, 2001.

Yienger, J. J., Galanter, M., Holloway, T. A., Phadnis, M. J., Guttikunda, S. K., Carmichael, G. R., Moxim, W. J., and Levy II, H.: The episodic nature of air pollution transport from Asia to North America, J. Geophys. Res., 105, 26931–26945, doi:10.1029/2000JD900309, 2000.

Yu, R., Xu, Y., Zhou, T., and Li, J.: Relation between rainfall duration and diurnal variation in the warm season precipitation over central eastern China, Geophys. Res. Lett., 34, Li3703, doi:10.1029/2007GL030315, 2007.

Yue, X. and Wang, H.: The springtime North Asia cyclone activity index and the Southern Annular Mode, Adv. Atmos. Sci., 25, 673–679, 2008.

Yurganov, L. N., McMillan, W. W., Dzhola, A. V., Grechko, E. I., Jones, N. B., and van derWerf, G.: Global AIRS and MOPITT CO measurements: validation, comparison, and links to biomass burning variations and carbon cycle, J. Geophys. Res., 113, D09301, doi:10.1029/2007JD009229, 2008.

Zhang, L., Jacob, D. J., Bowman, K. W., Logan, J. A., Turquety, S., Hudman, R. C., Li, Q. B., Beer, R., Worden, H. M., Worden, J. R., Rinsland, C. P., Kulawik, S. S., Lampel, M. C., Shephard, M. W., Fisher, B. M., Eldering, A., and Avery, M. A.: Ozone-CO correlations determined by the TES satellite instrument in continental outflow regions, Geophys. Res. Lett., 33, L18804, doi:10.1029/2006GL026399, 2006.

Zhang, Q., Streets, D. G., Carmichael, G. R., He, K. B., Huo, H., Kannari, A., Klimont, Z., Park, I. S., Reddy, S., Fu, J. S., Chen, D., Duan, L., Lei, Y., Wang, L. T., and Yao, Z. L.: Asian emissions in 2006 for the NASA INTEX-B mission, Atmos. Chem. Phys., 9, 5131–5153, doi:10.5194/acp-9-5131-2009, 2009.

Zhao, C., Wang, W., Yang, Y., Fu, R., Cunnold, D., and Choi, Y.: Impact of East Asian summer monsoon on the air quality over China: view from space, J. Geophys. Res., 115, D09301, doi:10.1029/2009JD012745, 2010.

Zhao, T. L., Gong, S. L., Zhang, X. Y., and Jaffe, D. A.: Asian dust storm influence on North American ambient PM levels: observational evidence and controlling factors, Atmos. Chem. Phys., 8, 2717–2728, doi:10.5194/acp-8-2717-2008, 2008.

Zhou, D., Ding, A., Mao, H., Fu, C., Wang, T., Chan, L. Y., Ding, K., Zhang, Y., Liu, J., Lu, A., and Hao, N.: Impacts of the East Asian monsoon on lower tropospheric ozone over coastal South China, Environ. Res. Lett., 8, 044011, doi:10.1088/1748-9326/8/4/044011, 2013.

Chemical and stable carbon isotopic composition of PM$_{2.5}$ from on-road vehicle emissions in the PRD region and implications for vehicle emission control policy

S. Dai[1,2], X. Bi[1], L. Y. Chan[1], J. He[1,3], B. Wang[3], X. Wang[1], P. Peng[1], G. Sheng[1], and J. Fu[1]

[1]State Key Laboratory of Organic Geochemistry and Guangdong Key Laboratory of Environmental Resources Utilization and Protection, Guangzhou Institute of Geochemistry, Chinese Academy of Sciences, Guangzhou 510640, China
[2]University of Chinese Academy of Sciences, Beijing 100049, China
[3]Institute of Atmospheric Environmental Safety and Pollution Control, Jinan University, Guangzhou 510632, China

Correspondence to: X. Bi (bixh@gig.ac.cn)

Abstract. Vehicle emissions are a major source of urban air pollution. In recent decade, the Chinese government has introduced a range of policies to reduce vehicle emissions. In order to understand the chemical characteristics of PM$_{2.5}$ from on-road vehicle emissions in the Pearl River Delta (PRD) region and to evaluate the effectiveness of control policies on vehicle emissions, the emission factors of PM$_{2.5}$ mass, elemental carbon (EC), organic carbon (OC), water-soluble organic carbon (WSOC), water-soluble inorganic ions (WSII), metal elements, organic compounds and stable carbon isotopic composition were measured in the Zhujiang tunnel of Guangzhou, in the PRD region of China in 2013. Emission factors of PM$_{2.5}$ mass, OC, EC and WSOC were 92.4, 16.7, 16.4 and 1.31 mg vehicle^{-1} km^{-1} respectively. Emission factors of WSII were 0.016 (F^{-}) $\sim$ 4.17 (Cl^{-}) mg vehicle^{-1} km^{-1}, contributing about 9.8 % to the PM$_{2.5}$ emissions. The sum of 27 measured metal elements accounted for 15.2 % of PM$_{2.5}$ emissions. Fe was the most abundant metal element, with an emission factor of 3.91 mg vehicle^{-1} km^{-1}. Emission factors of organic compounds including n-alkanes, polycyclic aromatic hydrocarbons, hopanes and steranes were 91.9, 5.02, 32.0 and 7.59 µg vehicle^{-1} km^{-1}, respectively. Stable carbon isotopic composition δ^{13}C value was -25.0‰ on average. An isotopic fractionation of 3.2‰ was found during fuel combustion. Compared to a previous study in Zhujiang tunnel in 2004, emission factors of PM$_{2.5}$ mass, EC, OC, WSII except Cl^{-} and organic compounds decreased by 16.0 $\sim$ 93.4 %, which could be attributed to emission control policy from 2004 to 2013. However, emission factors of most of the metal elements increased significantly, which could be partially attributed to the changes in motor oil additives and vehicle conditions. There are no mandatory national standards to limit metal content from vehicle emissions, which should be a concern of the government. A snapshot of the 2013 characteristic emissions of PM$_{2.5}$ and its constituents from the on-road vehicular fleet in the PRD region retrieved from our study would be helpful for the assessment of past and future implementations of vehicle emission control policy.

1 Introduction

Vehicle emissions are a major source of urban air pollution and account for approximately 14 $\sim$ 50 % of total fine particle mass in urban areas (Sheesley et al., 2007; Wang et al., 2008; Yu et al., 2013). The environmental and health effects of vehicle emissions have been our concern during the last decades. Numerous studies have been conducted to characterize vehicular particulate matter (PM) emissions in many countries with respect to emission factors, chemical composition and size distribution (Chiang and Huang, 2009; Laschober et al., 2004; Pio et al., 2013). The characteristics of vehicle emissions in China were studied by tunnel experiments, dynamometer tests and road monitoring (He et al., 2008; Jin et al., 2014; Song et al., 2012). Because of the differences in fuel qualities, engine conditions and operation practices, the PM emissions from vehicles varied over regions and time.

The Pearl River Delta (PRD) region, located on the southern coast of China, has experienced serious atmospheric pollution with its rapid urbanization and industrialization in the last few decades. Vehicle emissions account for approximately $25 \sim 30\%$ of total fine PM in the PRD region (http://epaper.southcn.com/nfdaily/html/2014-01/03/content_7261687.htm). Peer-reviewed papers have reported emission factors and chemical characteristic of $PM_{2.5}$ from vehicle emissions in the PRD region by means of tunnel studies in the Zhujiang (Guangzhou) and Wutong (Shenzhen) tunnels (He et al., 2006, 2008; X. F. Huang et al., 2006). However, the sampling in these studies was conducted in 2004. The Environmental Protection Agency of Guangdong revised the motor-vehicle-exhaust-pollution prevention and control regulations of Guangdong in 2008 and released the "PRD Regional Air Quality Management Plan" and "A Clean Air Plan" in 2010 to improve the relevant air quality through policies and measures. The emission standards for newly registered vehicles were tightened to China IV standards and better-quality gasoline and diesel were supplied in 2013. Therefore, the characteristics of PM emissions from vehicles in the PRD region might have changed throughout these years.

Tunnel experiments and chassis dynamometer tests were widely used to measure various pollutants emitted from vehicles (He et al., 2006; Heeb et al., 2003). However, dynamometer test is limited because it cannot account for vehicle fleet composition and emission characteristics related to break and tire wear and resuspension of road dust (Thorpe and Harrison, 2008). Tunnel studies have been demonstrated to be a suitable setup to measure PM emissions from on-road mixed fleets (Chiang and Huang, 2009; Laschober et al., 2004; Pio et al., 2013).

This study was carried out in a roadway tunnel located in the PRD region. We report here the emission factors of $PM_{2.5}$ mass, organic carbon (OC), elemental carbon (EC), water-soluble inorganic ions (WSII), metal elements, water-soluble organic carbon (WSOC), organic compounds and stable carbon isotopic composition. WSOC has the potential to modify the hygroscopicity of particles, PM size and cloud condensation nuclei activities (Shulman et al., 1996); however, it is often ignored in previous studies owing to the hydrophobic nature of the organic aerosol from primary vehicle emissions. Stable carbon isotope (δ^{13}C) is very useful for tracing sources (Lopez-Veneroni, 2009; Widory, 2006), and it was also less reported for vehicular exhaust emissions (Ancelet et al., 2011; Widory, 2006). The objectives of this study are (1) to obtain comprehensive information on the chemical and stable carbon isotopic composition of $PM_{2.5}$ emissions from on-road vehicles in the PRD region; (2) to compare our results with the previous study conducted in the same tunnel in 2004; (3) to evaluate the effectiveness of the implementation

of vehicle emission control policies from 2004 to 2013 in the PRD region. Although the fleet composition in this tunnel was probably different from the vehicle composition in the PRD region, it does not affect the conclusions in this paper.

2 Experimental

2.1 Tunnel sampling

$PM_{2.5}$ samples were collected from 10 to 14 August 2013 from the roadway tunnel (Zhujiang tunnel) located in Guangzhou, China. It has two bores, each of which has three lanes with traffic in the same direction, as shown in Fig. 1. Two high-volume $PM_{2.5}$ samplers (GUV-15HBL1, Thermo, USA) were placed at a distance of 75 m from the entrance and 75 m from the exit. The vehicle speed in the Zhujiang tunnel was 18 to 45 km h^{-1}, with an average vehicle speed of 33.4 km h^{-1} during the sampling. The PM samples were collected at about 1.13 m^3 min^{-1} through the quartz fiber filters (QFFs, 20.3 cm $\times$ 25.4 cm, Whatman). Other devices such as diffusion denuders and foam plugs were not used due to the difficulties in applying these devices. Consequently, volatilization losses or adsorption artifacts may occur on the filter for semi-volatile organic compounds, especially for the low molecular weight compounds due to their high volatility (Kavouras et al., 1999). However, the calculation of emission factors was based on the concentration differences between the exit and entrance of the tunnel; thus, the potential losses or adsorption artifacts of semi-volatile organic compounds would be partly deducted. Field blank samples were also collected by loading filters into the samplers but without pulling air through. The ventilation system of the tunnel was turned off during the sampling period; thus, the dispersion of air pollutants in the tunnel was mainly due to the piston effect arising from the traffic flow. The sampled filters were wrapped with annealed aluminum foil and stored in a refrigerator at $-40\,°$C until analysis. The meteorological parameters were synchronously recorded. A video camera was placed at the exit to record the passing vehicles during the sampling periods. The videotapes were then used to determine the vehicle counts and to classify the vehicles into three categories, namely, diesel vehicles (DV) (heavy-duty trucks, light-duty trucks and large passenger cars), gasoline vehicles (GV) (small cars and motorcycles) and liquefied petroleum gas vehicles (LPGV) (buses and taxis). The average traffic density during sampling was 1797 per hour with DV, GV and LPGV proportions of $13.7 \pm 2.7\%$, $59.8 \pm 8.8\%$ and $26.5 \pm 7.9\%$, respectively. More details of the vehicle counts and meteorological conditions are summarized in Table S1 in the Supplement.

2.2 Chemical analysis

The $PM_{2.5}$ mass concentrations were determined gravimetrically by weighing the quartz filters before and after sam-

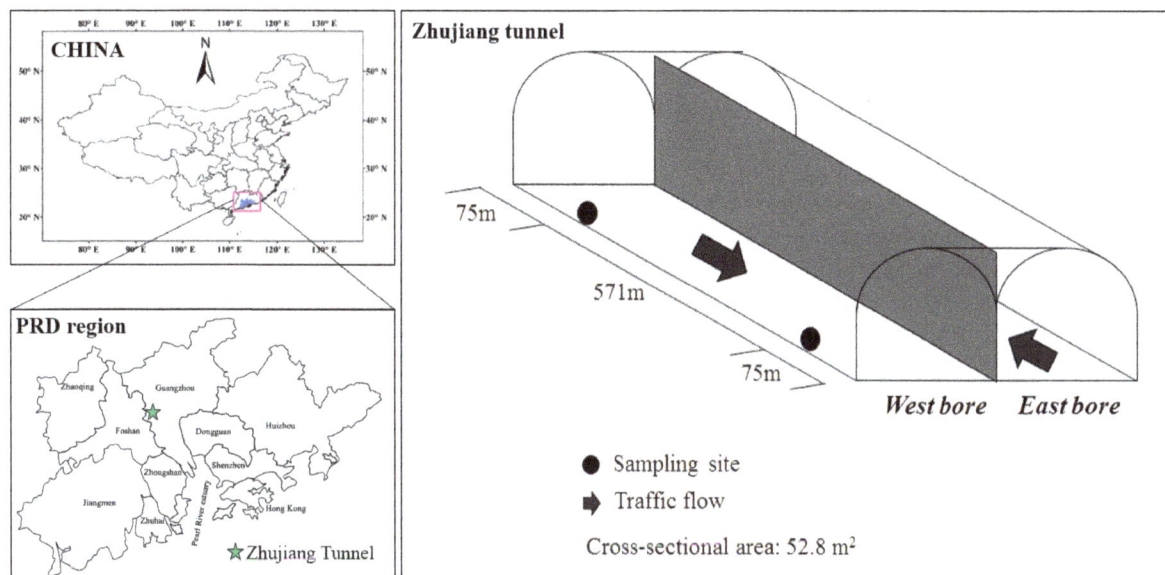

Figure 1. Sampling schematic diagram of the Zhujiang tunnel.

pling. The samples were conditioned in an electronic hygrothermostat for 24 h at 25 °C and 50 % relative humidity before weighing. Then, samples were analyzed for OC / EC, WSOC, WSII, metal elements, organic compounds and stable carbon isotope. The experimental methods of the chemical analysis are available in the Supplement.

2.3 Calculation of emission factor

Average emission factor was calculated for each sampling period on the basis of the concentration difference between the exit and entrance of the tunnel by the following equation (Handler et al., 2008):

$$EF = (C_{\text{out}} - C_{\text{in}})V/NL,$$

where EF is the emission factor of a species in unit of mg vehicle^{-1} km^{-1} or μg vehicle^{-1} km^{-1}, N is the number of vehicles passing through the tunnel, L is the distance between inlet and outlet sampling locations, C_{out} and C_{in} are the measured species concentration at the tunnel outlet and inlet, respectively, and V is the corresponding air volume calculated from the cross-sectional area of the tunnel, the average wind speed and the sampling duration of each filter. The average concentrations of all measured species at the inlet and outlet sampling locations and the corresponding emission factors in this study are presented in Tables S2–4.

3 Results and discussion

3.1 Characteristics of PM$_{2.5}$ emissions from vehicles in the PRD region

3.1.1 PM$_{2.5}$ mass, OC, EC, WSOC, WSII, metal elements

The PM$_{2.5}$ mass emission factors ranged from 79.8 to 107 mg vehicle^{-1} km^{-1}, with an average of 92.4 ± 8.9 mg vehicle^{-1} km^{-1}. Average OC and EC emission factors were 16.7 ± 1.9 and 16.4 ± 2.1 mg vehicle^{-1} km^{-1}, respectively, and they accounted for 18.1 ± 2.1 and 17.7 ± 2.2 % of PM$_{2.5}$ mass emissions. The ratio of OC to EC in the Zhujiang tunnel ranged from 0.77 to 1.35, with an average of 1.03. Previous studies have shown that the OC / EC ratio is useful to separate gasoline engine emissions from diesel emissions. Higher values (> 2) are associated with GV and LPGV exhaust, and lower values (0.3 to ∼ 0.9) are associated with DV exhaust (Cadle et al., 1999; Cheng et al., 2010; Gillies and Gertler, 2000). Therefore, the low OC / EC ratios in this study, which are closer to that from DV exhaust, indicate that diesel vehicles played an important role in the PM$_{2.5}$ emissions although the proportion of DV was only 13.7 % during the sampling. Additionally, it should be noted that emissions of EC from heavy-duty trucks are expected to be relatively low under the low-speed operating conditions in the tunnel (Kweon et al., 2002). Therefore, the ratio could be lower at the actual driving condition of vehicle fleet with a higher speed on the road. The concentration of WSOC at the inlet was 6.21 μg m^{-3} (Table S2) with a percentage of 31.1 % of OC, which is close to that of ambient air (Ding et al., 2008; Ho et al., 2006). At the outlet of the tunnel, the concentration of WSOC was $8.00 \mu g\,m^{-3}$, representing 17.9 % of OC. The WSOC had been reported to contribute on average 20 % to OC in the exit of Marseille roadway tunnel (El Haddad et al., 2009), in which background influence was included. The calculated emission factor of WSOC in this study ranged from 0.5 to $2.8\,mg\,vehicle^{-1}\,km^{-1}$ with an average of $1.31\,mg\,vehicle^{-1}\,km^{-1}$, which represents 7.84 % of that of OC. Such a WSOC fraction is considerably lower than that previously measured for biomass burning particles (71 %) (Mayol-Bracero et al., 2002). However, it could influence the hygroscopicity of particles and the formation of secondary aerosols (Ho et al., 2006; Rogge et al., 1993b; Weber et al., 2007) and is worthy of more attention and in-depth research.

The sum of WSII comprised about 9.8 % of the $PM_{2.5}$ emissions, with emission factors of 4.17, 0.104, 0.609, 2.88, 0.165, 0.177 and $0.953\,mg\,vehicle^{-1}\,km^{-1}$ for Cl^-, NO_3^-, SO_4^{2-}, Na^+, NH_4^+, Mg^{2+} and Ca^{2+}, respectively. The other WSII had a minor contribution ($<0.1\,mg\,vehicle^{-1}\,km^{-1}$). In total, 27 measured metal elements contributed 15.2 % to the $PM_{2.5}$ emissions. Fe was the most abundant element, with an emission factor of $3.91\,mg\,vehicle^{-1}\,km^{-1}$, followed by Na $3.53\,mg\,vehicle^{-1}\,km^{-1}$, Al $3.15\,mg\,vehicle^{-1}\,km^{-1}$, Ca $1.93\,mg\,vehicle^{-1}\,km^{-1}$, Mg $0.496\,mg\,vehicle^{-1}\,km^{-1}$ and K $0.338\,mg\,vehicle^{-1}\,km^{-1}$, which accounted for 4.2, 3.8, 3.4, 2.1, 0.5 and 0.4 % of $PM_{2.5}$ mass emissions, respectively. These six elements contributed 95.0 % to the total metal emissions. Emission factors of other metals ranged from 0.0001 (Ag) to 0.25 (Ba) $mg\,vehicle^{-1}\,km^{-1}$, with a sum of $0.71\,mg\,vehicle^{-1}\,km^{-1}$. It is worth noting that emission factors of elements Na, K, Mg and Ca were significantly higher than those of their corresponding water-soluble parts (Table S3). The differences can be attributed to the water-insoluble matter carrying these metal elements, such as calcium and magnesium carbonates and Na-, K- and Mg-bearing aluminosilicate species (Pio et al., 2013).

$PM_{2.5}$ mass was also obtained by summing OM, EC, geological component, sea salt and major water soluble inorganic ions (NH_4^+, SO_4^{2-}, NO_3^-). OC was multiplied by 1.4 to estimate mass of OM (He et al., 2008). The geological component of $35\,mg\,vehicle^{-1}\,km^{-1}$ was estimated based on the Al emission data as presented in Table S3. A typical road dust Al composition is 9 % on average (Tiittanen et al., 1999). Sea salt of $9\,mg\,vehicle^{-1}\,km^{-1}$ was estimated by Na, assuming sea salt contains 32 % of Na. Thus, the average $PM_{2.5}$ reconstructed mass was 91.8 % of the gravimetric value. This discrepancy can be attributed to the uncertainties in the weighing process, the estimation methods and uncalculated components.

3.1.2 Organic compounds

The average emission factors and abbreviated names of 67 individual organic compounds identified in the Zhujiang tunnel, including n-alkanes, polycyclic aromatic hydrocarbons (PAHs), hopanes and steranes are listed in Table S4. These organic compounds accounted for 0.59 % of the OM and 0.11 % of the $PM_{2.5}$ mass emissions. The distributions of organic molecular markers associated with $PM_{2.5}$ are known to be source indicative despite their small mass fractions (Schauer et al., 1996; Simoneit, 1986). n-Alkanes are an important class of organic compounds in atmospheric aerosols, and their homologue distribution may indicate different pollution sources (Rogge et al., 1993a). In this study, the n-alkane traces were dominated by C11–C36 with no odd–even carbon number predominance and the maximum was at C24, consistent with the characteristics of vehicle emissions reported by Simoneit (1984, 1985). The emission factors of individual n-alkanes were in the range of 0.22 (C13) ∼ 13.3 (C24) $\mu g\,vehicle^{-1}\,km^{-1}$ (Table S4).

There has been a worldwide concern to PAHs due to their known carcinogenic and mutagenic properties. PAHs are thought to be the result of incomplete combustion. In total, 15 priority PAHs (the results of naphthalene have not been discussed in this study due to its low recovery) were identified and quantified. The emission factor of total PAHs varied from 4.56 to $5.54 \mu g\,vehicle^{-1}\,km^{-1}$ in this study. The emission factor of benzo[a]pyrene (BaP), which is often used as an indicator of PAHs and regarded by the World Health Organization as a good index for whole PAH carcinogenicity, was in the range of 0.37 to $0.46 \mu g\,vehicle^{-1}\,km^{-1}$. The emission factors for other compounds ranged from 0.006 (acenaphthene) to 0.89 (pyrene) $\mu g\,vehicle^{-1}\,km^{-1}$ (Table S4). Pyrene (PYR) was the most abundant compound, followed by chrysene (CHR), benzo[ghi]perylene (BghiP) and benz[a]anthracene (BaA), which is different from biomass burning and coal combustion (Huang et al., 2014; Shen et al., 2012). PAHs diagnostic ratios have been used as a tool for identifying pollution emission sources including ANT / (ANT + PHE), FLA / (FLA + PYR), BaA / (BaA + CHR), BbF / (BbF + BkF), IcdP / (IcdP + BghiP) and BaP / (BaP + BghiP) (Tobiszewski and Namiesnik, 2012; Yunker et al., 2002; Zhang et al., 2005). We summarized PAH ratios mentioned above in Fig. 2 for three combustion sources including vehicle emissions, biomass burning and coal combustion. On the whole, the six ratios in this study are similar to the other tunnel experiments, though environmental conditions of tunnels are different to some extent. It is also suggested that the ratio of FLA / (FLA + PYR) and IcdP / (IcdP + BghiP) might be more suitable to distinguish vehicle emissions from biomass burning and coal combustion.

Hopanes and steranes are known molecular markers of aerosol emissions from fossil fuel utilization (Simoneit, 1985). Rogge et al. (1993a) and Schauer et al. (1996) showed that these petroleum biomarkers can be used to trace motor vehicle exhaust contributions to airborne PM in the southern Californian atmosphere. Fourteen major hopanes homologues with emission factors in the range

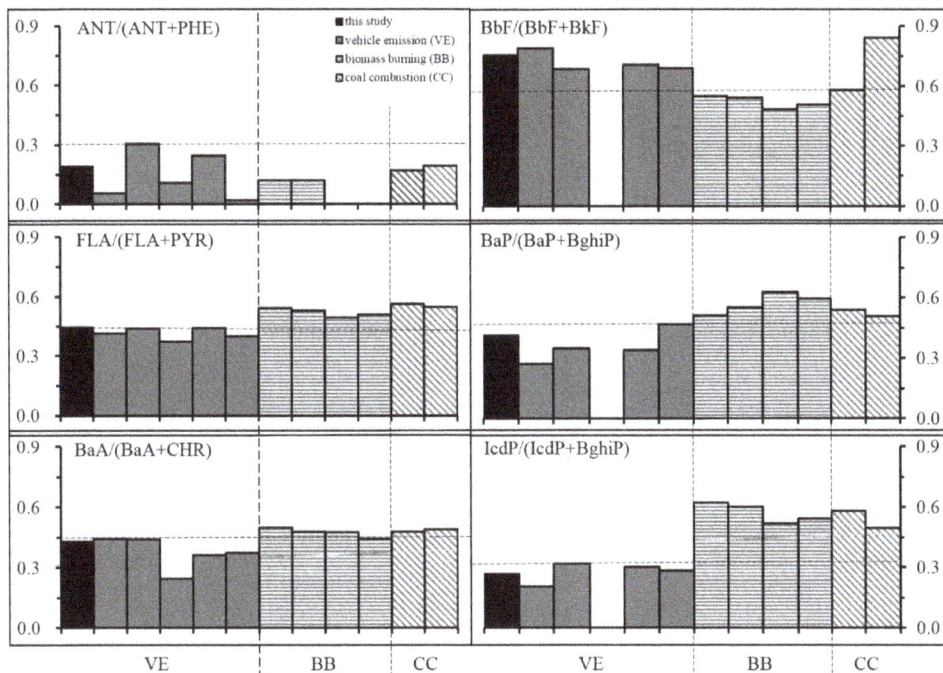

Figure 2. ANT / (ANT + PHE), FLA / (FLA + PYR), BaA / (BaA + CHR), BbF / (BbF + BkF), BaP / (BaP + BghiP) and IcdP / (IcdP + BghiP) ratios for three source emissions. ANT: anthracene, PHE: phenanthrene, FLA: fluoranthene, PYR: pyrene, BaA: benz[a]anthracene, CHR: chrysene, BbF: benzo[b]fluoranthene, BkF: benzo[k]fluoranthene, BaP: benzo[a]pyrene, BghiP: benzo[ghi]perylene, IcdP: indeno[cd]pyrene. The vehicle emission composition is from data collected in roadway tunnels (this study; He et al., 2006, 2008; Ancelet et al., 2011; Ho et al., 2009; Oda et al., 2001). The biomass burning profiles are obtained from nine straws (Shen et al., 2011), 26 firewood (Shen et al., 2012), three plant leaves and branches (Sheesley et al., 2003) and two biomass burning briquettes (Sheesley et al., 2003). The coal combustion profiles are obtained from the average value of PAH ratios from the combustion of five coals (Shen et al., 2011) and main coal-mining regions in China (Zhang et al., 2008).

of $0.46 \sim 9.14\,\mu g\,\text{vehicle}^{-1}\,\text{km}^{-1}$ and 12 steranes homologues in the range of $0.31 \sim 0.97\,\mu g\,\text{vehicle}^{-1}\,\text{km}^{-1}$ were identified in this study. $17\alpha(\text{H}),21\beta(\text{H})$-hopane (HP30) was the most abundant component with the emission factor of $9.14\,\mu g\,\text{vehicle}^{-1}\,\text{km}^{-1}$. The emission factor of total hopanes was $32.0\,\mu g\,\text{vehicle}^{-1}\,\text{km}^{-1}$. Emissions of the S hopanes for the extended $17\alpha(\text{H}),21\beta(\text{H})$-hopane homologues $>$C31 were always higher than those of the corresponding R pairs. All these characteristics of hopanes in the Zhujiang tunnel are consistent with those in gasoline and diesel exhausts (Rogge et al., 1993a; Simoneit, 1985) and in other tunnel studies (see Fig. S1 in the Supplement). Emission factors of individual sterane ranged from 0.31 to $0.97\,\text{ng}\,\text{vehicle}^{-1}\,\text{km}^{-1}$, and the sum of their emission factors was $7.58\,\mu g\,\text{vehicle}^{-1}\,\text{km}^{-1}$. The most abundant homologue was C29$\alpha\beta\beta$-stigmastane (20R) (29$\alpha\beta\beta$R), followed by 29$\alpha\alpha\alpha$S and 29$\alpha\beta\beta$S.

3.1.3 Stable carbon isotope

Stable carbon isotope analysis of vehicle emissions in Zhujiang tunnel yielded δ^{13}C values ranging from -25.5 to -24.7% with an average value of $-25.0 \pm 0.2\%$ and is comparable to previously reported ranges of -29 to -24.6% (Table 1) for vehicular fuel emissions. Generally, the variation in $\delta^{13}\text{C}_{\text{Fuel}}$ could affect the δ^{13}C of hydrocarbons (Keppler et al., 2004; Yamada et al., 2009). In the PRD region, the δ^{13}C values of gasoline and diesel were on average $-28.6 \pm 0.6\%$ and $-27.8 \pm 0.2\%$; small variations of fuel δ^{13}C were observed (Hu et al., 2014). We calculated the isotopic differences between $\delta^{13}\text{C}_{\text{PM2.5}}$ and $\delta^{13}\text{C}_{\text{Fuel}}$, which represent the apparent isotopic fractionation occurring during fuel burning. It expressed as Δ^{13}C ($\%$) and is defined by the following equation (Yamada et al., 2009).

$$\Delta^{13}\text{C}_{\text{PM2.5}-\text{Fuel}} = \left(\frac{\delta^{13}\text{C}_{\text{PM2.5}} + 1000}{\delta^{13}\text{C}_{\text{Fuel}} + 1000} - 1 \right) \times 1000$$

In this study, the value of $\Delta^{13}\text{C}_{\text{PM2.5}-\text{Fuel}}$ ranged from 2.7 to 3.5% with an average of 3.2%, indicating that an isotopic fractionation occurred during fuel combustion. Comparing the stable isotopic carbon value of vehicular fuel emissions with other particulate emission sources (see Table 1), different emission sources showed different stable carbon isotopic composition. For total carbon in PM2.5 samples, δ^{13}C ($\%$) of coal and fuel oil combustion are -23.9% and -26.0%, respectively, while that of vehicle emissions is $-25.9 \sim -25.0\%$. Obviously, the δ^{13}C ($\%$) of vehicle

Table 1. δ^{13}C values (‰) of PM from vehicle emissions in this study and other emission sources.

Emission sources and sampling site	Particle types	δ^{13}C values	Sampling time	Reference
Vehicular fuel emissions				
Vehicle emissions (Zhujiang tunnel, China)	$PM_{2.5}$ / TC	-25.0 ± 0.3	Aug 2013	This study
Vehicle emissions (tunnel of Rio de Janeiro, Brazil)	PM / OC	-25.4	Apr 1985	Tanner and Miguel (1989)
Vehicle emissions (tunnel of Rio de Janeiro, Brazil)	PM / EC	-24.8	Apr 1985	Tanner and Miguel (1989)
Complete combustion of diesel	PM / TC	-29	N/A	Widory (2006)
Complete combustion of gasoline	PM / TC	-27	N/A	Widory (2006)
Vehicle emissions (Cassier tunnel, Canada)	$PM_{2.5}$ / OC	-27.1	N/A	L. Huang et al. (2006)
Vehicle emissions (Cassier tunnel, Canada)	$PM_{2.5}$ / EC	-26.9	N/A	L. Huang et al. (2006)
Diesel vehicle emissions (central Camionera del Norte, Mexico)	$PM_{2.5}$ / TC	-24.6 ± 0.3	Mar 2002	Lopez-Veneroni (2009)
Gasoline vehicle emissions (tunnel of Avenida Chapultepec, Mexico)	$PM_{2.5}$ / TC	-25.5 ± 0.1	Mar 2002	Lopez-Veneroni (2009)
Vehicle emissions (Mount Victoria tunnel, New Zealand)	$PM_{2.5}$ / TC	-25.9 ± 0.8	Dec 2008 to Mar 2009	Ancelet et al. (2011)
Non-vehicular fuel sources				
Coal combustion (Paris, France)	$PM_{2.5}$ / TC	-23.9 ± 0.5	May to Sep 2002	Widory et al. (2004)
Coal combustion (Yurihonjo, Japan)	$PM_{2.5}$ / EC	-23.3	N/A	Kawashima and Haneishi (2012)
Charcoal combustion (Yurihonjo, Japan)	$PM_{2.5}$ / EC	-27.4 ± 1.7	N/A	Kawashima and Haneishi (2012)
Fireplace soot (Yurihonjo, Japan)	PM / EC	-26.5 ± 0.1	N/A	Kawashima and Haneishi (2012)
Fuel oil combustion (Paris, France)	$PM_{2.5}$ / TC	-26.0 ± 0.5	May to Sep 2002	Widory et al. (2004)
Dust particles				
Street dust (Mexico City, Mexico)	$PM_{2.5}$ / TC	-21 ± 0.2	Mar 2002	Lopez-Veneroni (2009)
Street dust (Yurihonjo, Japan)	$PM_{2.5}$ / EC	$-18.4 \sim -16.4$	Nov 2009	Kawashima and Haneishi (2012)
Biomass burning				
C4 plant	PM / TC	-13 ± 4	N/A	Boutton (1991)
C4 plant (Yurihonjo, Japan)	$PM_{2.5}$ / EC	$-19.3 \sim -16.1$	Apr to Nov 2009	Kawashima and Haneishi (2012)
C3 plant	PM / TC	-27 ± 6	N/A	Boutton (1991)
C3 plant (Yurihonjo, Japan)	$PM_{2.5}$ / EC	$-34.7 \sim -28.0$	Apr to Nov 2009	Kawashima and Haneishi (2012)

Table 2. Vehicle emission standards and limits for PM and NO_x implemented in Guangzhou after 2000.

Emission standard	Year[a]	Limit for PM		Limit for NO_x	
		$g\,km^{-1}$[b]	$g\,kWh^{-1}$[c]	$g\,km^{-1}$[b]	$g\,kWh^{-1}$[c]
China I	2001	$0.14 \sim 0.40$	$0.40 \sim 0.68$	–	$8.0 \sim 9.0$
China II	2004	$0.08 \sim 0.20$	0.15	–	7.0
China III	2007	$0.05 \sim 0.10$	$0.10 \sim 0.21$	$0.15 \sim 0.78$	5.0
China IV	2010	$0.025 \sim 0.060$	$0.02 \sim 0.03$	$0.08 \sim 0.39$	3.5

[a] Year of implementation; [b] for light-duty vehicles; [c] for compression ignition and gas-fueled positive ignition engines of vehicles.

emissions is not significantly different from that of coal and fuel oil combustion. However, they are obviously different from other sources, like dust particles ($-21 \sim -18.4$‰), C3 plants ($-19.3 \sim -13$‰) and C4 plants ($-34.7 \sim -27$‰). Therefore, δ^{13}C might be used to distinguish the fossil fuel combustion from other sources.

3.2 Comparison to previous studies conducted in the same tunnel

To investigate the variation of chemical emission characteristics from vehicles in the PRD region over the past decade, we compared the chemical emission characteristics of this study to that of previous study (He et al., 2008) for the same tunnel in 2004 (see Figs. 3 and 4). Figure 3 shows that $PM_{2.5}$ mass, OC and EC decreased significantly from 2004 to 2013. The

reason can be partly attributed to the implementation of pollution control measures for Chinese vehicle emissions. During this 9-year period, vehicle emission standards have raised two levels (from China II in 2004 to China IV in 2013) (Table 2). Additionally, comparing the fleet composition of 2013 to 2004 in Zhujiang tunnel, we found that the proportion of DV and GV decreased while LPGV increased. LPG is a type of clean energy, and LPGV is known to emit much less PM mass than GV and DV (Allen et al., 2001; Myung et al., 2014; Yang et al., 2007). LPG could be combusted more completely than gasoline and diesel. Changes mentioned above contributed greatly to the decrease of emission factors of OC and EC (31.3 and 66.9 %) and $PM_{2.5}$ mass (16.0 %) from 2004 to 2013. However, the emissions of $PM_{2.5}$ mass, OC and EC are still significantly higher than those measured in

Table 3. Vehicle fuel standards and limits for sulfur content ($mg\,kg^{-1}$) implemented in Guangzhou after 2000.

Standard	China I		China II		China III		China IV	
	limit	year[a]	limit	year	limit	year	limit	year
Gasoline	1000	2001	500	2005	150	2006	50	2010
Diesel	2000	2002	500	2003	350	2010	50	2013
LPG	–	–	270[b]	2003	–	–	50	2013

[a] Year of implementation; [b] unit: $mg\,m^{-3}$.

other countries (see Table S5). The implication of these high emission levels is that both the fuel quality and engine technologies in the PRD region need to be further improved.

It is also found from Fig. 3 that emission factors of NO_3^-, SO_4^{2-} and NH_4^+ decreased from 2004 to 2013. Improvement of fuel quality resulted in decreasing of sulfate emission factor from 3.87 to 0.61 $mg\,vehicle^{-1}\,km^{-1}$, since the amount of sulfur in fuel is slashed by $81.5 \sim 95\%$ in China IV (2013) compared to that in China II (2004) (Table 3). The emission levels of nitrate and ammonium were about one-tenth of those observed in 2004, possibly because NO_x emission standards tightened from 2004 to 2013 (Table 2), leading to lower production of ammonium nitrate. The emission factor of chloride is significantly higher than that obtained from Zhujiang tunnel in 2004 and other tunnels. Chloride was found up to 74 $mg\,vehicle^{-1}\,km^{-1}$ in PM_{10} in the Howell tunnel due to the application of salt to melt ice on roadways in the winter (Lough et al., 2005). However, it is not applicable in Guangzhou. The good correlation between Cl^- and Na^+ ($r^2 = 0.992$) indicates that resuspension of sea salt particles combined with vehicle emission PM might be a major source (He et al., 2008).

Emission factors of most of the metal elements increased in Zhujiang tunnel from 2004 to 2013 with the exception of Cd and Pb. Na emissions increased 3.16 $mg\,vehicle^{-1}\,km^{-1}$ in 2013 from 2004. Na correlated weakly with Cl^- ($r^2 = 0.374$) and Na^+ ($r^2 = 0.429$). This indicates that Na emissions had other sources and was not only from the resuspension of sea salt particles. The other four most abundant elements including Fe, Ca, Mg and K increased 1 to 3 times, probably because of resuspended road dust. However, the wind speed in 2013 was not found to be significantly higher than that in 2004 (3.8 $m\,s^{-1}$ in 2013 vs. 3.0 $m\,s^{-1}$ in 2004). This minor difference in wind speed could not account for the large increase. Furthermore, examination of the number of vehicles per hour in 2013 and 2004 suggests that there were fewer vehicles per hour in 2013. Therefore, a more plausible explanation is that there was a lot more dust on the road in 2013. Other sources would also cause the increased emissions of these elements, such as oil additives (Mg, Ca, Cu, Zn) (Cadle et al., 1997), the wear of engines (Fe) (Cadle et al., 1997; Garg et al., 2000) and brakes and tires (Al, Fe,Cu, Mn, Cd, Ni, Pb and Zn) (Garg et al., 2000; Pio et al., 2013).

Additionally, emissions of Zn, Cu, Mn, Cr, Ni, V, As, Co, U and Tl increased 0.5 to 4.5 times. Although the sum of these elements did not exceed 0.5 % of $PM_{2.5}$ mass, they are important for health effects. Lower emission factors of Pb ($0.01 \pm 0.0007\,mg\,vehicle^{-1}\,km^{-1}$) in 2013 than in 2004 could be a result of the phasing out of leaded gasoline across China in the late 1990s.

Figure 4 shows a comparison of organic compound emissions in Zhujiang tunnel between 2004 and 2013. The n-alkane homologues exhibited a smooth hump-like distribution with the most abundance at C24, as shown in Fig. 4a. Such a distribution pattern was similar to patterns observed in Zhujiang tunnel in 2004. However, there are some differences. Firstly, the highest abundant n-alkane shifted from C23 in 2004 to C24 in 2013. This difference might be explained by the shift of gas–particle partitioning as alkanes of < C26 are semi-volatile. However, the t test showed that the temperatures were not significantly different ($p = 0.14$) between this study ($33.0 \pm 2.3\,°C$) and that in 2004 ($31.8 \pm 1.0\,°C$). Thus, the differences of C_{max} cannot be regarded as a result of temperature differences. Furthermore, C_{max} was found to be C24 in every test of this study although the temperature ranged from 28.6 to 36.1 °C. It was reported that the n-alkane in the highest abundance was C20 for DV and C25 or C26 for GV in dynamometer tests (Rogge et al., 1993a; Schauer et al., 1999, 2002). As the emissions collected in tunnel studies present a composite result of emissions from a mixed vehicle fleet, the lower fraction of DV in 2013 was more likely the cause of the shift of C_{max}. Secondly, emission factors of C16–C26 in 2013 were significantly lower than those in 2004, while this trend reversed gradually after C27. Emission factors of the PAHs decreased by $67.6 \sim 93.4\%$. BaP equivalent emission factors decreased by 88.1 % from 2004 to 2013 (Table S6). This could be attributed to the variation of fleet composition between 2004 and 2013. PAHs emitted from LPGV are about one-third of that from GV (Yang et al., 2007), while DV emit more PAHs than GV (Phuleria et al., 2006). Therefore, the higher proportion of LPGV and lower proportion of DV resulted in the lower emission factor of PAHs in 2013 than that in 2004. Emission factors of hopanes also decreased from 2004 to 2013; the percentage of decrease ranged from 56.2 to 68.7 %. However, the distributions of hopane series derived from dif-

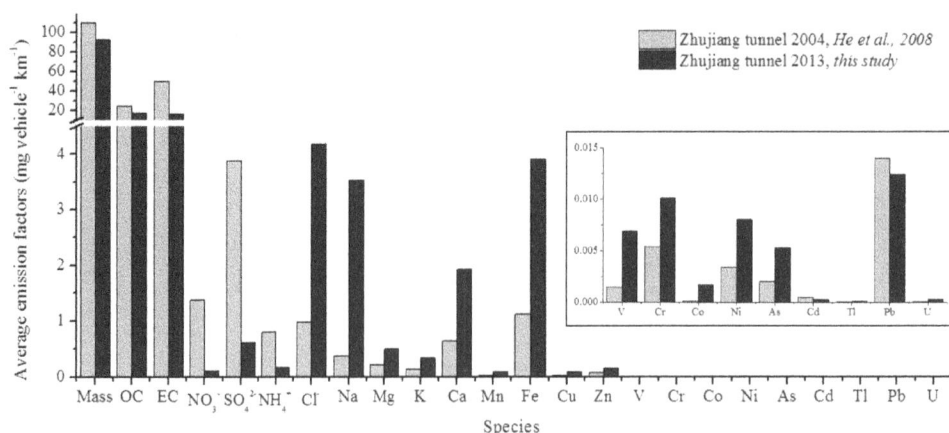

Figure 3. Comparison of PM$_{2.5}$, OC, EC, WSII and metal emissions in Zhujiang tunnel sampling in 2004 and 2013.

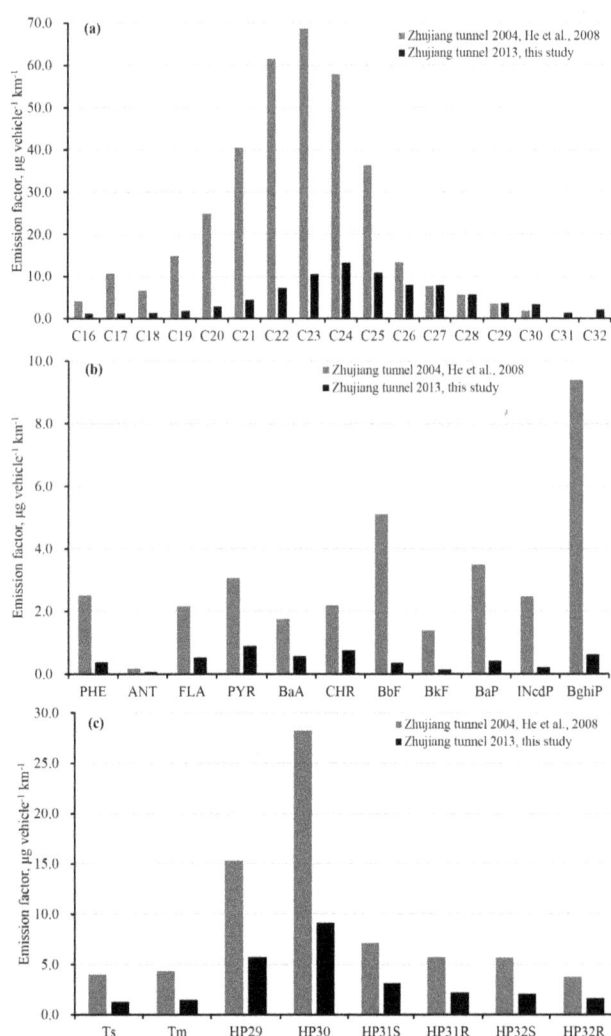

Figure 4. Comparison of organic compounds emissions in Zhujiang tunnel sampling in 2004 and 2013.

ferent tunnel studies were very similar (see Fig. S1). This suggests that the hopane emission characteristics might be independent of the fleet composition. This is a reasonable result given that hopanes originate from the lubricating oil used in DV, GV and LPGV rather than from the fuel (He et al., 2008; Phuleria et al., 2006). Owing to the fact that more units in heavy-duty vehicles need lubrication, emission factors of hopanes attributable to heavy-duty vehicles were higher than those attributable to light-duty vehicles (Phuleria et al., 2006). Reduction of the proportion of heavy-duty vehicles (buses, heavy-duty trucks, large passenger cars) in fleet composition in 2013 (11.3 %) compared to that in 2004 (20 %) might be the reason that emission factors of hopanes decreased.

3.3 Implications for vehicle emission control policy

Vehicle emission control strategies and policies adopted by Guangdong province can be classified as emission control on vehicles, fuel-quality improvements and alternative fuel utilization. PM emission standards for newly registered vehicles were tightened from China II in 2004 to China IV in 2013 (Table 2). The reduction of on-road high-PM-emitting vehicles, the phasing in of lower-PM-emitting vehicles and more environmentally friendly on-road vehicles with more advanced engines following the implementation of these emission standards were effective for decreasing PM emissions. Emission factors of PM decreased by 16 % from 2004 to 2013. Also for NO$_x$, the emission limit was reduced to about half from 2004 to 2013. This change in emission standards that limit NO$_x$ emissions is a major factor in the decrease of emission factors of nitrate and ammonium by about 90 %. Additionally, the national standards have been revised several times to improve fuel quality to adapt to stringent vehicle emission standards (Table 3). Sulfur content, for example, showed a sharp decrease by over 90 % from 2004 to 2013, resulting in the decrease of the emission factor of sulfate by 70 %. LPG and liquefied natural gas have grad-

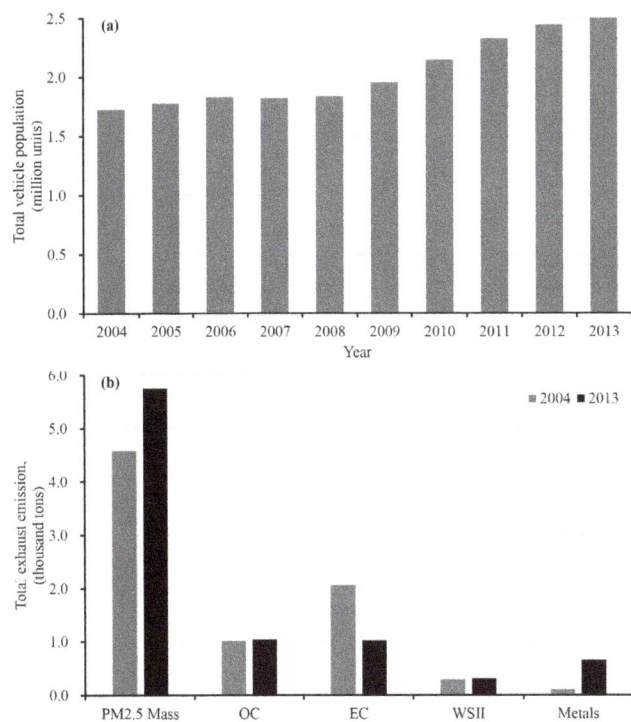

Figure 5. (a) Growth in total vehicle population in Guangzhou during 2004–2013. (b) Total exhaust emissions of $PM_{2.5}$ mass, OC, EC, WSII and metal in 2004 and 2013.

ually taken the place of diesel and gasoline as the fuel of taxis and buses after 2004; these vehicles now seldom use diesel and gasoline as fuel (http://www.southcn.com/news/gdnews/nanyuedadi/200707040173.htm). The application of clean fuel led to nearly complete combustion and resulted in much lower emissions from taxis and buses. In general, our results suggest that these strategies are effective to reduce emission factors of $PM_{2.5}$ mass, as well as OC, EC, WSII and organic compounds in $PM_{2.5}$. However, the total vehicle population increased year by year. As shown in Fig. 5a, the total vehicle population increased by 49.1 % from 2004 to 2013. Total emissions of vehicle exhaust of $PM_{2.5}$ mass (calculated as emission factors multiply by annual average driving distance per car and vehicle population; Wu et al., 2012) increased by 25.2 % from 2004 to 2013 (Fig. 5b). Consequently, it is demonstrated that more stringent emission standards are higher quality of fuel and more utilization of clean fuels are necessary to offset the impacts induced by the growth in vehicle population and to improve air quality in the PRD region. Additionally, owing to a lack of mandatory national standards limiting metal content in vehicle emissions, the emissions of the majority of metals increased from 2004 to 2013 (Figs. 3 and 5b). In China, heavy metals, including As, Cr, Cu, Ni and Tl, are listed as key substances to be preferentially monitored in the atmospheric environment (SEPA, 2003); thus, the increase of metal elements should raise the awareness of the government due to their health concern.

4 Conclusions

$PM_{2.5}$ samples were collected between 10 and 14 August 2013 in Zhujiang tunnel, Guangzhou, to acquire a comprehensive snapshot of the chemical characteristics of vehicle emissions. The average emission factors of $PM_{2.5}$ mass, EC, OC, WSOC, WSII, metal elements, organic compounds and stable carbon isotope were measured. Stable carbon isotope $\delta^{13}C$ values indicate that an isotopic fractionation of 3.2 ‰ occurred during fuel combustion. Compared to a previous study in Zhujiang tunnel in 2004, emission factors of $PM_{2.5}$ mass, EC, OC and major WSII decreased due to control-policy-induced changes from 2004 to 2013: change of fleet composition, implementation of more stringent gasoline and diesel emission standards, improvement in fuel quality and clean fuel used for taxis and buses. The shift in n-alkanes distribution and decreased PAHs emissions were due to the lower proportion of DV in 2013 than in 2004, and the decrease in emission factors of hopanes was due to the reduction of the proportion of heavy-duty vehicles. Our study shows that control polices for vehicles emissions by the government were effective to decrease the emission factors of $PM_{2.5}$, EC, OC and WSII from on-road vehicular fleets. However, the increase in emissions of most metal elements should raise the awareness of the government, since metal elements, especially heavy metals, could affect human health. Also, in order to offset the impacts of the growth of the vehicle population and to improve air quality in the PRD region, more stringent and aggressive emission control policies are necessary.

Acknowledgements. This work was supported by the "Strategic Priority Research Program (B)" of the Chinese Academy of Sciences (XDB05020205), the Foundation for Leading Talents from Guangdong province government and the National Nature Science Foundation of China (No. 41405131). J. Fu is grateful for the support of the "Team Project of the Guangdong Natural Science Foundation, China" (S2012030006604). This is contribution no. 2042 from GIGCAS.

Edited by: B. N. Duncan

References

Allen, J. O., Mayo, P. R., Hughes, L. S., Salmon, L. G., and Cass, G. R.: Emissions of size-segregated aerosols from on-road vehicles in the Caldecott Tunnel, Environ. Sci. Technol., 35, 4189–4197, 2001.

Ancelet, T., Davy, P. K., Trompetter, W. J., Markwitz, A., and Weatherburn, D. C.: Carbonaceous aerosols in an urban tunnel, Atmos. Environ., 45, 4463–4469, 2011.

Boutton, T. W.: Stable carbon isotope ratios of natural materials II, Atmospheric, terrestrial, marine, and freshwater environments, in: Carbon Isotope Techniques, edited by: Coleman D. C. and Fry B., Academic Press, San Diego, 173–186, 1991.

Cadle, S. H., Mulawa, P. A., Ball, J., Donase, C., Weibel, A., Sagebiel, J. C., Knapp, K. T., and Snow, R.: Particulate emission rates from in use high emitting vehicles recruited in Orange County, California, Environ. Sci. Technol., 31, 3405–3412, 1997.

Cadle, S. H., Mulawa, P. A., Hunsanger, E. C., Nelson, K., Ragazzi, R. A., Barrett, R., Gallagher, G. L., Lawson, D. R., Knapp, K. T., and Snow, R.: Composition of light-duty motor vehicle exhaust particulate matter in the Denver, Colorado area, Environ. Sci. Technol., 33, 2328–2339, 1999.

Cheng, Y., Lee, S. C., Ho, K. F., Chow, J. C., Watson, J. G., Louie, P. K. K., Cao, J. J., and Hai, X.: Chemically-speciated on-road $PM_{2.5}$ motor vehicle emission factors in Hong Kong, Sci. Total Environ., 408, 1621–1627, 2010.

Chiang, H. L. and Huang, Y. S.: Particulate matter emissions from on-road vehicles in a freeway tunnel study, Atmos. Environ., 43, 4014–4022, 2009.

Ding, X., Zheng, M., Yu, L. P., Zhang, X. L., Weber, R. J., Yan, B., Russell, A. G., Edgerton, E. S., and Wang, X. M.: Spatial and seasonal trends in biogenic secondary organic aerosol tracers and water-soluble organic carbon in the southeastern United States, Environ. Sci. Technol., 42, 5171–5176, 2008.

El Haddad, I., Marchand, N., Dron, J., Temime-Roussel, B., Quivet, E., Wortham, H., Jaffrezo, J. L., Baduel, C., Voisin, D., Besombes, J. L., and Gille, G.: Comprehensive primary particulate organic characterization of vehicular exhaust emissions in France, Atmos. Environ., 43, 6190–6198, 2009.

Garg, B. D., Cadle, S. H., Mulawa, P. A., Groblicki, P. J., Laroo, C., and Parr, G. A.: Brake wear particulate matter emissions, Environ. Sci. Technol., 34, 4463–4469, 2000.

Gillies, J. A., and Gertler, A. W.: Comparison and evaluation of chemically speciated mobile source $PM_{2.5}$ particulate matter profiles, J. Air Waste Manage., 50, 1459–1480, 2000.

Handler, M., Puls, C., Zbiral, J., Marr, I., Puxbaum, H., and Limbeck, A.: Size and composition of particulate emissions from motor vehicles in the Kaisermuhlen-Tunnel, Vienna, Atmos. Environ., 42, 2173–2186, 2008.

He, L. Y., Hu, M., Huang, X. F., Zhang, Y. H., Yu, B. D., and Liu, D. Q.: Chemical characterization of fine particles from on-road vehicles in the Wutong tunnel in Shenzhen, China, Chemosphere, 62, 1565–1573, 2006.

He, L. Y., Hu, M., Zhang, Y. H., Huang, X. F., and Yao, T. T.: Fine particle emissions from on-road vehicles in the Zhujiang Tunnel, China, Environ. Sci. Technol., 42, 4461–4466, 2008.

Heeb, N. V., Forss, A. M., Saxer, C. J., and Wilhelm, P.: Methane, benzene and alkyl benzene cold start emission data of gasoline-driven passenger cars representing the vehicle technology of the last two decades, Atmos. Environ., 37, 5185–5195, 2003.

Ho, K. F., Lee, S. C., Cao, J. J., Li, Y. S., Chow, J. C., Watson, J. G., and Fung, K.: Variability of organic and elemental carbon, water soluble organic carbon, and isotopes in Hong Kong, Atmos. Chem. Phys., 6, 4569–4576, doi:10.5194/acp-6-4569-2006, 2006.

Ho, K. F., Ho, S. S. H., Lee, S. C., Cheng, Y., Chow, J. C., Watson, J. G., Louie, P. K. K., and Tian, L.: Emissions of gas- and particle-phase polycyclic aromatic hydrocarbons (PAHs) in the Shing Mun Tunnel, Hong Kong, Atmos. Environ., 43, 6343–6351, 2009.

Hu, P., Wen, S., Liu, Y. L., Bi, X. H., Chan, L. Y., Feng, J. L., Wang, X. M., Sheng, G. Y., and Fu, J. M.: Carbon isotopic characterization of formaldehyde emitted by vehicles in Guangzhou, China, Atmos. Environ., 86, 148–154, 2014.

Huang, L., Brook, J. R., Zhang, W., Li, S. M., Graham, L., Ernst, D., Chivulescu, A., and Lu, G.: Stable isotope measurements of carbon fractions (OC / EC) in airborne particulate: A new dimension for source characterization and apportionment, Atmos. Environ., 40, 2690–2705, 2006.

Huang, W., Huang, B., Bi, X. H., Lin, Q. H., Liu, M., Ren, Z. F., Zhang, G. H., Wang, X. M., Sheng, G. Y., and Fu, J. M.: Emission of PAHs, NPAHs and OPAHs from residential honeycomb coal briquette combustion, Energ. Fuel., 28, 636–642, 2014.

Huang, X. F., Yu, J. Z., He, L. Y., and Hu, M.: Size distribution characteristics of elemental carbon emitted from Chinese vehicles: Results of a tunnel study and atmospheric implications, Environ. Sci. Technol., 40, 5355–5360, 2006.

Jin, T. S., Qu, L., Liu, S. X., Gao, J. J., Wang, J., Wang, F., Zhang, P. F., Bai, Z. P., and Xu, X. H.: Chemical characteristics of particulate matter emitted from a heavy duty diesel engine and correlation among inorganic and PAH components, Fuel, 116, 655–661, 2014.

Kavouras, I. G., Lawrence, J., Koutrakis, P., Stephanou, E. G., and Oyola, P.: Measurement of particulate aliphatic and polynuclear aromatic hydrocarbons in Santiago de Chile: source reconciliation and evaluation of sampling artifacts, Atmos. Environ., 33, 4977–4986, 1999.

Kawashima, H., and Haneishi, Y.: Effects of combustion emissions from the Eurasian continent in winter on seasonal delta C-13 of elemental carbon in aerosols in Japan, Atmos. Environ., 46, 568–579, 2012.

Keppler, F., Kalin, R. M., Harper, D. B., McRoberts, W. C., and Hamilton, J. T. G.: Carbon isotope anomaly in the major plant C_1 pool and its global biogeochemical implications, Biogeosciences, 1, 123-131, doi:10.5194/bg-1-123-2004, 2004.

Kweon, C. B., Foster, D. E., Schauer, J. J., and Okada, S.: Detailed chemical composition and particle size assessment of diesel engine exhaust, SAE Technical Paper Series, No. 2002-01-2670, doi:10.4271/2002-01-2670, 2002.

Laschober, C., Limbeck, A., Rendl, J., and Puxbaum, H.: Particulate emissions from on-road vehicles in the Kaisermühlen-tunnel (Vienna, Austria), Atmos. Environ., 38, 2187–2195, 2004.

Lopez-Veneroni, D.: The stable carbon isotope composition of $PM_{2.5}$ and PM_{10} in Mexico City Metropolitan Area air, Atmos. Environ., 43, 4491–4502, 2009.

Lough, G. C., Schauer, J. J., Park, J. S., Shafer, M. M., Deminter, J. T., and Weinstein, J. P.: Emissions of metals associated with motor vehicle roadways, Environ. Sci. Technol., 39, 826–836, 2005.

Mayol-Bracero, O. L., Guyon, P., Graham, B., Roberts, G., Andreae, M. O., Decesari, S., Facchini, M. C., Fuzzi, S., and Artaxo, P.: Water-soluble organic compounds in biomass burning aerosols over Amazonia – 2. Apportionment of the chemical composition and importance of the polyacidic fraction, J. Geophys. Res., 107, 8091, doi:10.1029/2001JD000522, 2002.

Myung, C. L., Ko, A., Lim, Y., Kim, S., Lee, J., Choi, K., and Park, S.: Mobile source air toxic emissions from direct injection spark ignition gasoline and LPG passenger car under various in-use vehicle driving modes in Korea, Fuel Process. Technol., 119, 19–31, 2014.

Oda, J., Nomura, S., Yasuhara, A., and Shibamoto, T.: Mobile sources of atmospheric polycyclic aromatic hydrocarbons in a roadway tunnel, Atmos. Environ., 35, 4819–4827, 2001.

Phuleria, H. C., Geller, M. D., Fine, P. M., and Sioutas, C.: Size-resolved emissions of organic tracers from light-and heavy-duty vehicles measured in a California roadway tunnel, Environ. Sci. Technol., 40, 4109–4118, 2006.

Pio, C., Mirante, F., Oliveira, C., Matos, M., Caseiro, A., Oliveira, C., Querol, X., Alves, C., Martins, N., Cerqueira, M., Camoes, F., Silva, H., and Plana, F.: Size-segregated chemical composition of aerosol emissions in an urban road tunnel in Portugal, Atmos. Environ., 71, 15–25, 2013.

Rogge, W. F., Hildemann, L. M., Mazurek, M. A., Cass, G. R., and Simoneit, B. R. T.: Sources of Fine Organic Aerosol .2. Noncatalyst and Catalyst-Equipped Automobiles and Heavy-Duty Diesel Trucks, Environ. Sci. Technol., 27, 636–651, 1993a.

Rogge, W. F., Mazurek, M. A., Hildemann, L. M., Cass, G. R., and Simoneit, B. R. T.: Quantification of Urban Organic Aerosols at a Molecular-Level – Identification, Abundance and Seasonal-Variation, Atmos. Environ. A-Gen., 27, 1309–1330, 1993b.

Schauer, J. J., Rogge, W. F., Hildemann, L. M., Mazurek, M. A., Cass, G. R., and Simoneit, B. R. T.: Source apportionment of airborne particulate matter using organic compounds as tracers, Atmos. Environ., 30, 3837–3855, 1996.

Schauer, J. J., Kleeman, M. J., Cass, G. R., and Simoneit, B. R. T.: Measurement of emissions from air pollution sources. 2. C-1 through C-30 organic compounds from medium duty diesel trucks, Environ. Sci. Technol., 33, 1578–1587, 1999.

Schauer, J. J., Kleeman, M. J., Cass, G. R., and Simoneit, B. R. T.: Measurement of emissions from air pollution sources. 5. C-1-C-32 organic compounds from gasoline-powered motor vehicles, Environ. Sci. Technol., 36, 1169–1180, 2002.

SEPA: State Environmental Protection Administration, Air And Waste Gas Monitor Analysis Method, China Environmental Science Press, Beijing, 4–12, 2003.

Sheesley, R. J., Schauer, J. J., Chowdhury, Z., Cass, G. R., and Simoneit, B. R. T.: Characterization of organic aerosols emitted from the combustion of biomass indigenous to South Asia, J. Geophys. Res., 108, 4285, doi:10.1029/2002JD002981, 2003.

Sheesley, R. J., Schauer, J. J., Zheng, M., and Wang, B.: Sensitivity of molecular marker-based CMB models to biomass burning source profiles, Atmos. Environ., 41, 9050–9063, 2007.

Shen, G. F., Tao, S., Wang, W., Yang, Y. F., Ding, J. N., Xue, M. A., Min, Y. J., Zhu, C., Shen, H. Z., Li, W., Wang, B., Wang, R., Wang, W. T., Wang, X. L., and Russell, A. G.: Emission of Oxygenated Polycyclic Aromatic Hydrocarbons from Indoor Solid Fuel Combustion, Environ. Sci. Technol., 45, 3459–3465, 2011.

Shen, G. F., Tao, S., Wei, S. Y., Zhang, Y. Y., Wang, R., Wang, B., Li, W., Shen, H. Z., Huang, Y., Chen, Y. C., Chen, H., Yang, Y. F., Wang, W., Wang, X. L., Liu, W. X., and Simonich, S. L. M.: Emissions of Parent, Nitro, and Oxygenated Polycyclic Aromatic Hydrocarbons from Residential Wood Combustion in Rural China, Environ. Sci. Technol., 46, 8123–8130, 2012.

Shulman, M. L., Jacobson, M. C., Carlson, R. J., Synovec, R. E., and Young, T. E.: Dissolution behavior and surface tension effects of organic compounds in nucleating cloud droplets, Geophys. Res. Lett., 23, 277–280, 1996.

Simoneit, B. R. T.: Organic-Matter of the Troposphere .3. Characterization and Sources of Petroleum and Pyrogenic Residues in Aerosols over the Western United-States, Atmos. Environ., 18, 51–67, 1984.

Simoneit, B. R. T.: Application of Molecular Marker Analysis to Vehicular Exhaust for Source Reconciliations, Int. J. Environ. Anal. Chem., 22, 203–233, 1985.

Simoneit, B. R. T.: Characterization of Organic-Constituents in Aerosols in Relation to Their Origin and Transport - a Review, Int. J. Environ. Anal. Chem., 23, 207–237, 1986.

Song, S. J., Wu, Y., Jiang, J. K., Yang, L., Cheng, Y., and Hao, J. M.: Chemical characteristics of size-resolved PM$_{2.5}$ at a roadside environment in Beijing, China, Environ. Pollut., 161, 215–221, 2012.

Tanner, R. L. and Miguel, A. H.: Carbonaceous Aerosol Sources in Rio De Janeiro, Aerosol Sci. Technol., 10, 213–223, 1989.

Thorpe, A. and Harrison, R. M.: Sources and properties of non-exhaust particulate matter from road traffic: A review, Sci. Total Environ., 400, 270–282, 2008.

Tiittanen, P., Timonen, K. L., Ruuskanen, J., Mirme, A., and Pekkanen, J.: Fine particulate air pollution, resuspended road dust and respiratory health among symptomatic children, Eur Respir J, 13, 266–273, 1999.

Tobiszewski, M. and Namiesnik, J.: PAH diagnostic ratios for the identification of pollution emission sources, Environ. Pollut., 162, 110–119, 2012.

Wang, H. L., Zhuang, Y. H., Wang, Y., Sun, Y., Yuan, H., Zhuang, G. S., and Hao, Z. P.: Long-term monitoring and source apportionment of PM$_{2.5}$/PM$_{10}$ in Beijing, China, J. Environ. Sci-China, 20, 1323–1327, 2008.

Weber, R. J., Sullivan, A. P., Peltier, R. E., Russell, A., Yan, B., Zheng, M., de Gouw, J., Warneke, C., Brock, C., Holloway, J. S., Atlas, E. L., and Edgerton, E.: A study of secondary organic aerosol formation in the anthropogenic-influenced southeastern United States, J. Geophys. Res., 112, D13302, doi:10.1029/2007JD008408, 2007.

Widory, D.: Combustibles, fuels and their combustion products: A view through carbon isotopes, Combust. Theor. Model., 10, 831–841, 2006.

Widory, D., Roy, S., Le Moullec, Y., Goupil, G., Cocherie, A., and Guerrot, C.: The origin of atmospheric particles in Paris: a view through carbon and lead isotopes, Atmos. Environ., 38, 953–961, 2004.

Wu, Y. Y., Zhao, P., Zhang, H. W., Wang, Y., and Mao, G. Z.: Assessment for Fuel Consumption and Exhaust Emissions of China's Vehicles: Future Trends and Policy Implications, Sci. World J., 591343, 1–8, 2012.

Yamada, K., Hattori, R., Ito, Y., Shibata, H., and Yoshida, N.: Carbon isotopic signatures of methanol and acetaldehyde emitted from biomass burning source, Geophys. Res. Lett., 36, L18807, doi:10.1029/2009GL038962, 2009.

Yang, H. H., Chien, S. M., Cheng, M. T., and Peng, C. Y.: Comparative study of regulated and unregulated air pollutant emissions before and after conversion of automobiles from gasoline power to liquefied petroleum gas/gasoline dual-fuel retrofits, Environ. Sci. Technol., 41, 8471–8476, 2007.

Yu, L. D., Wang, G. F., Zhang, R. J., Zhang, L. M., Song, Y., Wu, B. B., Li, X. F., An, K., and Chu, J. H.: Characterization and Source Apportionment of $PM_{2.5}$ in an Urban Environment in Beijing, Aerosol Air Qual. Res., 13, 574–583, 2013.

Yunker, M. B., Macdonald, R. W., Vingarzan, R., Mitchell, R. H., Goyette, D., and Sylvestre, S.: PAHs in the Fraser River basin: a critical appraisal of PAH ratios as indicators of PAH source and composition, Org. Geochem., 33, 489–515, 2002.

Zhang, X. L., Tao, S., Liu, W. X., Yang, Y., Zuo, Q., and Liu, S. Z.: Source diagnostics of polycyclic aromatic hydrocarbons based on species ratios: A multimedia approach, Environ. Sci. Technol., 39, 9109–9114, 2005.

Zhang, Y. X., Schauer, J. J., Zhang, Y. H., Zeng, L. M., Wei, Y. J., Liu, Y., and Shao, M.: Characteristics of particulate carbon emissions from real-world Chinese coal combustion, Environ. Sci. Technol., 42, 5068–5073, 2008.

4

Evidence for tropospheric wind shear excitation of high-phase-speed gravity waves reaching the mesosphere using the ray-tracing technique

M. Pramitha[1]**, M. Venkat Ratnam**[1]**, A. Taori**[1]**, B. V. Krishna Murthy**[2]**, D. Pallamraju**[3]**, and S. Vijaya Bhaskar Rao**[4]

[1]National Atmospheric Research Laboratory (NARL), Gadanki, India
[2]B1, CEBROS, Chennai, India
[3]Physical Research Laboratory (PRL), Ahmadabad, India
[4]Department of Physics, Sri Venkateswara University, Tirupati, India

Correspondence to: M. Venkat Ratnam (vratnam@narl.gov.in)

Abstract. Sources and propagation characteristics of high-frequency gravity waves observed in the mesosphere using airglow emissions from Gadanki (13.5° N, 79.2° E) and Hyderabad (17.5° N, 78.5° E) are investigated using reverse ray tracing. Wave amplitudes are also traced back, including both radiative and diffusive damping. The ray tracing is performed using background temperature and wind data obtained from the MSISE-90 and HWM-07 models, respectively. For the Gadanki region, the suitability of these models is tested. Further, a climatological model of the background atmosphere for the Gadanki region has been developed using nearly 30 years of observations available from a variety of ground-based (MST radar, radiosondes, MF radar) and rocket- and satellite-borne measurements. ERA-Interim products are utilized for constructing background parameters corresponding to the meteorological conditions of the observations. With the reverse ray-tracing method, the source locations for nine wave events could be identified to be in the upper troposphere, whereas for five other events the waves terminated in the mesosphere itself. Uncertainty in locating the terminal points of wave events in the horizontal direction is estimated to be within 50–100 km and 150–300 km for Gadanki and Hyderabad wave events, respectively. This uncertainty arises mainly due to non-consideration of the day-to-day variability in the tidal amplitudes. Prevailing conditions at the terminal points for each of the 14 events are provided. As no convection in and around the terminal points is noticed, convection is unlikely to be the source. Interestingly, large ($\sim 9\,\mathrm{m\,s^{-1}}$ km^{-1}) vertical shears in the horizontal wind are noticed near the ray terminal points (at 10–12 km altitude) and are thus identified to be the source for generating the observed high-phase-speed, high-frequency gravity waves.

1 Introduction

Atmospheric gravity waves (GWs) play an important role in the middle atmospheric structure and dynamics. They transport energy and momentum from the source region (mainly troposphere) to the upper atmosphere. The waves are dissipated upon reaching a critical level, transferring energy and momentum to the mean flow and leading to changes in the thermal structure of the atmosphere (Fritts and Alexander 2003). Several sources are identified for the generation of GWs, which include deep convection, orographic effect, vertical shear of horizontal wind and geostrophic adjustment. For GW generation from deep convection, mainly three mechanisms are considered (Fritts and Alexander 2003). These are (i) pure thermal forcing (e.g., Salby and Garcia 1987; Alexander et al., 1995; Piani et al., 2000; Fritts and Alexander 2003; Fritts et al., 2006), (ii) the mechanical oscillator effect (e.g., Clark et al., 1986; Fovell et al., 1992) and (iii) the obstacle effect (e.g., Clark et al., 1986; Pfister et al., 1993; Vincent and Alexander 2000). The importance of these depends upon the vertical profile of local shear and time dependence of latent heat release. GWs from the convection source can have a wide range of phase speeds, frequencies and wavelengths, unlike those from orography, which are generally confined to low ground-based frequencies and phase speeds (e.g., Queney., 1948; Lilly and Kennedy 1973; Nastrom and Fritts 1992; Eckermann and Preusse 1999; Alexander et al., 2010). In the shear excitation mechanism, mainly two processes, namely sub-harmonic interaction and envelope radiation (Fritts and Alexander 2003), are considered. The latter process can yield horizontal scales of a few tens of kilometers and phase speeds comparable to the mean wind. The geostrophic adjustment source is effective mainly in high latitudes (e.g., O'Sullivan and Dunkerton 1995; Shin Suzuki et al., 2013; Plugonven and Zhang 2014).

In general, significant progress has been made in the understanding of the physical processes leading to the spectrum of GWs through both observations and modeling. However, identification of the exact sources for the generation of GWs and their parameterization in the models remains a challenge (Geller et al., 2013). In order to identify the gravity wave sources, hodograph analysis has been widely used. Hodograph analysis can be used to identify the gravity wave parameters, which can be used as input parameters to the ray tracing. With the hodograph, the direction of propagation of the wave and hence the location of the source can be ascertained. However, this method is applicable only for medium- and low-frequency waves, as for the high-frequency GWs the hodograph will not be an ellipse but nearly a straight line. Further, as it assumes monochromatic waves, it is not always applicable in the real atmosphere. Notwithstanding this limitation, with this method, convection and vertical shear have been identified as the possible sources of the observed medium and low-frequency GWs in the troposphere and lower stratosphere over many places (e.g., Venkat Ratnam et al., 2008). It becomes difficult to apply this method for GWs that are observed in the mesosphere and lower thermosphere (MLT) region, where simultaneous measurements of temperatures (with wind) would not be available.

A more appropriate method in such cases is ray tracing (Marks and Eckermann 1995), which is widely used to identify the sources of GWs observed at mesospheric altitudes. Several studies (Hecht et al., 1994; Taylor et al., 1997; Nakamura et al., 2003; Gerrard et al., 2004; Brown et al., 2004; Wrasse et al., 2006; Vadas et al., 2009; and references therein) have been carried out to identify the sources for the GWs observed in the mesosphere using airglow images and in the stratosphere using radiosonde and lidar data (Guest et al., 2000; Hertzog et al., 2001). In mesospheric studies, important GW parameters, such as periodicities and horizontal wavelengths (and sometimes vertical wavelengths when two imagers are simultaneously used), are directly derived. A major limitation to the ray-tracing method is the non-availability of realistic information on the background atmosphere. This is difficult to obtain with the available suite of instrumentation for rendering identification of the source of the waves. Nevertheless, possible errors involved in identifying the terminal point of the waves with and without realistic background atmosphere have been estimated (e.g., Wrasse et al., 2006; Vadas et al., 2009).

Over the Indian region, several studies (Venkat Ratnam et al., 2008, and references therein) have been carried out for extracting GW parameters using various instruments (MST radar, radiosondes, lidar and satellite observations). In a few studies (Kumar, 2006, 2007; Dhaka et al., 2002; Venkat Ratnam et al., 2008; Debashis Nath et al., 2009; Dutta et al., 2009; Leena et al., 2012a, b), possible sources in the troposphere for their generation are identified which include convection, wind shear and topography.

In the present investigation, a reverse ray-tracing method is implemented to identify the sources of the GWs at mesospheric altitudes observed from an airglow imager located at Gadanki (13.5° N, 79.2° E) and from a balloon experiment which carried an ultraviolet imaging spectrograph from Hyderabad (17.5° N, 78.5° E). Wave amplitudes are also traced back, including both radiative and diffusive damping. In Sect. 2 we described the instrumentation, in Sect. 3 the theory behind ray tracing, in Sect. 4 the background atmosphere used for ray tracing, in Sect. 5 application of the ray-tracing method and in Sect. 6 identification of the sources of the observed waves.

2   Database

2.1   Airglow imager observations at Gadanki and methodology for extracting GW characteristics

The NARL Airglow Imager (NAI) located at Gadanki is equipped with a 24 mm Mamiya fisheye lens. It can monitor OH, $O(^1S)$ and $O(^1D)$ emissions and has a 1024×1024 pixel CCD as the detector, and has a field of view of 90°, avoiding nonlinearity arising at higher zenith angles. In the present study, only observations of $O(^1S)$ emission which originate at $\sim 93-100$ km (with a peak emission altitude of ~ 97 km) are used. The exposure time used to measure the intensities of emissions was 70 s. After the image was captured, it was analyzed and corrected for the background brightness, star brightness and actual coordinates. The area covered in the image is 200 km $\times 200$ km with a spatial resolution of 0.76 km near zenith and 0.79 km at the edges. More details on the NAI are discussed by Taori et al. (2013).

We observed three wave events between 14:29 and 14:51, 15:44 and 15:50, and 20:45 and 21:17 UTC, respectively, on 17 March 2012 (Fig. 1), as well as two wave events between 15:47 and 16:27 and between 16:31 and 16:54 UTC, respectively, on 19 March 2012 in the $O(^1S)$ airglow emission intensities. In these images, crests of the waves are emphasized by yellow freehand lines, and the motion of the waves is apparent in the successive images shown one below the other. Red arrows indicate the direction of the propagation of the waves. Horizontal wavelengths of the GWs are determined

Table 1. GW characteristics (direction of propagation, φ; horizontal wavelength, λ_h; vertical wavelength, λ_z; period, T; phase speed, C; and intrinsic frequency, ω_{ir}) for events observed over Gadanki (G) and Hyderabad (H). The terminal point locations (latitude, longitude and altitude) are also shown for each event. Conditions leading to the termination for each wave event are also shown. Events for which ray paths terminated at mesospheric altitude are indicated with an asterisk.

Events	ϕ (degrees)	λ_h (km)	T (min)	C (m s^{-1})	Longitude (degrees)	Latitude (degrees)	Altitude (km)	ω_{ir} (rad s^{-1})	Termination condition
Gadanki location									
G1	102	85(26)	18	78	79.9	10.8	13	0.00058	$m^2 < 0$
G2	98	34(28.9)	9	63	79.4	12.3	17	0.0116	WKB > 1
G3	132	12(13.6)	6	33	79.2	13.37	96.9*	0.0006	Intrinsic frequency approaching zero
G4	62	134(40)	12	186	79.14	13.2	92.9*	0.0093	WKB > 1
G5	142	16(2)	8	33	79.9	12.7	66.9*	0.0156	WKB > 1 and $m^2 < 0$
Hyderabad location									
H1	11	39(23.7)	16	41	70.2	15.8	10.5	0.0028	WKB > 1
H2	16	57(31.5)	16	59	75.3	16.4	13.5	0.0046	WKB > 1
H3	21	74(39)	16	77	75.9	16.3	14.5	0.0049	WKB > 1
H4	11	39(19.6)	20	32.5	76.3	17.1	67.6*	0.00083	$m^2 <$ limiting condition and intrinsic frequency approaching zero
H5	16	57(25)	20	48	72.7	15.7	12.5	0.0029	WKB > 1
H6	21	74(31)	20	61.7	74.7	15.8	13.5	0.0035	WKB > 1
H7	11	39(17.7)	23	28	75.8	16.9	68.5*	0.00087	$m^2 <$ limiting condition and intrinsic frequency approaching zero
H8	16	57(23)	23	41	68.3	14.8	11.5	0.0022	WKB > 1
H9	21	74(27)	23	54	73.4	15.4	13.5	0.0032	WKB > 1

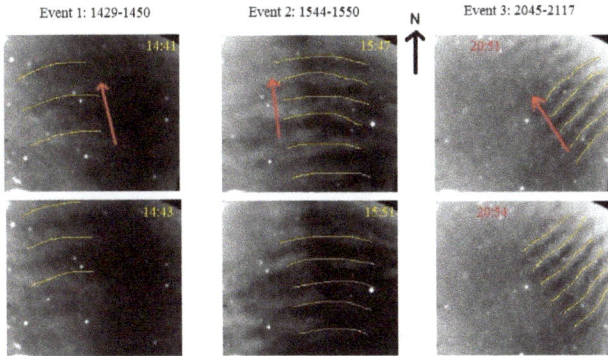

Event 1: 1429–1450   Event 2: 1544–1550   Event 3: 2045–2117

Figure 1. Identification of three wave events (left to right) obtained from the airglow emission intensities originating from O(^{1}S) emissions from Gadanki. The wave crests are emphasized by yellow freehand lines. Motion of waves can be obtained by successive images and the direction of propagation is shown by red arrows. Time of occurrence of events is shown in each image in UT (hh:mm).

by applying 2-D FFT (fast Fourier transform) to the observed airglow images. The periods of the GWs are estimated by applying 1-D FFT in time to the complex 2-D FFT in space. Direction of propagation and phase speed of GWs are identified using successive images. More details of the method-

ology for estimating the GW parameters from NAI observations are provided in Taori et al. (2013). Table 1 summarizes the GW parameters extracted from the five wave events (G1 to G5) mentioned above. In general, the waves corresponding to these events are moving in the NNW direction. Zonal (k) and meridional (l) wave numbers are calculated using the relations $k = k_h \cos\phi$ and $l = k_h \sin\phi$, where k_h is the horizontal wave number and φ is the horizontal direction of propagation observed from the airglow imager. The vertical wavelengths are calculated using the GW dispersion relation

$$\omega_{ir}^2 = \frac{N^2(k^2 + l^2) + f^2(m^2 + \alpha^2)}{k^2 + l^2 + m^2 + \alpha^2}, \tag{1}$$

where ω_{ir} is the intrinsic frequency of the wave, N is the Brunt–Väisälä frequency, f is the Coriolis frequency and m is the vertical wave number. Zonal, meridional and vertical wavelengths can be derived from the parameters given in Table 1. The background atmosphere used for ray tracing is developed using 30 years of observations from various sources and will be discussed more in Sect. 4.

2.2 Daytime GW observations at Hyderabad obtained through optical emissions

A multi-wavelength imaging echelle spectrograph (MISE) is used to obtain daytime emission intensities of oxygen emissions at 557.7, 630.0 and 777.4 nm in the MLT region at Hyderabad. MISE obtains high-resolution spectra of daytime skies which are compared with the reference solar spectrum. The difference obtained between the two yields information on the airglow emissions. The details of the emission extraction process and calibration procedures of the emission intensities and the salient results obtained in terms of wave coupling of atmospheric regions demonstrating the capability of this technique have been described elsewhere (Pallamraju et al., 2013; Laskar et al., 2013). In the present experiment, the slit oriented along the magnetic meridian enabled information on the meridional scale size of waves (λ_y) at O(^{1}S) emission altitude of ~ 100 km (in the daytime). An ultraviolet imaging spectrograph with its slit oriented in the east–west direction was flown on a high-altitude balloon (on 8 March 2010), which provided information on the zonal scale sizes of waves (λ_x) using the OI 297.2 nm emissions that originate at ~ 120 km. Both MISE and UVIS are slit spectrographs with array detectors providing 2-D information, with one direction yielding high-spectral-resolution spectrum (0.012 nm at 589.3 nm and 0.2 nm at 297.2 nm for MISE and UVIS, respectively) and the orthogonal direction yielding information on the dynamics over 330 km (in the y direction for OI 557.7 nm emission) and 170 km (in the x directions for the OI 297.2 nm emission). The spatial resolutions of these measurements are around 50 and 11 km, respectively. The details of the experiment and the wave characteristics in terms of λ_x, λ_y, λ_H (horizontal scale sizes), time periods (τ), propagation speeds (c_H) and propagation direction (θ_H) obtained by this instrument at a representative altitude of 100 km are described in detail in Pallamraju et al. (2014). Nine events from this experiment which occurred on 8 March 2010 are considered in the present study for investigating their source regions and are marked as H1 to H9 in Table 1. All wave events observed at Gadanki and Hyderabad correspond to high-frequency, high-phase-speed gravity waves as seen from their large vertical wavelengths, small periods and high phase speeds (Table 1).

2.3 Outgoing long-wave radiation (OLR) and brightness temperature in the infrared band (IR BT)

Satellite data of OLR/IR BT are used as a proxy for tropical deep convection. In general, the daily NOAA interpolated OLR can be used to obtain information on the synoptic-scale convection. However, for local convection on smaller spatial and temporal scales, the IR BT data merged from all available geostationary satellites (GOES-8/10, METEOSAT-7/5 GMS) are obtained from the National

Center for Environment Prediction (NCEP) Climate Prediction Center (source: ftp://disc2.nascom.nasa.gov/data/s4pa/TRMM_ANCILLARY/MERG/). The merged IR BT with a pixel resolution of 4 km is available for 60° N to 60° S (geostationary). The data in the east–west direction begin from 0.082° E with a grid increment of 0.03637° longitude and those in the north–south direction from 59.982° N with a grid increment of 0.03638° latitude (Janowiak et al., 2001). The BT data set is retrieved for every half-hour interval over regions of ±5° around Gadanki and Hyderabad on 17 March 2012 and 8 March 2010, respectively, to see whether any convective sources were present in these locations. Since the waves under study are high-frequency waves propagating at high phase speeds with smaller horizontal wavelengths, a maximum 5° × 5° grid is considered to be adequate. In general, the regions with OLR < 240 W/m^2 are treated as convective areas.

3 Reverse ray-tracing method

We followed the treatment of ray tracing given by Marks and Eckermann (1995). Note that the ray-tracing theory is applicable only when WKB (Wentzel–Kramers–Brillouin) approximation is valid. When the WKB parameter δ given by

$$\delta = \frac{1}{m^2}\left|\frac{\partial m}{\partial z}\right| \approx \left|\frac{1}{C_{gz}m^2}\frac{dm}{dt}\right| \qquad (2)$$

(where C_{gz} is the vertical group velocity, m is the vertical wave number, t is the time and z is the altitude) is less than unity, the approximation is taken to be valid.

In order to calculate the wave amplitude, we used the wave action equation of the form

$$\frac{\partial A}{\partial t} + \nabla \cdot (C_g A) = -\frac{2A}{\tau}, \qquad (3)$$

where $A = E/\omega_{ir}$ represents the wave action density; C_g represents the group velocity vector; and $E = \frac{\rho_0}{2}[\overline{u'^2} + \overline{v'^2} + \overline{w'^2} + N^2\overline{\zeta'^2}]$ represents the wave energy density, which is the sum of kinetic and potential energy components, as described by wave perturbations in zonal, meridional and vertical velocities (u', v', w'), as well as vertical displacement (ζ'). Here ρ_0 is the background density and τ is the damping timescale (Marks and Eckermann, 1995). Using the peak horizontal velocity amplitude along the horizontal wave vector we can calculate the wave action density using the equation

$$A = \frac{1}{4}\frac{\rho_0|\hat{u}_{\backslash\backslash}|^2}{\omega_{ir}}\left\{1 + \frac{f^2}{\omega_{ir}^2} + \frac{N^2 + \omega_{ir}^2}{N^2 - \omega_{ir}^2}\left(1 - \frac{f^2}{\omega_{ir}^2}\right)\right\}. \qquad (4)$$

In order to avoid spatial integration in the wave action equation we can write Eq. (3) in terms of the vertical flux of wave action $F = C_{gz}A$, where F is the vertical flux of wave action

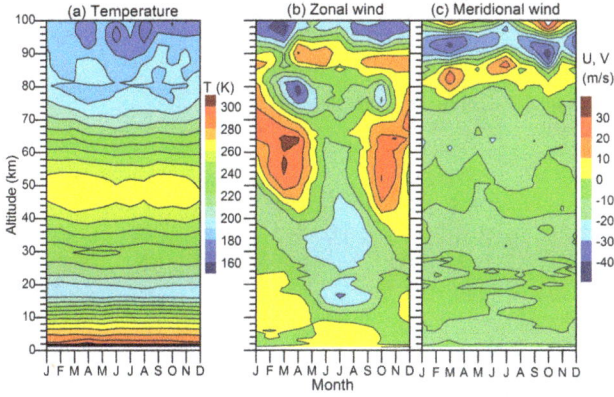

Figure 2. Climatological monthly mean contours of (a) temperature, (b) zonal wind and (c) meridional wind obtained over Gadanki region combining a variety of instruments listed in Table 2.

and C_{gz} is vertical component of the group velocity. Assuming negligible contribution from higher-order terms, Eq. (4) can be written as

$$\frac{dF}{dt} = -\frac{2}{\tau}F. \qquad (5)$$

As the wave moves through the atmosphere, amplitude damping takes place, which is mainly due to eddy diffusion and infrared radiative cooling by CO_2 and O_3. At higher altitudes (above about 100 km), molecular diffusion becomes important as compared to the eddy diffusion. We can calculate the damping rate due to diffusion using

$$\tau_D^{-1} = D(k^2 + l^2 + m^2 + \alpha^2), \qquad (6)$$

where $D = D_{Eddy} + D_{molecular}$ represents the sum of eddy and molecular diffusivities. Details of the calculation of D are given in the next section. In order to calculate the infrared radiative damping from 20 to 100 km, we used the damping rate calculation method given by Zhu (1993). The total damping rate is calculated using the following equation:

$$\tau^{-1} =$$

$$\frac{\tau_r^{-1}\left(\frac{1-f^2/\omega_{ir}^2}{1-\omega_{ir}^2/N^2}\right) + \tau_D^{-1}\left(1 + \frac{f^2}{\omega_{ir}^2} + \frac{1-f^2/\omega_{ir}^2}{N^2/\omega_{ir}^2 - 1} + Pr^{-1}\frac{1-f^2/\omega_{ir}^2}{1-\omega_{ir}^2/N^2}\right)}{\left\{1 + \frac{f^2}{\omega_{ir}^2} + \frac{N^2+\omega_{ir}^2}{N^2-\omega_{ir}^2}\left(1-\frac{f^2}{\omega_{ir}^2}\right)\right\}},$$

$$(7)$$

where Pr is Prandtl number. Note that the damping effect will be less for high-frequency wave.

4   Background atmosphere

In order to carry out reverse ray tracing, information on background atmospheric parameters (U, V and T) is required

right from the initial point (mesosphere) to the termination point (frequently in the troposphere). In general, there is no single instrument which can probe the troposphere, stratosphere and mesosphere simultaneously. Note that in order to trace the ray we require atmospheric parameters for a specified latitude–longitude grid. Since the observed wave events belong to high frequencies (GWs with short horizontal wavelengths and high vertical wavelengths), we require the background information for grid sizes of at least $5° \times 5°$ around Gadanki and Hyderabad. For the information on temperature and density at the required grids, we used Extended Mass Spectrometer and Incoherent Scatter Empirical (MSISE-90) model data (Hedin, 1991) from the surface to 100 km with an altitude resolution of 0.1 km for a $0.1° \times 0.1°$ grid around these locations. Note that the MSISE-90 model is an empirical model which provides temperature and density data from the surface to the thermosphere. For horizontal winds at the required grids, we used the outputs from the Horizontal Wind Model (HWM-07) (Drob et al., 2008) data. This model was developed by using a total of 60×10^6 observations available from 35 different instruments spanning 50 years. Further, long-term data available from a variety of instruments (MST radar, MF radar, rocketsondes, radiosondes, HRDI/UARS and SABER/TIMED satellites) in and around ($\pm 5°$) Gadanki have been used to develop a background climatological model profiles of U, V and T on a monthly basis. Details of the data used to develop the background temperature, and horizontal winds are provided in Table 2. Monthly mean contours of temperature, zonal and meridional winds obtained from the climatological model (from now on referred to as the Gadanki model) are shown in Fig. 2. In general, significant features of the background atmospheric structure for a typical tropical region can be noticed from this figure. Tropopause, stratopause and mesopause altitudes are located at around 16–18, 48–52 and 98–100 km with temperatures of 190–200, 260–270 and 160–170 K, respectively. Mesospheric semiannual oscillation around 80–85 km is also seen (Fig. 2a). The Tropical Easterly Jet at around 16 km during the Indian summer monsoon season (June-July-August) and semiannual oscillation near the stratopause (and at 80 km with different phase) are also clearly visible in the zonal wind (Fig. 2b). Meridional winds do not exhibit any significant seasonal variation in the troposphere and stratosphere but show large variability in the mesosphere (Fig. 2c). These overall features in the background temperature and wind match well with those reported considering data from different instruments by Kishore Kumar et al. (2008a, b).

The profiles of T obtained from the MSISE-90 model and U and V from HWM-07 for 17 March 2012 are shown in Fig. 3a and 3b and c, respectively. The Gadanki model mean temperature profile for the month of March and the temperature profile obtained from TIMED/SABER and mean temperature obtained from ERA-Interim for the month of March 2012 are also superimposed in Fig. 3a for comparison. Very

Table 2. Details of instruments and parameters measured, altitude range in which data are available, and the duration of the data considered for developing the Gadanki atmospheric model.

Instrument (parameter(s) obtained)	Altitude range covered	Duration of the data considered
Indian MST radar (U, V)	4–21 and 65–85 km	1996–2012
Radiosonde (U, V, T)	1–30 km	2006–2012
Lidar (T)	30–75 km	1998–2012
Rocket (U, V, T)	22–80 km	1970–1991, 2002–2007
HALOE, HRDI/UARS (T, U, V)	65–110 km	1991–2000
SABER/TIMED (T)	30–110 km	2002–2012

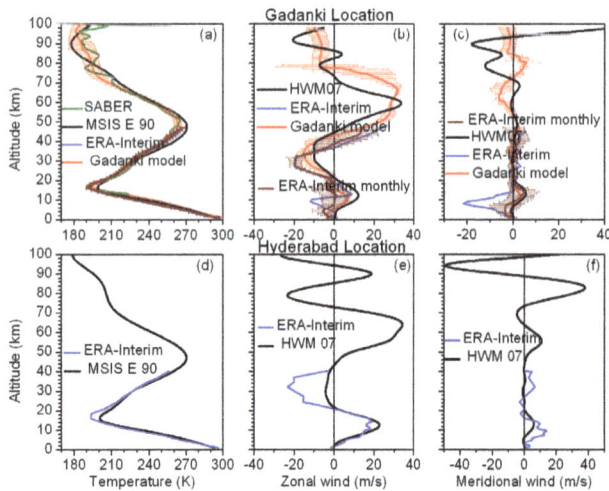

Figure 3. Profiles of **(a)** temperature **(b)** zonal wind and **(c)** meridional wind obtained using ERA-Interim data products for 17 March 2012, 12:00 UTC, over the Gadanki region. Profiles obtained from a variety of sources over Gadanki (Gadanki model) listed in Table 2 are also superimposed in the respective panels for comparison. Plots **(d–e)** are same as **(a–c)** but obtained for Hyderabad on 8 March 2010. The temperature profile obtained from MSISE-90 and the zonal and meridional winds obtained from HWM-07 for the same day are also provided in the respective panels.

good agreement between the profiles can be noticed. The profiles of U and V obtained from the Gadanki model for the month of March and also the monthly mean of the ERA-Interim are also superimposed in Fig. 3b and c, respectively. In general, a good match is seen between the Gadanki model and ERA-Interim and HWM-07 models up to the altitude of the stratopause. The differences between the two above the stratopause could be due to tidal winds, which have large amplitudes at mesospheric altitudes. Though tidal amplitudes are already included in the HWM-07 model, their day-to-day variability may be contributing to these differences. In order to avoid any bias due to day-to-day variability of the tides at mesospheric altitudes, we have considered tidal amplitudes of 5, 10 and 15 K in temperature and 10, 20 and 30 m s^{-1} in wind at 97 km to represent day-to-day variability.

In general, the troposphere is a highly dynamic region, though the amplitudes of tides are considerably small. In order to consider more realistic horizontal winds in the troposphere and stratosphere, we further considered the ERA-Interim products (Dee et al., 2011). These data are available at 6 h intervals with a $1.5° \times 1.5°$ grid resolution at 37 pressure levels covering from the surface (1000 hpa) to the stratopause (~ 1 hPa). The profiles of T, U and V from ERA-Interim for 17 March 2012 for 12:00 UTC are also superimposed in Fig. 3a, b and c, respectively. In general, good agreement between the other models and ERA-Interim model can be noticed, particularly in V in the lower and upper levels, except between 10 and 20 km. To summarize, we have considered the following wind models: (1) ERA-Interim (from the surface to 40 km) and HWM-07 models from 40 to 100 km, (2) the Gadanki model and (3) zero wind ($U = 0$ and $V = 0$). Using these background atmosphere profiles, we calculated the relevant atmospheric parameters like N^2 and H. Profiles of T, U and V obtained using ERA-Interim data products for 8 March 2010, 06:00 UTC, over the Hyderabad region are shown in Fig. 3d–f, respectively. T, U and V profiles as obtained from MSISE-90 and HWM-07 for the same day are also provided in the respective panels. The background atmosphere information for wave events over Hyderabad is obtained in a manner similar to that mentioned above for Gadanki.

In order to calculate diffusive damping, we used eddy diffusivity profiles for the troposphere and lower stratosphere and mesosphere which are obtained using MST radar (Narayana Rao et al., 2001) at Gadanki as shown in Fig. 4a. In the altitude regions where there are data gaps, we extrapolated/interpolated the diffusivity profiles, and the approximated profile with different analytical exponential functions are also shown in Fig. 4a. The eddy diffusivity profile of Hocking (Hocking, 1991) that is presented in Marks and Eckermann (1995) is also superimposed for comparison. Note that Hocking's profile corresponds mainly to midlatitudes. In general, eddy diffusivity is relatively higher in Hocking's profile than in the Gadanki profile. This same (Gadanki) profile is also used for Hyderabad events. In Fig. 4b, molecular diffusivity is shown. It is seen that the molecular diffusivity exceeds the eddy diffusivity at altitudes > 80 km.

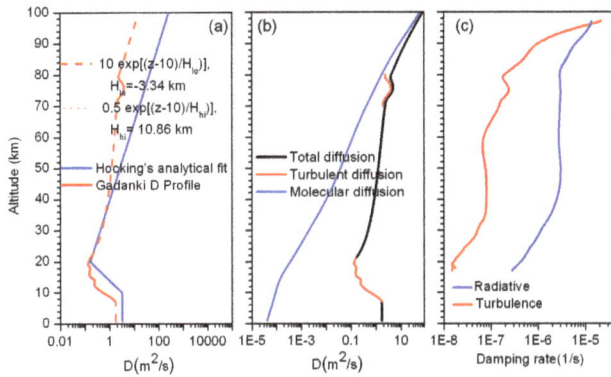

Figure 4. (a) Profile of eddy diffusivity (thick red line) obtained from Gadanki MST radar (Rao et al., 2001) in the troposphere, lower stratosphere and mesosphere. A fitted profile (dotted line) with exponential function is also shown. Hocking's (Hocking 1991) analytical curve (extrapolated) is also superimposed for comparison. **(b)** Profiles of eddy, molecular and total diffusivity. **(c)** Radiative and diffusive damping rates.

We have also taken into account molecular diffusivity in the ray-tracing calculation while considering the total diffusivity above 80 km, and the total diffusivity profile is shown in Fig. 4b. Radiative and diffusive damping rates corresponding to event G1 observed over Gadanki are shown in Fig. 4c for illustration. It is seen that radiative damping rate is higher than the diffusive damping rate below 95 km. This is the case for the other 13 events (G2–G5 and H1–H9) as well.

5 Application of reverse ray tracing for the wave events

By using the background parameters and the ray-tracing equations, we trace back the ray path(s) to identify the GW source region(s). We used a Runge–Kutta fourth-order method for numerical integration at the time step of $\delta t = 100\,\mathrm{m}\,C_{\mathrm{gz}}^{-1}$, where 100 m is the height step downwards from 97 km (the peak altitude of the airglow layer) and C_{gz} is the vertical group velocity. As the ray-tracing treatment is valid only when WKB approximation holds well, the ray integration is terminated whenever the WKB approximation is violated. We terminated the ray (1) when m^2 became negative, which means that the wave cannot propagate vertically; (2) when intrinsic frequency was less than zero or approaching zero, which means that the wave reached a critical layer and is likely to break beyond this; (3) when the WKB parameter was approaching values greater than 1 (beyond which WKB approximation breaks); and (4) when square of vertical wave number was becoming greater than $1 \times 10^{-6}\,\mathrm{cyc^2\,m^{-2}}$ (approaching critical level) (Wrasse et al., 2006). We calculated the wave action and thus the amplitude along the ray path by including the damping. As the information on wave amplitudes cannot be unambiguously determined from the optical emission intensity measurements, we assumed the GW am-

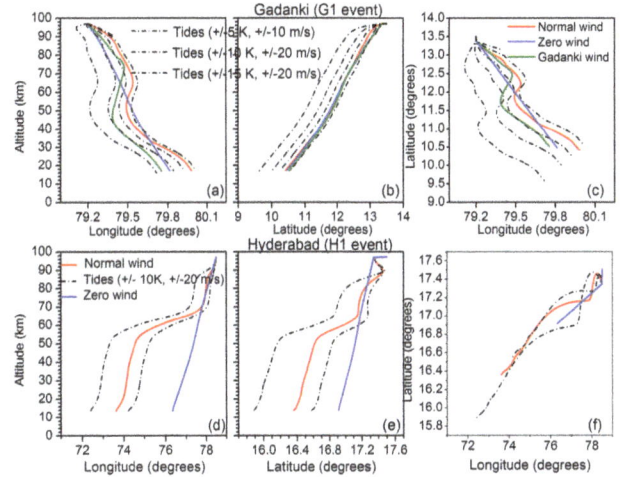

Figure 5. Ray paths for wave event G1 (starting at 97 km) in the **(a)** longitude–altitude, **(b)** latitude–altitude and **(c)** longitude–latitude cross sections. Ray paths obtained while considering different background wind conditions (normal wind, zero wind and Gadanki model wind) and the day-to-day variability of tides are also superimposed (dotted lines). Panels **(d–f)** are the same as **(a–c)** but for wave event H1. Note that the Gadanki atmospheric model wind is not used for the wave events over Hyderabad.

plitude as unity (at 97 km) and traced back the relative amplitudes along the ray path. Further, as we have not considered the local time variation of the background parameters, the ground-based wave frequency will be a constant. However, note that the intrinsic frequency still varies with altitude because of the varying background horizontal winds.

The observed and calculated GW parameters (intrinsic frequency; wave period; and zonal, meridional and vertical wave numbers) for all the wave events measured at the peak airglow emission altitudes as described in Sect. 2.1 and 2.2 are given as initial parameters to the ray-tracing code. We considered all the different combinations of observed wave parameters including the errors in the observations for obtaining the ray paths and the uncertainties in them. Note that atmospheric tides have large amplitudes in the MLT region which, at times, can be comparable to those of the background wind. As mentioned earlier, though tidal amplitudes are considered in the HWM-07 model, their day-to-day variability is not taken into account in the model. Amplitudes of the tides may reach values as high as $20\,\mathrm{m\,s^{-1}}$ over equatorial latitudes (Tsuda et al., 1999). As already mentioned, we have included day-to-day variability of tidal amplitudes into temperature and winds. In general, above the stratopause, tidal amplitudes are large and increase exponentially with altitude. It is interesting to note that (figure not shown) the variability in the background atmospheric parameters developed using data from a suite of instruments as mentioned above lies within the variability due to tides. Ray path calculations are also carried out for these background profiles.

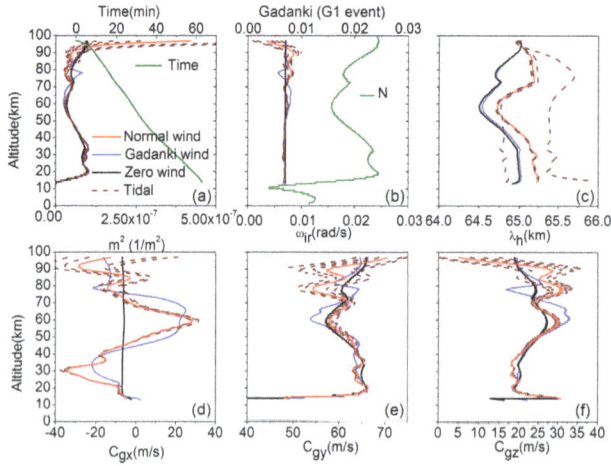

Figure 6. Profiles of (**a**) the square of vertical wave number (m^2), (**b**) intrinsic frequency (ω_{ir}) and Brunt–Väisälä frequency (N) (green); (**c**) horizontal wavelength; and (**d**) zonal, (**e**) meridional and (**f**) vertical group velocities for wave event G1. Profiles of the same obtained while considering the three different background winds (different colored lines) and the day-to-day variability of tides are also superimposed (dotted lines) in the respective panels. The observation time at the ray start and corresponding times along the ray time is also shown in (**a**) with the axis on the top.

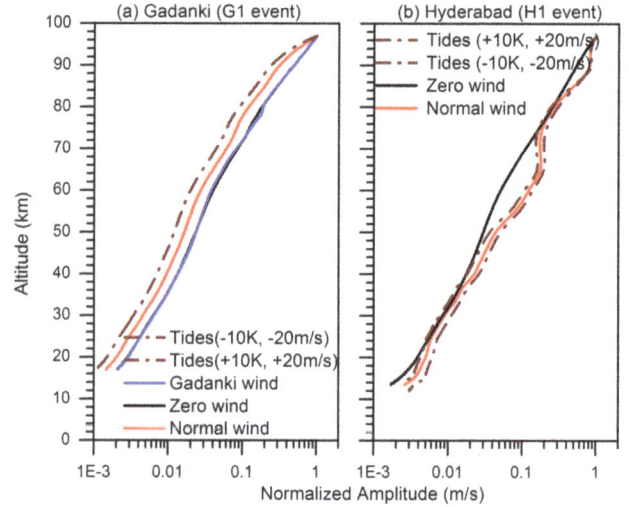

Figure 7. Normalized amplitudes of gravity waves observed for wave events (**a**) G1, and (**b**) H1, over Gadanki and Hyderabad, respectively. Amplitudes with three different background wind conditions along with different tidal amplitudes are also shown.

We traced the ray path using the above initial parameters from the initial latitude (13.5° N/17.5° N), longitude (79.2° E/78.5° E) and altitude (97 km). The ray paths for wave event G1 with the longitude–altitude, latitude–altitude and longitude–latitude are shown in Fig. 5a–c, respectively, for Gadanki and in Fig. 5d–f for (H1) Hyderabad. Ray paths obtained while considering different background conditions (normal wind, zero wind and Gadanki model wind) and the day-to-day variability of tides are also superimposed with dotted lines. When we considered zero (Gadanki) wind, a shift of 71 km (25 km) in the horizontal position of the terminal point is observed with respect to that for normal wind for wave event G1. The shift reduced to 19 km and increased to 47 and 97 km when we considered the tidal variability of $+5$ K, $+10\,\mathrm{m\,s^{-1}}$; $+10$ K, $+20\,\mathrm{m\,s^{-1}}$; and $+15$ K, $+30\,\mathrm{m\,s^{-1}}$, respectively, with respect to the normal wind. The shift is ~ 15 km for the tidal variability of -5 K, $-10\,\mathrm{m\,s^{-1}}$. The ray terminated in the mesosphere itself for tidal variability of -10 K, $-20\,\mathrm{m\,s^{-1}}$ and -15 K, $-30\,\mathrm{m\,s^{-1}}$ (figure not shown).

Over Hyderabad, for wave event H1, shown in Fig. 5d–f, the shifts in the horizontal location of the terminal point are 305.6 km (148.7 km) for tidal variability of $+10$, $+20\,\mathrm{m\,s^{-1}}$ (-10 K, $-20\,\mathrm{m\,s^{-1}}$) with reference to zero wind. This difference is only 59.5 km for tidal variability of -10 K, $-20\,\mathrm{m\,s^{-1}}$ with respect to the normal wind. The terminal point locations for the rest of the wave events for normal winds are listed in Table 1. Note that, out of the five wave events over Gadanki, two wave events (G3 and G4) were

terminated in the upper mesosphere itself and one (G5) was terminated at 67 km. Over Hyderabad, out of the nine wave events, two wave events (H4 and H7) were terminated at ~ 67 km. In general, all the wave events which propagated down to the upper troposphere terminated between 10 and 14.5 km, except for case G2, which was terminated at 17 km due to violation of the WKB approximation. The violation of the WKB approximation at 17 km could be due to sharp temperature gradients near tropopause.

Profiles of the square of the vertical wave number (m^2); intrinsic frequency (ω_{ir}) and Brunt–Väiäsälä frequency (N); horizontal wavelength (λ_h); and zonal, meridional and vertical group speed for the event G1 are shown in Fig. 6a–f, respectively. Profiles of these parameters obtained for different background wind conditions (normal wind, zero wind and Gadanki model wind) and the day-to-day variability of tides are also superimposed in the respective panels. The differences with and without the variability of tides in the parameters mentioned above are small below the stratopause and quite high above. Note that the effect of Doppler shifting of the wave frequency is larger at higher altitudes due to higher wind amplitudes. Around 13 km, the Brunt–Väisälä frequency is less than that of the intrinsic frequency, and thus the square of the vertical wave number is negative there (Fig. 6b). The variation in the horizontal wavelength with height (Fig. 6c) is small. Zonal group speed shows (Fig. 6d) nearly the same behavior as that of the zonal wind. The intrinsic frequency, ω_{ir}, exceeded N at 13 km altitude, and due to this m^2 became negative and the ray path was terminated there. The observation time at the ray start and the times along the ray time shown in Fig. 6a reveal that it took 63 min for the ray propagation.

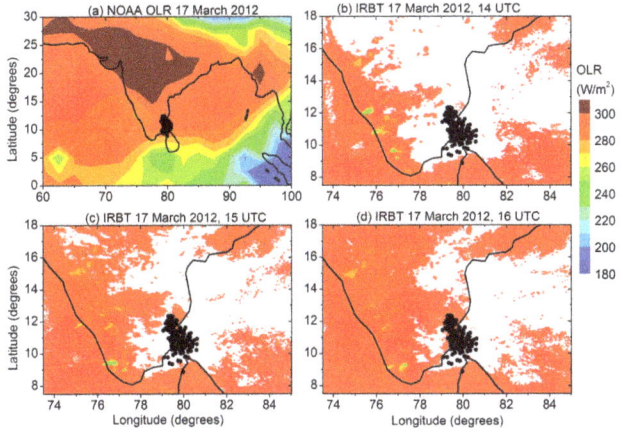

Figure 8. Daily mean latitude–longitude section of (**a**) OLR observed using NOAA products over the Indian region on 17 March 2012. Panels (**b–d**) are the same as (**a**) but for IR BT observed at 14:00, 15:00 and 20:00 UTC, respectively. Open (closed) circles in (**a**) (**b–d**) depict the terminal points of the ray paths shown in Fig. 4.

As mentioned earlier, the information on the wave amplitudes is not available from the observations. Therefore we used the GW amplitude as unity (at the altitude of observation) and traced back the relative amplitudes along the ray path. Profiles of amplitudes of GWs observed for wave events G1 and H1 over Gadanki and Hyderabad are shown in Fig. 7a and b, respectively. Amplitudes with three different background wind conditions along with different tidal amplitudes are also shown in the respective panels. Unity wave amplitude at the observed region translates to an amplitude of 10^{-3} near the source region. Amplitude growth is found to be higher when either Gadanki or zero wind models are considered and slightly lower for the normal wind. The growth is highly reduced when tidal variability in the background wind is considered. However, higher-amplitude growth rates were obtained over Hyderabad when we considered normal wind along with tidal variability as opposed to zero wind. Similar growth rates are also obtained for other wave events (not shown). Thus, background winds play an important role in the growth rates of GWs.

6 Discussion on the potential source(s) of the GW events

The geographical locations of the terminal points for different combinations of background winds along with different combinations of tidal variability are shown in Figs. 8 and 9 for the Gadanki and Hyderabad wave events, respectively. In these figures, the contour encircling all the points (not drawn in the panels of the figure) represents the horizontal spread of uncertainty due to background conditions (including tidal variability). The terminal point of the ray (in the troposphere) is expected to be the location of the GW source. Since 9 out

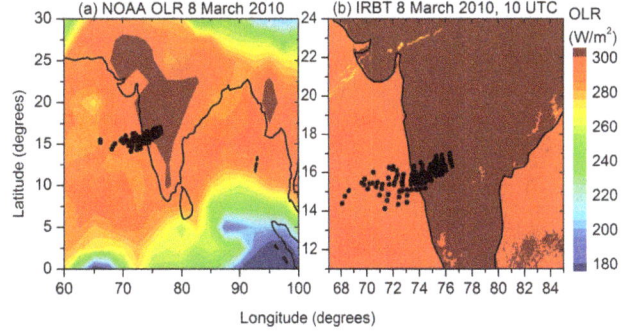

Figure 9. Same as Fig. 8 but for wave events observed over Hyderabad on 8 March 2010. Note that IR BT is shown only for 10:00 UTC.

of 14 wave events were terminated between 10 and 17 km, we search for the possible sources around this altitude at the location.

In general, primary sources for the GW generation over tropics are orography, convection and vertical shear in the horizontal winds. In the present case, GWs are unlikely to be the orographic origin as the observed waves have phase speeds much greater than zero. Tropical deep convection is assumed to be a primary source of the generation of a wide spectrum of GWs in the tropical latitudes. As mentioned earlier, OLR/IR BT is considered to be the proxy for the tropical deep convection. The lower the OLR/BT values, the higher the cloud top and hence the deeper the convection. OLR (IR BT) $< 240\,\mathrm{W\,m^{-2}}$ (K) is taken to represent deep convection. However, convection may exist at locations away from the observational site, and waves generated at those locations can propagate to the mesospheric altitudes over the site. In order to see the presence or otherwise of convection in the vicinity of the termination location, a latitude–longitude cross section of NOAA-interpolated OLR obtained for 17 March 2012 (8 March 2010) is shown in Fig. 8a (Fig. 9a) for the Gadanki (Hyderabad) region. The terminal points of the rays for wave events G1 and G2 (H1–H9 except H4 and H7) with different background wind conditions and different combinations of variability of the tides are also shown in the figure. There is no convection in and around the Gadanki (Hyderabad) region, as can be noticed from this figure. Note that this plot is with a coarse grid ($2.5° \times 2.5°$ latitude–longitude) averaged for a day. The observed GWs could be generated due to localized sources having shorter temporal and spatial scales than those seen from the NOAA OLR data used. In order to examine this, we have used IR BT data which are available at $4\,\mathrm{km} \times 4\,\mathrm{km}$ grid size and on a half-hour basis. Latitude and longitude section of hourly IR BT at 14:00 (10:00), 15:00 and 16:00 UTC is shown in Fig. 8b–d (Fig. 9b), respectively. The terminal points with and without variability of the tides are also shown. Interestingly no cloud patches are seen at any of the times mentioned above. Thus, convection can be ruled out as a possible source of the observed wave events.

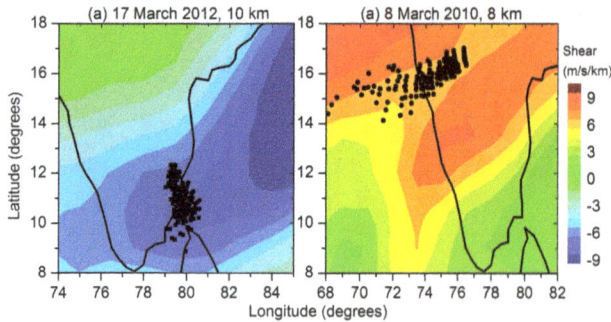

Figure 10. Latitude–longitude section of vertical shear in the horizontal wind observed using ERA-Interim data products on (**a**) 17 March 2012 at 10 km and (**b**) 8 March 2010 at 8 km. Filled circles show the terminal points of the ray paths estimated using three different wind conditions and tidal amplitudes.

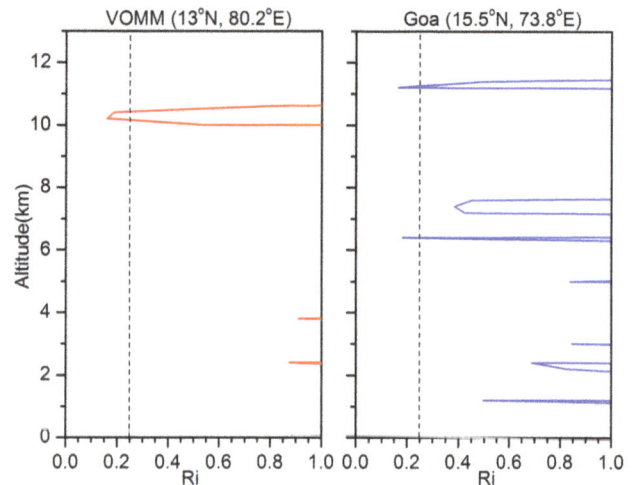

Figure 11. Profiles of Richardson number calculated close to the termination point using radiosonde data for the (**a**) Gadanki and (**b**) Hyderabad locations.

The other possible source of GW generation is the vertical shear of the horizontal wind. The vertical shear in horizontal winds at an altitude of 10 km (8 km) on 17 March 2012 (8 March 2010) as a function of latitude–longitude is shown in Fig. 10a (Fig. 10b). The terminal points of the rays for both the wave events with and without the day-to-day variability of the tides are also shown in the figure. Interestingly, at all the terminal points (in the troposphere), strong vertical shear in the horizontal wind is seen, which is quite high (8–9 m s^{-1} km^{-1}). In order to see whether these waves could be generated due to nonlinear interaction (through Kelvin–Helmholtz instability, KHI), the Richardson number ($Ri = \frac{N^2}{(dU/dz)^2}$) for the nearest location is calculated (using nearby radiosonde data) and is shown in Fig. 11. From the figure it can be noticed that Ri is < 0.25, showing that Ri satisfies the condition for instability for the observed waves at both the stations. Thus, the shear is unstable and hence conducive for the excitation of KHI, leading to the generation of the propagating GWs through nonlinear interaction. Note that shear excitation of the GWs has been examined theoretically using both linear and nonlinear approaches (e.g., Fritts, 1982, 1984; McIntyre, 1978). For the excitation of radiating GWs by KHIs at a shear layer, the two mechanisms that are examined are the vortex pairing (sub-harmonic interaction) and envelope radiation (Fritts, 1984). The vortex pairing is found to be highly dependent on the minimum Ri, whereas the envelope radiation mechanism is found to provide efficient radiating wave excitation in the absence of propagating unstable modes (Fritts, 1984). Theoretical and numerical simulation work needs to be carried out to examine which of these mechanisms is adequate for the observed events in the present study. This aspect is beyond the scope of the present study and is planned to be taken up in the future.

Note that five wave events terminated at mesospheric altitudes. We examined the background atmospheric condition which can lead to the termination of these wave events at such high altitudes. The ray paths for two wave events observed on the same day over Gadanki could propagate down below with the same background atmosphere. When wave parameters related to this event are examined (Table 1), it can be seen that the phase speeds are small when compared to the other two wave events. We have introduced a wave at around 15 km, with all the wave parameters similar to that observed at 97 km for this event, and applied forward ray tracing. It is seen that the ray propagated up to 50 km and was terminated. Note that strong vertical shear in the background wind is seen at this altitude (Fig. 3). To investigate the role of shear in the process of propagation of waves, the shear is reduced to almost 0 in the 50–80 km altitude region. Under such conditions this wave event also could propagate to ~ 16 km in the reverse ray tracing. This reveals that the background wind shear is obstructing the ray path. It is quite likely that the wave was ducted between 50 and 80 km, and similar results are obtained for the other cases which were terminated in the mesosphere. This indicates that wind shears at mesospheric altitudes are responsible for termination at mesospheric altitudes for these events.

7 Summary and conclusions

Identification of the GW sources for the 14 wave events observed over Gadanki and Hyderabad using optical airglow measurements is presented. A reverse ray-tracing method is developed to obtain the location of the source regions of the GWs in the troposphere/mesosphere. We made use of the MSISE-90 model for temperature and the HWM-07 model for the zonal and meridional winds in addition to the ERA-Interim products in the lower atmosphere (1000 to 1 hPa pressure levels) and the Gadanki climatological model and zero wind model for the background atmosphere. We have also incorporated the expected variability of tidal amplitudes of 5, 10 and 15 K in temperature and 10, 20 and 30 m s^{-1} in wind, respectively. The terminal points lie in the range of 50–100 and 60–300 km for Gadanki and Hyderabad, respectively, when different wind and tidal variabilities are used. Wave action is implemented taking into account the radiative and diffusive damping. Considering the wave amplitude as unity at 97 km, the amplitude of the wave is traced back to the source region for different wind models. Out of the 14 events examined, 9 ray paths terminated in the troposphere. The remaining 5 events were terminated in the mesosphere itself. We examined for possible sources for the 9 events where the ray paths terminated in the troposphere.

Orography as the possible source was ruled out, as wave events have high phase speeds. No tropical deep convection in and around Gadanki and Hyderabad was noticed near the ray terminal points. Interestingly, strong vertical shear in the horizontal wind is observed near the terminal points, and these large shears are considered to be the source of the GW events observed at the mesospheric altitudes. Preusse et al. (2008) discussed the transparency of the waves to the atmosphere in different seasons. They reported that, during equinox times, the atmosphere is more transparent to the high phase speed and shorter horizontal wavelength waves than it is in the solstices. Waves with shorter (< 10 km) horizontal wavelengths tend to be removed by vertical reflection or evanescence at the source, and slower phase speeds are more prone to the critical-level removal. This leads to a preference for waves with longer horizontal wavelengths and faster ground-based phase speeds to reach the MLT. However, they observed that many rays penetrated into the MLT at the tropical latitude where wind speed is low in comparison to the mid- and high latitudes. In our case, whenever phase speed was low for short horizontal wavelength waves, the wave did not reach the troposphere and was stopped at mesospheric altitude itself. While there is strong evidence for convectively generated gravity waves, evidence of tropospheric wind-shear-generated GWs is rather sparse (Mastrantonio 1976; Fritts and Alexander 2003). The present study clearly demonstrated that high-frequency, high-phase-speed GWs observed in the mesosphere can be generated by tropospheric wind shear. Examination of the background wind conditions and wave parameters for the events that were terminated in the mesosphere revealed that the phase speeds were quite small for these strong vertical shears in the 50–80 km region (and at 95 km), resulting in termination of the ray paths. A potential explanation is that the waves generated in the troposphere are ducted between 50 and 80 km and the waves observed above this region are due to leakage of waves from the duct. It is also likely that the observed GWs in these cases (G3, G4, G5, H5 and H7) are from secondary wave generation due to wave breaking at the termination region. While secondary wave generation due to convectively generated waves has been investigated (e.g., Zhou et al., 2002; Chun and Kim, 2008), such investigations have not yet been

carried out for GWs of shear origin. This aspect needs further investigation. Note that we have tested the reverse ray-tracing method successfully for 14 wave events. Further, wave action is also implemented successfully by assuming the wave amplitudes as unity, as information on the same is not available from optical observations. However, a greater number of cases need to be examined, particularly for the events that occur during the Indian summer monsoon season, when convection and strong vertical shears in the horizontal winds co-exist due to the prevailing Tropical Easterly Jet (Venkat Ratnam et al., 2008). A few experiments are planned to be conducted at Gadanki with simultaneous operation of MST radar, radiosondes, Rayleigh lidar, airglow imagers and meteor radar that can together provide information right from the troposphere to the MLT region. Note that such a study on the vertical propagation of mesoscale gravity wave from the lower to upper atmosphere was performed recently by Shin Suzuki et al. (2013) using an airglow imager and lidar over the Arctic region.

Acknowledgements. This work was done as a part of the SAFAR and CAWSES India phase II programs. We thank the NARL staff for providing data used in the present study. We are deeply grateful to NOAA and ECMWRF for providing OLR and ERA-Interim data, respectively, used in the present study through their ftp sites. HWM-07 model data are obtained from http://nssdcftp.gsfc.nasa.gov/models/atmospheric/hwm07/   This work is supported by the Department of Space, Government of India.

Edited by: W. Ward

References

Alexander, M. J., Holton, J. R., and Durran, D. R.: The gravity wave response above deep convection in a squall line simulation, J. Atmos. Sci., 52, 2212–2226, 1995.

Alexander , M. J., Geller, M., McLandress, C., Polavarapu, S., Preusse, P., Sassi, F., Sato, K., Eckermann, S., Ern, M., Hertzog, A., Kawatani, Y., Pulido, M., Shaw, T. A., Sigmond, M., Vincent, R., and Watanabe, S.: Recent developments in gravity-wave effects in climate models, and the global distribution of gravity-wave momentum flux from observations and models, Q. J. R. Meteorol. Soc., 136, 1103–1124, 2010.

Brown, L. B., Gerrard, A. J., Meriwether, J. W., and Makela, J. J.: All-sky imaging observations of mesospheric fronts in OI 557.7 nm and broadband OH airglow emissions: Analysis of frontal structure, atmospheric background conditions, and potential sourcing mechanisms, J. Geophys. Res., 109, D19104, doi:10.1029/2003JD004223, 2004.

Chun, H. Y. and Kim, Y. H.: Secondary waves generated by breaking of convective gravity waves in the mesosphere and their influence in the wave momentum flux, J. Geophys. Res., 113, D23107, doi:10.1029/2008JD009792, 2008.

Clark, T. L., Hauf, T., and Kuettner, J. P.: Convectively forced internal gravity waves: Results from two-dimensional numerical experiments, Q. J. R. Meteorol. Soc.,112, 899–925, doi:10.1002/qj.49711247402,1986.

Debashis Nath., Venkat Ratnam, M., Jagannadha Rao, V. V. M., Krishna Murthy, B. V., and Vijaya Bhaskara Rao, S.: Gravity wave characteristics observed over a tropical station using high-resolution GPS radiosonde soundings, J. Geophys. Res., 114, D06117, doi:10.1029/2008JD011056, 2009

Dee, D. P., Uppala, S. M., Simmons, A. J., Berrisford, P., Poli, P., Kobayashi, S., Andrae, U., Balmaseda, M. A., Balsamo, G., Bauer, P., Bechtold, P., Beljaars, A. C. M., van de Berg, L., Bidlot, J., Bormann, N., Delsol, C., Dragani, R., Fuentes, M., Geer, A. J., Haimberger, L., Healy, S. B., Hersbach, H., Hólm, E. V., Isaksen, L., Kållberg, P., Köhler, M., Matricardi, M., McNally, A. P., Monge-Sanz, B. M., Morcrette, J.-J., Park, B. K., Peubey, C., de Rosnay, P., Tavolato, C., Thépaut J.-N., and Vitart , F.: The ERA-Interim-reanalysis: configuration and performance of the data assimilation system, Q. J. R. Meteorol. Soc., 137, 553–597, 2011.

Dhaka, S. K., Choudhary, R. K., Malik, S., Shibagaki, Y., Yamanaka, M. D., and Fukao, S.: Observable signatures of a convectively generated wave field over the tropics using Indian MST radar at Gadanki (13.5° N, 79.2° E), Geophys. Res. Lett., 29, 1872, doi:10.1029/2002GL014745, 2002.

Drob, D.P., Emmert, J. T., Crowley, G., Picone, J. M., Shepherd, G. G., Skinner, W., Hays, P., Niciejewski, R. J., Larsen, M., She, C. Y., Meriwether, J. W., Hernandez, G., Jarvis, M. J., Sipler, D. P., Tepley, C. A., O'Brien, M. S., Bowman, J. R., Wu, Q., Murayama, Y., Kawamura, S., Reid, I. M., and Vincent, R. A.: An empirical model of the Earth's horizontal wind fields: HWM07, J. Geophys. Res., 113, A12304, doi:10.1029/2008JA013668, 2008.

Dutta, G., Ajay Kumar, M. C., Vinay Kumar, P., Venkat Ratnam, M.,Chandrashekar, M., Shibagaki, Y., Salauddin, M., and Basha, H. A.: Characteristics of high-frequency gravity waves generated by tropical deep convection: Case studies, J. Geophys. Res., 114, D18109, doi:10.1029/2008JD011332, 2009.

Eckermann, S. D.: Ray-tracing simulation of the global propagation of inertia gravity waves through the zonally averaged middle atmosphere, J. Geophys. Res., 97, 15849–15866, 1992.

Fovell, R., Durran, D., and Holton, J. R.: Numerical simulations of convectively generated stratospheric gravity waves, J. Atmos. Sci., 49, 1427–1442, 1992.

Fritts, D. C.: Shear Excitation of Atmospheric Gravity Waves, J. Atmos. Sci., 39, 1936–1952, 1982.

Fritts, D. C.: Shear Excitation of Atmospheric Gravity Waves, Part II: Nonlinear Radiation from a Free Shear Layer, J. Atmos. Sci., 41, 524–537, 1984.

Fritts, D. C. and Alexander, M. J.: Gravity wave dynamics and effects in the middle atmosphere, Rev. Geophys., 41, doi:10.1029/2001RG000106, 2003.

Fritts, D. C., Sharon, L., Vadas, K. W., and Werne, J. A.: Mean and variable forcing of the middle atmosphere by gravity waves, J. Atmos. Sol. Terr. Phys., 68, 247–265, 2006.

Geller, M. A., Alexander, M. J., Love, P. T., Bacmeister, J., Ern, M., Hertzog, A., Manzini, E., Preusse, P., Sato, K., Scaife, A., and Zhou, T.: A comparison between gravity wave momentum fluxes in observations and climate models, J. Climate, 26, 6383–6405, doi:10.1175/JCLI-D-12-00545.1, 2013.

Gerrard, A. J., Kane, T. J., Eckermann, S. D., and Thayer, J. P.: Gravity waves and mesospheric clouds in the summer middle atmosphere: A comparison of lidar measurements and ray modeling of gravity waves over Sondrestrom, Greenland, J. Geophys. Res., 109, D10103, doi:10.1029/2002JD002783, 2004.

Guest, F. M., Reeder, M. J., Marks, C. J., and Karoly, D. J.: Inertia–Gravity Waves Observed in the Lower Stratosphere over Macquarie Island, J. Atmos. Sci., 57, 2000.

Hecht, J. H., Walterscheid, R. L., and Ross, M. N.: First measurements of the two-dimensional horizontal wave number spectrum from CCD images of the nightglow, J. Geophys. Res., 99, 11449–11460, 1994.

Hedin, A. E.: Extension of thc MSIS Thermosphere model into the middle and Lower atmosphere, J. Geophys. Res., 96, 1159–117, 1991.

Hertzog, A., Souprayen, C., and Hauchecorne, A.: Observation and backward trajectory of an inertio-gravity wave in the lower stratosphere, Ann. Geophys., 19, 1141–1155, 2001, http://www.ann-geophys.net/19/1141/2001/.

Hocking, W. K.: The effects of middle atmosphere turbulence on coupling between atmospheric regions, J. Geomag., Geoelc., 43, 621–636, 1991.

Janowiak, J. E., Joyce, R. J., and Yarosh, Y.: A real-time global half-hourly pixel-resolution IR dataset and its applications, Bull. Amer. Meteor. Soc., 82, 205–217, 2001.

Jones, W. L.: Ray tracing for internal gravity waves, J. Geophys. Res., 74, 2028–2033, 1969.

Kumar, K. K.: VHF radar observations of convectively generated gravity waves: Some new insights, Geophys. Res. Lett., 33, L01815, doi:10.1029/2005GL024109, 2006.

Kumar, K. K.: VHF radar investigations on the role of mechanical oscillator effect in existing convectively generated gravity waves, Geophys. Res. Lett., 34, L01803, doi:10.1029/2006GL027404, 2007.

Kishore Kumar, G., Venkat Ratnam, M., Patra, A. K., Vijaya Bhaskara Rao, S., and Russell, J.: Mean thermal structure of the low-latitude middle atmosphere studied using Gadanki Rayleigh lidar, Rocket, and SABER/TIMED observations, J. Geophys. Res., 113, D23106, doi:10.1029/2008JD010511, 2008a.

Kishore Kumar, G., Venkat Ratnam, M., Patra, A. K., Jagannadha Rao, V. V. M., Vijaya Bhaskar Rao, S., Kishore Kumar, K., Gurubaran, S., Ramkumar, G., and Narayana Rao, D.: Low-latitude mesospheric mean winds observed by Gadanki mesosphere-stratosphere- troposphere (MST) radar and comparison with Rocket, High Resolution Doppler Imager (HRDI), and MF radar measurements and HWM93, J. Geophys. Res., 113, D19117, doi:10.1029/2008JD009862, 2008b.

Laskar, F. I., Pallamraju, D., Vijaya Lakshmi, T., Anji Reddy, M., Pathan, B. M., and Chakrabarti, S.: Investigations on vertical coupling of atmospheric regions using combined multiwavelength optical dayglow, magnetic, and radio measurements, J. Geophys. Res. Space Physics., 118, 4618–4627, doi:10.1002/jgra.50426, 2013.

Leena, P. P., Venkat Ratnam, M., KrishnaMurthy, B. V., and VijayaBhaskaraRao, S.: Detection of high frequency gravity waves using high resolution radiosonde observations, J. Atmos. Sol. Terr. Phys., 77, 254–259, 2012a.

Leena, P. P., Venkat Ratnam, M., and Krishna Murthy, B. V.: Inertia gravity wave characteristics and associated fluxes observed using five years of radiosonde measurements over a tropical station, J. Atmos. Sol. Terr. Phys., 84–85, 37–44, 2012b.

Lilly, D. K. and Kennedy, P. J.: Observations of a stationary mountain wave and its associated momentum flux and energy dissipation, J. Atmos. Sci., 30, 1135–1152, 1973.

Mastrantonio, G., Einaudi, F., Fua, D., and Lalas, D. P.: Generation of gravity waves by jet streams in the at-mosphere, J. Atmos. Sci., 33, 1730–1738, 1976.

Marks, C. J. and Eckermann, S. D.: A three-dimensional non-hydrostatic ray-tracing model for gravity waves: formulation and preliminary results for the middle atmosphere, J. Atmos. Sci., 52, 1959–1984, 1995.

McIntyre, M. E. and Weissman, M. A.: On Radiating Instabilities and resonant over-reflection, J. Atmos. Sci., 35, 1190–1196, 1978.

Nakamura, T., Aono, T., Tsuda, T., Admiranto, A. G., and Achmad Suranto, E.: Mesospheric gravity waves over a tropical convective region observed by OH airglow imaging in Indonesia, Geophys. Res. Lett., 30, 1882–1885, 2003.

Narayana Rao, D., Ratnam, M. V., Rao, T. N., and Rao, S. V. B.: Seasonal variation of vertical eddy diffusivity in the troposphere, lower stratosphere and mesosphere over a tropical station, Ann. Geophys., 19, 975–984, 2001,
http://www.ann-geophys.net/19/975/2001/.

Nastrom, G. D. and Fritts, D. C.: Sources of mesoscale variability of gravity waves I: topographic excitation, J. Atmos. Sci., 49, 101–110, 1992.

O'Sullivan, D. and Dunkerton, T. J.: Generation of Inertia Gravity waves in a simulated life cycle of Baroclinic Instability, J. Atmos. Sci., 52, 1995.

Pallamraju, D., Laskar, F. I., Singh, R. P., Baumgardner, J., and Chakrabarti, S.: MISE: A Multiwavelength Imaging Spectrograph using Echelle grating for daytime optical agronomy investigations, J. Atmos. Sol-Terr. Phys., 103, 176–183, 2013.

Pallamraju, D., Baumgardner, J., Singh, R. P., Laskar, F. I., Mendillo, C., Cook, T., Lockwood, S., Narayanan, R., Pant, T. K., and Chakrabarti, S.: Daytime wave characteristics in the mesosphere lower thermosphere region: results from the balloon-borne investigations of regional-atmospheric dynamics experiment, J. Geophys. Res., 119, 2229–2242, doi:10.1002/2013JA019368, 2014.

Pfister, L., Chan, K. R., Bui, T.P., Bowen, S., Legg, M., Gary, B., Kelly, K., Proffitt, M. and Starr, W.: Gravity Waves Generated by a Tropical Cyclone During the STEP Tropical Field Program: A Case Study, J. Geophys. Res., 98, D5, 1993.

Piani, C., Durran, D., Alexander, M. J., and Holton, J. R.: A numerical study of three-dimensional gravity waves triggered by deep tropical convection and their role in the dynamics of the QBO, J. Atmos. Sci., 57, 3689–3702, doi:10.1175/1520-0469(2000)057<3689, 2000.

Plougonven, R. and Zhang, F.: Internal gravity waves from atmospheric jets and fronts, Rev. Geophys., 52, 33–76, doi:10.1002/2012RG000419, 2014.

Preusse, P., Eckermann, S. D., and Ern, M.: Transparency of the atmosphere to short horizontal wavelength gravity waves, J. Geophys. Res., 113, D24104, doi:10.1029/2007JD009682, 2008.

Queney, P.: The problem of air flow over mountains: A summary of theoretical results, Bull. Am. Meterol. Soc., 29, 16–26, 1948.

Salby, M. L. and Garcia, R. R.: Transient response to localized episodic heating in the Tropics, Part 1: Excitation and short-time Near field behavior, J. Atmos. Sci., 44, 458–498, 1987.

Schoeberl, M. R.: A ray tracing model of gravity wave propagation and breakdown in the middle atmosphere, J. Geophys. Res., 90, 7999–8010, doi:10.1029/JD090iD05p07999, 1985.

Suzuki, S., Lubkena, F.-J., Baumgarten, G., Kaifler, N., Eixmann, R., Williams, B. P., and Nakamura, T.: Vertical propagation of a mesoscale gravity wave from the lower to the upper atmosphere, J. Atmos. Sol. Terr. Phys., 97, 29–36, 2013.

Taori, A., Jayaraman, A., and Kamalakar, V.: Imaging of mesosphere–thermosphere airglow emissions over Gadanki (13.5° N, 79.2° E) – first results, J. Atmos. Sol. Terres. Phys., 93, 21–28, 2013.

Taylor, M. J., Pendleton, W. R. J., Clark, S., Takahashi, H., Gobbi, D., and Goldberg, R. A.: Image measurements of short-period gravity waves at equatorial latitudes, J. Geophys. Res., 102, 26283–26299, 1997.

Tsuda, T., Ohnishi, K., Isoda, F., Nakamura, T., Vincent, R. A., Reid, I. M., Harijono, S. W. B., Sribimawati, T., Nuryanto, A., and Wiryosumarto, H.: Coordinated radar observations of atmospheric diurnal tides in equatorial regions, Earth Planets Space, 51, 579–592, 1999.

Vadas, S. L., Taylor, M. J., Pautet, P.-D., Stamus, P. A., Fritts, D. C., Liu, H.-L., São Sabbas, F. T., Rampinelli, V. T., Batista, P., and Takahashi, H.: Convection: the likely source of the medium-scale gravity waves observed in the OH airglow layer near Brasilia, Brazil, during the SpreadFEx campaign, Ann. Geophys., 27, 231–259, doi:10.5194/angeo-27-231-2009, 2009.

Venkat Ratnam, M., Narendra Babu, A., Jagannadha Rao, V. V. M., Vijaya Baskar Rao, S., and Narayana Rao, D.: MST radar and radiosonde observations of inertia-gravity wave climatology over tropical stations: source mechanisms, J. Geophys. Res., 113, D07109, doi:10.1029/2007JD008986, 2008.

Vincent, R. A. and Alexander, M. J.: Gravity waves in the tropical lower stratosphere: an observational study of seasonal and interannual variability, J. Geophys. Res., 105, 17971–17982, 2000.

Wrasse, C. M., Nakamura, T., Tsuda, T., Takahashi, H., Medeiros, A. F., Taylor, M. J., Gobbi, D., Salatun, A., Suratno, Achmad, E., and Admiranto, A. G.: Reverse ray tracing of the mesospheric gravity waves observed at 23° S (Brazil) and 7° S (Indonesia) in airglow imagers, J. Atmos. Sol. Terr. Phys., 68, 163–181, 2006.

Zhou, X. L., Holton, J. R., and Mullendore, G. L.: Forcing of secondary waves by breaking of gravity waves in the mesosphere, J. Geophys. Res., 107, D7, 4058, 10.1029/2001JD001204, 2002.

Zhu, X.: Radiative damping revisited: parameterization of damping rate in the middle atmosphere, J. Atmos. Sci., 50, 3008–3012, 1993.

5

Dust aerosol radiative effects during summer 2012 simulated with a coupled regional aerosol–atmosphere–ocean model over the Mediterranean

P. Nabat[1], **S. Somot**[1], **M. Mallet**[2], **M. Michou**[1], **F. Sevault**[1], **F. Driouech**[3], **D. Meloni**[4], **A. di Sarra**[4], **C. Di Biagio**[5], **P. Formenti**[5], **M. Sicard**[6], **J.-F. Léon**[2], and **M.-N. Bouin**[7]

[1]Météo-France, CNRM-GAME, Centre national de recherches météorologiques, UMR3589, Toulouse, France
[2]Laboratoire d'Aérologie, Toulouse, France
[3]Direction de la Météorologie Nationale, Casablanca, Morocco
[4]Laboratory for Earth Observations and Analyses, ENEA, Rome, Italy
[5]Laboratoire interuniversitaire des systèmes atmosphériques (LISA), UMR7583 – CNRS, Créteil, France
[6]Universitat Politechnica de Catalunya, Barcelona, Spain
[7]Météo-France, CMM, Centre de Météorologie Marine, Brest, France

Correspondence to: P. Nabat (pierre.nabat@meteo.fr)

Abstract. The present study investigates the radiative effects of dust aerosols in the Mediterranean region during summer 2012 using a coupled regional aerosol–atmosphere–ocean model (CNRM-RCSM5). A prognostic aerosol scheme, including desert dust, sea salt, organic, black-carbon and sulphate particles, has been integrated to CNRM-RCSM5 in addition to the atmosphere, land surface and ocean components. An evaluation of this aerosol scheme of CNRM-RCSM5, and especially of the dust aerosols, has been performed against in situ and satellite measurements, showing its ability to reproduce the spatial and temporal variability of aerosol optical depth (AOD) over the Mediterranean region in summer 2012. The dust vertical and size distributions have also been evaluated against observations from the TRAQA/ChArMEx campaign. Three simulations have been carried out for summer 2012 with CNRM-RCSM5, including the full prognostic aerosol scheme, only monthly-averaged AOD means from the aerosol scheme or no aerosols at all, in order to focus on the radiative effects of dust particles and the role of the prognostic scheme. Surface short-wave aerosol radiative forcing variability is found to be more than twice as high over regions affected by dust aerosols, when using a prognostic aerosol scheme instead of monthly AOD means. In this case downward surface solar radiation is also found to be better reproduced according to a comparison with several stations across the Mediterranean. A composite study over 14 stations across the Mediterranean, designed to identify days with high dust AOD, also reveals the improvement of the representation of surface temperature brought by the use of the prognostic aerosol scheme. Indeed the surface receives less radiation during dusty days, but only the simulation using the prognostic aerosol scheme is found to reproduce the observed intensity of the dimming and warming on dusty days. Moreover, the radiation and temperature averages over summer 2012 are also modified by the use of prognostic aerosols, mainly because of the differences brought in short-wave aerosol radiative forcing variability. Therefore this first comparison over summer 2012 highlights the importance of the choice of the representation of aerosols in climate models.

1   Introduction

Numerous and various aerosols affect the Mediterranean basin (Lelieveld et al., 2002), located at the crossroads of air masses carrying both natural (desertic particles, sea salt, volcanic ashes, etc.) and anthropogenic (black carbon, sulphate, etc.) particles. Because of their microphysical and optical properties, these aerosols can have strong effects on the regional radiative budget (e.g. Bergamo et al., 2008), with ensuing impact on climate (Zanis et al., 2012; Spyrou et al., 2013; Nabat et al., 2015) and ecosystems of the Mediterranean (Guieu et al., 2010). Among these aerosols, the Saharan desert dust particles represent an important contribution of aerosols for this region (Barnaba and Gobbi, 2004; Nabat et al., 2013). Indeed, dust particles coming from suspension, saltation and creeping processes associated with wind erosion (Knippertz and Todd, 2012) can move from northern Africa to the Mediterranean Sea and Europe (Moulin et al., 1997; Papadimas et al., 2008; Gkikas et al., 2013). These dust outbreaks are mainly driven by the synoptic meteorological conditions (Gkikas et al., 2012): they are more frequent in the eastern basin in winter and spring, in the central basin in spring and in the western basin in summer (Moulin et al., 1998). The ChArMEx initiative (Chemistry-Aerosol Mediterranean Experiment, http://charmex.lsce.ipsl.fr) has been launched for a few years in the framework of the MISTRALS (Mediterranean Integrated STudies at Regional And Local Scales) programme in order to improve our knowledge of aerosols and their impacts on climate in the Mediterranean. Thus, in early summer 2012, the ChArMEx/TRAQA (TRansport and Air QuAlity) campaign focused on the characterization of the polluted air masses over the Mediterranean basin through the study of representative case studies. A particularly intense dust event has been measured at the end of June with different observation means (balloons, aircraft, surface and remote-sensing measurements) and consequently represents a documented case to evaluate the aerosol schemes of regional climate models. Indeed the analysis of study cases is made possible by the use of a reanalysis as lateral boundary forcing which provides the real chronology of these events.

The aim of the present work is consequently to evaluate the direct and semi-direct effects of dust particles during summer 2012 both at the daily time scale and at the summer scale. We consider here a modelling approach with the following requirements. First of all, in order to simulate dust outbreaks, models need prognostic dust schemes (emission, transport, deposition) to uplift dust particles from arid areas and transport them in the atmosphere. Many climate models indeed use only monthly aerosol climatology (e.g. Tanré et al., 1984; Tegen et al., 1997) that cannot correspond to this kind of study. However, disregarding the chemistry-transport models (e.g. CHIMERE, MOCAGE) that do not have aerosol–climate interactions, several aerosol schemes already exist in different climate models (e.g. MACC, ECHAM-HAM, IPSL), evaluated in different intercomparison exercises (e.g. AEROCOM, Schulz et al., 2006, ACCMIP, Lamarque et al., 2013). With regards to dust aerosols, most of the climate models can simulate the main patterns of dust emission and transport (Woodage et al., 2010), but large uncertainties remain in the characterization of dust properties and the resulting impact on climate (Huneeus et al., 2011; Mahowald et al., 2013) notably because of differences in dust

emission parameterizations (Todd et al., 2008). Over the Euro-Mediterranean region, several studies have considered the effects of aerosols on climate using simulations with a prognostic scheme, both for anthropogenic aerosols (Zanis, 2009; Vogel et al., 2009; Meier et al., 2012) and dust particles (Santese et al., 2010; Spyrou et al., 2013). Moreover, the role of the Mediterranean Sea is essential in climate feedbacks (Somot et al., 2008; Artale et al., 2010; Herrmann et al., 2011), so that ocean–atmosphere coupled regional models have recently been developed (Krzic et al., 2011; Herrmann et al., 2011; Mariotti and Dell'Aquila, 2012; L'Hévéder et al., 2012; Turuncoglu et al., 2013; Nabat et al., 2015). The importance of this coupling in the aerosol–climate interactions in the Mediterranean has even been recently highlighted (Nabat et al., 2015). However, up to now, aerosol–climate studies with prognostic aerosol schemes have been achieved either with the COSMO (Vogel et al., 2009) or with the RegCM model (Giorgi et al., 2012) and have not included an ocean–atmosphere coupling yet, even if an ocean–atmosphere coupling is currently developed between RegCM and ROMS (Turuncoglu et al., 2013).

In addition, as the Mediterranean is also characterized by local winds, complex coastlines and orography, high-resolution modelling is needed to correctly reproduce the atmospheric circulation (Gibelin and Déqué, 2003; Gao et al., 2006; Giorgi and Lionello, 2008).

From our knowledge, none of these regional models can have simultaneous ocean–atmosphere coupling and prognostic aerosol schemes. In the present study, a new version of the coupled regional climate model system (RCSM) of the CNRM, called CNRM-RCSM5, has been developed, including an aerosol prognostic scheme derived from the GEMS/MACC project (Morcrette et al., 2009; Michou et al., 2014) in addition to the atmosphere, ocean and land-surface components. This new model tool thus complies with all the criteria mentioned above and should be able to help us to evaluate the direct and semi-direct effects of dust aerosols at the daily time scale. The data brought by the TRAQA campaign provide the opportunity to a first evaluation of the dust aerosol scheme before assessing the radiative aerosol effects. Additionally, including the other aerosol species allows a comparison of total aerosol optical depth (AOD) with remote-sensing measurements. Thus the present work aims at studying the radiative effects of dust aerosols in the Mediterranean area during summer 2012. The question of the difference between the use of climatological and prognostic aerosols in this model will also be raised, notably to study the consequences of this choice both on the daily and seasonal (for summer) variability of different meteorological parameters (radiation, temperature, cloud cover).

After a description of the aerosol scheme in Sect. 2 and its evaluation in Sect. 3, the radiative effects of aerosols are studied in Sect. 4 before the concluding remarks in Sect. 5.

2 Methodology

2.1 The CNRM-RCSM5 model

Four different components are included in this regional climate model system: the atmosphere with the regional climate model ALADIN-Climate (Déqué and Somot, 2008; Colin et al., 2010), the ocean with the regional model NEMOMED8 (Beuvier et al., 2010), the land-surface with the model ISBA (Noilhan and Mahfouf, 1996) and the aerosols, simulated interactively within ALADIN-Climate (see details in 2.2). ALADIN-Climate is a bi-spectral semi-implicit semi-Lagrangian regional model with a 50 km horizontal resolution and 31 vertical levels in the present work. The version 5.3 is used here bringing some improvements compared to the previous version 5.2 used in Nabat et al. (2015). As in the version used in Lucas-Picher et al. (2013), the long-wave (LW) radiation scheme is now based on the rapid radiation transfer model (RRTM, Mlawer et al., 1997), while the short-wave (SW) scheme initially developed by Morcrette (1989) has a finer spectral resolution (six bands). We also use here a spectral nudging method described in Radu et al. (2008), which enables us to keep large scales from the boundary forcing and thus impose the true natural climate variability that is essential to represent dust events notably. Here the wind vorticity and divergence, the surface pressure, the temperature and the specific humidity are nudged. The function used imposes a constant rate above 700 hPa and a relaxation zone between 700 and 850 hPa, while the levels below 850 hPa are free. The spatial wavelengths are similarly nudged beyond 400 km, with a relaxation zone between 200 and 400 km. Thus this method gives the model enough freedom to generate the aerosols at the surface while keeping the large scale conditions that are essential to simulate the true chronology.

The ocean model NEMOMED8 and the land surface model ISBA are the same models as used in Nabat et al. (2015). The ocean–atmosphere coupling is achieved by the OASIS3 coupler (Valcke, 2013) at a 3 h frequency, which represents an improvement compared to CNRM-RCSM4 described in Nabat et al. (2015). Note finally that contrary to CNRM-RCSM4, the coupling to the river routine scheme is not included in the present version of CNRM-RCSM5.

2.2 The aerosol scheme in ALADIN-Climate

Until the version 5.2 of ALADIN-Climate aerosols were represented in this model through monthly climatologies of aerosol optical depth for five aerosol types (desert dust, sea salt, black carbon, organic matter and sulphate) distributed vertically according to constant profiles. In the version 5.3 used here, a prognostic aerosol scheme has been included, adapted from the GEMS/MACC aerosol scheme (Morcrette et al., 2009; Benedetti et al., 2011; Michou et al., 2014). It includes the same five aerosol species that can be directly emit-

ted from the surface for dust and sea-salt particles or from external emission data sets for black carbon, organic matter and sulphate precursors. The spatial domain of our simulations has consequently been extended compared to the previous study of Nabat et al. (2015) in order to include all the sources generating aerosols that can be transported over the Mediterranean basin. As far as dust particles are concerned (Middleton and Goudie, 2001; Israelevich et al., 2012), the following sources are notably included in the domain: North African sources (Morocco, Algeria, Tunisia), the Hoggar mountains, the Tibesti Mountains, the Bodélé depression, Libya, Egypt and sources near the Red Sea (northeast Sudan, Djibouti). No aerosol is included in the lateral boundary forcing.

Sea-salt aerosols are generated by wind stress on ocean surface either because of air bubbles bursting at the sea surface or from spume droplets directly torn off the wave crests by the wind. Guelle et al. (2001) have reviewed different approaches to model these processes. The current formulation used in ALADIN-Climate is based on the studies of Guelle et al. (2001) and Schulz et al. (2004) that provide surface mass fluxes at 80 % relative humidity depending on 10 m wind, integrated for the three size bins defined in the scheme: 0.03 to 0.5, 0.5 to 5 and 5 to 20 μm. Note that the size distribution of emitted sea salt also depends on other factors, such as the sea surface temperature (Jaeglé et al., 2011), which are not taken into account in this current version. Dust emission processes depend on several factors such as soil characteristics (chemical composition, humidity, roughness) and surface wind speed. In the GEMS/MACC scheme, the dust parameterization follows Ginoux et al. (2001), who propose a simplified formulation of dust emission based on the wind speed and thresholds according to the fraction of bare soil and soil moisture. In ALADIN-Climate, this function has been replaced by the Marticorena and Bergametti (1995) parameterization that takes into account more soil characteristics coming from the ECOCLIMAP database (Masson et al., 2003), which provides information on the erodible fraction and the sand and clay fractions, allowing a classification of the soil textures. After the determination of an erosion threshold based on the soil distribution, the soil moisture and the roughness caused by nonerodible elements, the horizontal saltation flux is calculated proportionally to the third power of the wind friction velocity. The vertical flux is then inferred from this saltation flux, according to an empirical relationship given by Marticorena and Bergametti (1995), which notably depends on the soil clay content. The emitted dust size distribution is based on the work of Kok (2011). More details about this dust emission parameterization can be found in Nabat et al. (2012), who have used the same dust emission scheme in RegCM4. Once emitted dust particles are integrated in the three dust size bins of the scheme: 0.01 to 1.0, 1.0 to 2.5 and 2.5 to 20 μm.

The external emission data sets for the three other aerosol types come from Lamarque et al. (2010), who have provided inventories at 0.5° resolution of different species for climate

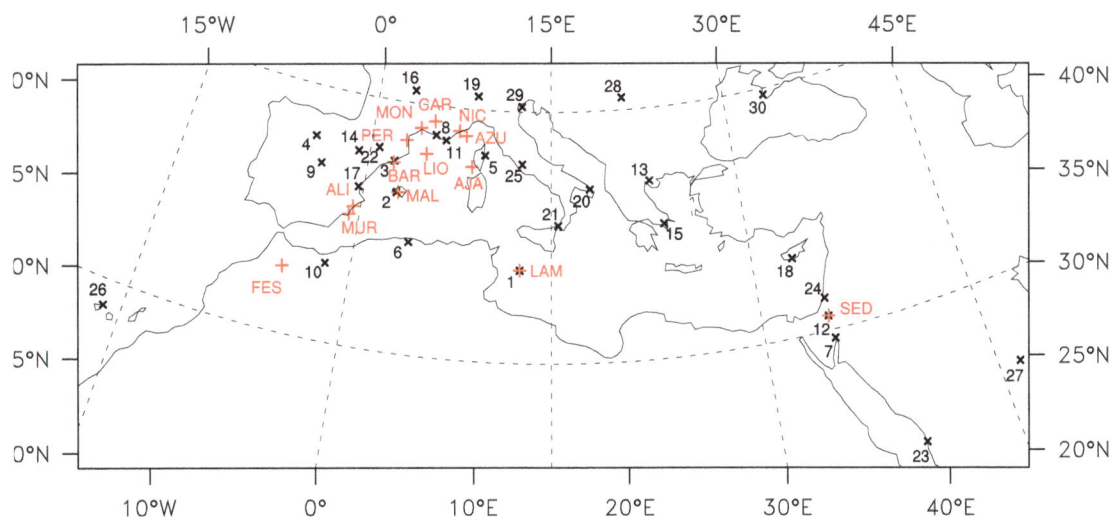

Figure 1. Stations of the AERONET network (black crosses, see the list of the corresponding numbers in Fig. 4). Red crosses indicate the stations providing measurements of surface radiation and temperature (see the list in Table 1).

models. These inventories include numerous sectors such as energy production, industries, domestic activities, agriculture, transport and fires. Organic and black carbon particles are separated between hydrophile and hydrophobic particles. SO_2 emitted particles can be transformed in SO_4, but 5 % of them are directly emitted as SO_4 aerosols (Benkovitz et al., 1996). Volcanic sulfur emissions are also included, as well as dimethylsulfide particles from oceans (see Michou et al., 2014).

All these aerosols gathered in 12 bins are then transported in the atmosphere before possible dry or wet deposition. More details about transport and deposition can be found in Morcrette et al. (2009). Optical properties (single scattering albedo and asymmetry factor) are fixed for each aerosol type, as defined in Nabat et al. (2013). The complexity of this aerosol scheme is similar to the one used in RegCM, but it does not include detailed chemical processes that can be found in COSMO-ART (Vogel et al., 2009). However, it enables our model to keep a low cost of calculations so that multi-annual simulations could be carried out for aerosol–climate studies. Note also that nitrate aerosols are not considered in this model.

2.3 Simulations

Three simulations have been carried out with CNRM-RCSM5, driven by the ERA-Interim reanalysis (Dee et al., 2011) as initial and lateral boundary forcing. First of all, the PROG simulation includes the whole aerosol prognostic scheme described previously. Secondly, in order to estimate the effect of aerosols on meteorological variables such as temperature and radiation, a simulation without aerosols is needed: the NO simulation does not include any aerosols. Thirdly, as the objective of this study is also to discuss the

choice of using climatological or prognostic aerosols, another simulation, called PROG-M, uses monthly AOD provided by PROG so that PROG and PROG-M share the same average aerosol content at the monthly scale. Comparisons between these simulations will enable us to estimate the aerosol effects on the radiative budget and regional climate and the implications of using a prognostic aerosol scheme instead of monthly climatologies. While an improvement on daily SW radiation variability is expected with the use of prognostic aerosols, it is more difficult to answer a priori for other daily parameters, 2 m temperature (T2m) and sea surface temperature (SST), and more generally for consequences on the summer average. The three simulations cover the summer 2012 period from 1 June to 31 August. A 1-month spin-up period has been performed for each simulation in order to have realist aerosol concentrations on 1 June.

2.4 Observation data

For the evaluation of the aerosols and their direct radiative effects, different observed data sets are used in the present work.

Simulated AOD is compared to satellite data from the MODerate resolution Imaging Spectroradiometer (MODIS, collection 5.1, standard and Deep Blue algorithms, 1° resolution; Tanré et al., 1997; Levy et al., 2007), the Multiangle Imaging SpectroRadiometer (MISR, Level3; Kahn et al., 2005, 2010) and the SEVIRI radiometer onboard the geostationary satellite Meteosat Second Generation. For the latter instrument, we use the algorithm of Carrer et al. (2010), which provides high-resolution AOD over both ocean and land surfaces. Nowadays, this algorithm is being implemented on the production chain of the ICARE thematic centre (http://www.icare.univ-lille1.fr) under the name of

Table 1. Stations used for the composite study. The total number of days when observations are available and among them the number of dusty days have been indicated.

Short name	Station	Lat	Long	Available days	Dusty days
MUR	Murcia	37.8	−0.8	83	23
BAR	Barcelona	41.3	2.1	85	10
MAL	Palma de Mallorca	39.6	2.6	74	13
ALI	Alicante	38.3	−0.6	90	15
AJA	Ajaccio	41.6	8.5	88	7
CAR	Carpentras	44.1	5.1	84	4
MON	Montpellier	43.6	4.0	75	7
NIC	Nice	43.7	7.2	88	4
PER	Perpignan	42.7	2.9	80	6
FES	Fès	33.9	−5.0	61	36
LIO	Gulf of Lions (buoy)	42.1	4.6	83	9
AZU	Azur (buoy)	43.4	7.8	78	5
LAM	Lampedusa	35.5	12.6	89	24
SED	Sde Boker	30.9	34.8	92	5

AERUS-GEO (Aerosol and surface albEdo Retrieval Using a directional Splitting method; application to GEO data by Carrer et al., 2014), a daytime averaged product.

Ground-based observations from 30 stations of the AErosol RObotic NETwork (AERONET, Holben et al., 1998, 2001) will also be considered (Fig. 1). These sunphotometer observations provide high-quality data (Level 2.0), which have been downloaded from the AERONET website (http://aeronet.gsfc.nasa.gov). All AOD data have been calculated at 550 nm using the Ångstrom coefficient when necessary to make comparisons and evaluation easier.

The TRAQA campaign has also provided interesting observations for dust aerosols, namely vertical profiles from lidar instruments in Barcelona and San Giuliano (Corsica). The Barcelona lidar system is part of the AC-TRIS/EARLINET network (Aerosols, Clouds, and Trace gases Research InfraStructure Network/European Aerosol Research Lidar Network, Pappalardo et al., 2014). The extinction coefficient profiles were retrieved by means of the two-component elastic lidar inversion algorithm constrained with the AERONET sun-photometer-derived AOD (Reba et al., 2010). In San Giuliano (42.28° N, 9.51° E), aerosol vertical profiles were acquired with a 355 nm backscattering lidar. The aerosol extinction coefficient profiles are estimated using the Klett's method and a fixed lidar ratio (Léon et al., 2015) from hourly averaged attenuated range-corrected lidar signals. Additionally, an ATR-42 research flight operated by SAFIRE (Service des avions français instrumentés pour la recherche en environnement) has also been realized during the TRAQA campaign. This study uses the airborne data from the Passive Cavity Aerosol Spectrometer Probe (PCASP), which measures particles between 0.1 and 3.2 μm.

In addition, the Météo-France and AEMET networks have provided daily radiation and 2 m temperature measurements (see Fig. 1 and Table 1). Radiation measurements have been

completed by the stations of Sde Boker (SED, SolRad-Net network, AERONET website), Lampedusa (LAM, coll. ENEA) and two Météo-France buoys located in the Gulf of Lions (LIO) and near the French Riviera (AZU). Lampedusa and the two buoys also provide SST measurements. All 14 stations providing surface radiation and temperature have been added in Fig. 1 (red crosses). It is worth mentioning that available data are provided by stations that are located for most of them in the western Mediterranean. However, in summer most of the dust outbreaks occur in this region because of frequent low-pressure systems over Morocco that favour the dust export over the western Mediterranean (Moulin et al., 1998; Gkikas et al., 2012).

Additionally, the MACC reanalysis (Morcrette et al., 2009) is also used in the present work as a means of evaluating the CNRM-RCSM5 simulations. This reanalysis includes data assimilation of AOD from the MODIS instrument.

3   Evaluation of the simulated aerosols

In this section, an evaluation of the simulated aerosols during summer 2012 is carried out against different available observations and climatologies. Depending on the parameter, several types of data sets are indeed required.

3.1   Total AOD: spatial evaluation

The AOD spatial distribution is firstly evaluated against different satellite products (MODIS, MISR and AERUS-GEO). The average total AOD in summer 2012 for each data set is shown in Fig. 2. The general spatial pattern shows a good agreement between satellites and CNRM-RCSM5. The highest values (up to 1.5) are indeed found over northern Africa and Arabian peninsula while the Mediterranean Sea is af-

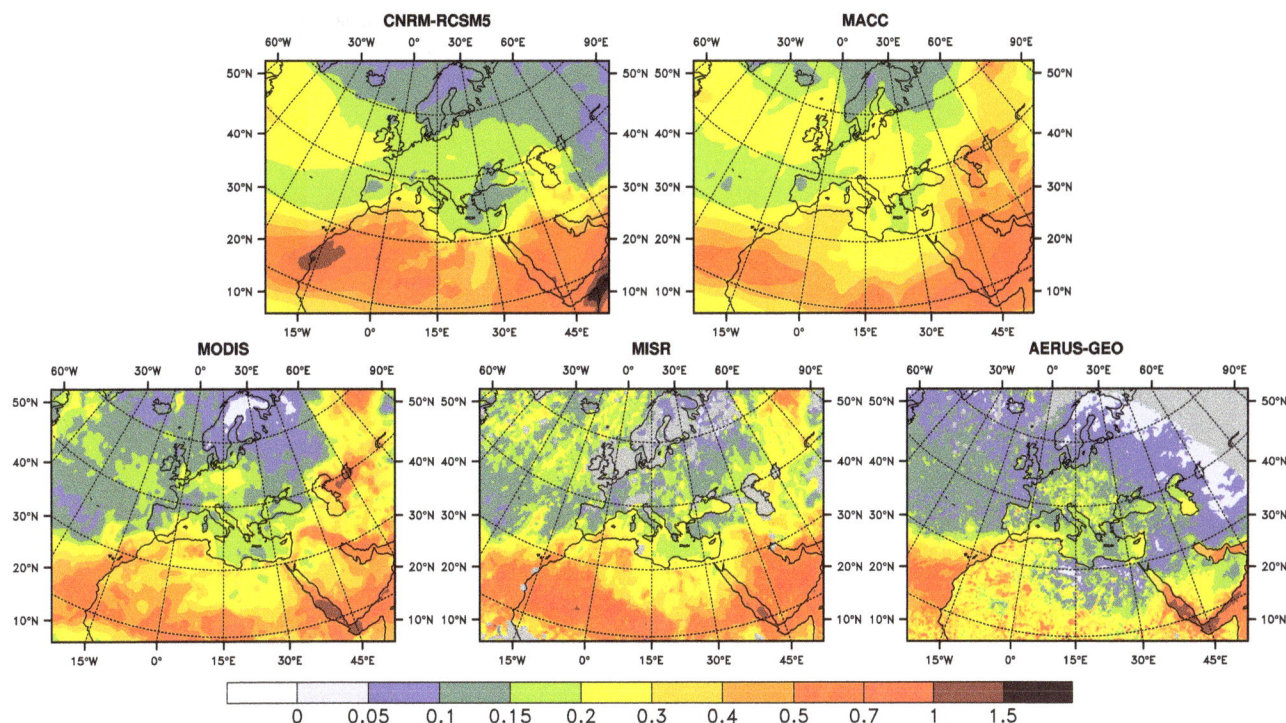

Figure 2. Mean aerosol optical depth at 550 nm in summer 2012 (JJA) simulated by CNRM-RCSM5 and MACC (top) and measured by three satellite instruments (MODIS, MISR and AERUS-GEO, bottom).

Table 2. Spatial correlation coefficients between AOD of the different data sets presented in Fig. 2.

Data sets	MODIS	MISR	AERUS-GEO	MACC
CNRM-RCSM5	0.64	0.77	0.65	0.74
MODIS		0.81	0.69	0.84
MISR			0.68	0.84
AERUS-GEO				0.61

fected by moderate AOD, ranging from 0.15 to 0.3, from the north-east to the south-west.

In greater detail, some differences can be noted between the model and satellite data. CNRM-RCSM5 AOD is closer to MISR over northern Africa, where a large zone of AOD higher than 0.5 can be identified in both data sets, while MODIS and especially AERUS-GEO show lower AOD. Similar conclusions can be drawn for the Arabian peninsula. Dust export over the Atlantic Ocean is, on the contrary, in very good agreement between the five products (AOD between 0.5 and 0.7). Over western and central Europe, MISR AOD is lower than MODIS, AERUS-GEO and CNRM-RCSM5. Large differences in AOD are also present in Eastern Europe and Russia, where MODIS shows higher AOD than the other data sets. However, this region is in the limit of the domain seen by SEVIRI (lower values in AERUS-GEO) and is also close to the border of the domain used in CNRM-

RCSM5, so that aerosols over this region may come from outside the domain. Finally, AOD over the northern Atlantic Ocean is higher in CNRM-RCSM5 than in satellite products, but the presence of numerous clouds in this area limits the quality of the satellite data there.

In summary, Table 2 presents the spatial correlations between these four products. All the correlations are higher than 0.6, confirming the general agreement and the ability of CNRM-RCSM5 to reproduce the main spatial patterns of AOD.

3.2 Total AOD: temporal evaluation

As far as the temporal dimension is concerned, an evaluation has been realized against ground-based measurements from the AERONET network in the Mediterranean area in terms of daily means. Indeed, AERONET measurements benefit from a higher temporal resolution than data from moving satellites and their accuracy is generally higher, about ± 0.01 (Holben et al., 1998) compared to ± 0.05 for satellites (Kahn et al., 2010; Levy et al., 2010). Figure 3 shows four temporal series across the Mediterranean basin at Oujda (a, Morocco, number 10 in Fig. 1), Mallorca (b, Spain, 2), Frioul (c, France, 8) and Lampedusa (d, Italy, 1). All these series show high daily variability because of frequent dust outbreaks in this season. The spectral nudging technique used in CNRM-RCSM5 enables the model to reproduce the true chronology of the synoptic meteorological conditions as shown in Herrmann

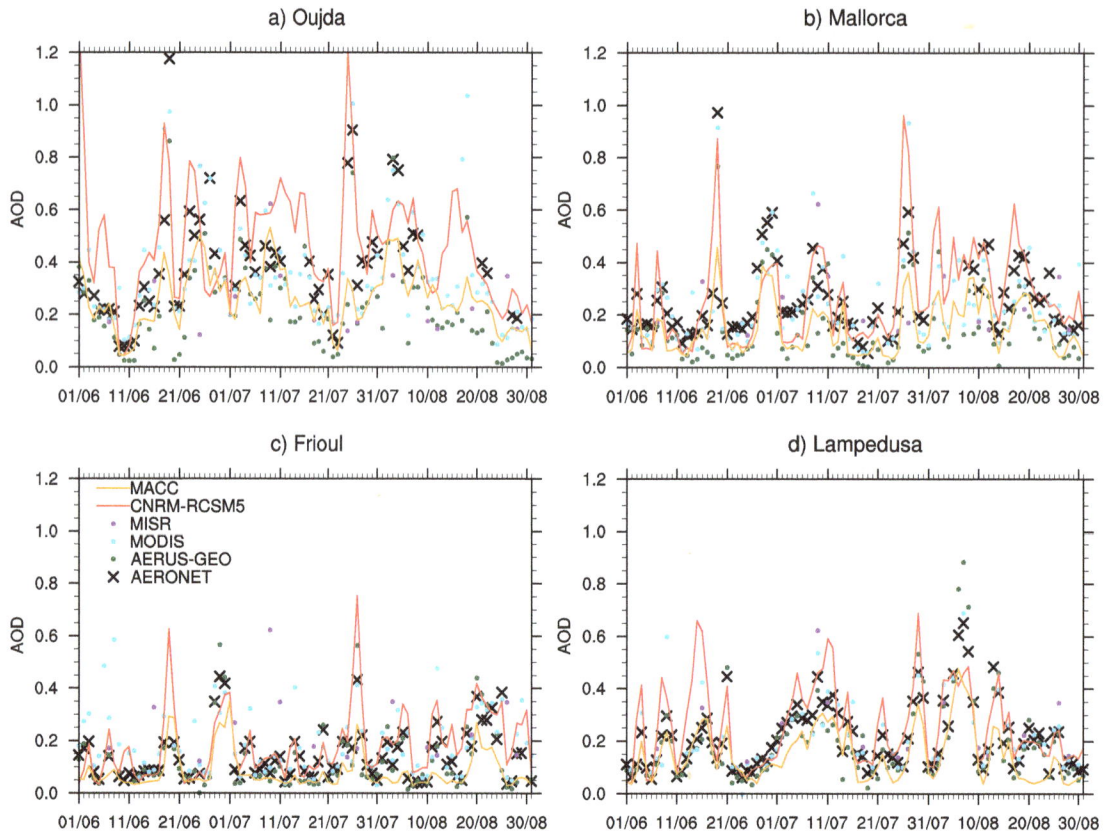

Figure 3. AOD (at 550 nm) temporal series between 1 June and 31 August 2012 simulated by CNRM-RCSM5 (red lines) and MACC (orange lines) and observed by AERONET sun photometers (black crosses), MODIS (blue points) , MISR (purple points) and AERUS-GEO (green points) at four stations of the AERONET network: Oujda (**a**, number 10 in Fig. 1), Mallorca (**b**, 2), Frioul (**c**, 8) and Lampedusa (**d**, 1).

et al. (2011), which is useful for driving dust emission in the present work. As a result, the model is able to reproduce the intensity and the chronology of most AOD peaks, such as those observed in Oujda (18 June, 25 July) in Mallorca (19 June, 9 July, 10 August), Frioul (28 June, 19 August) and Lampedusa (21 June, 13 August). However, CNRM-RCSM5 overestimate a few dust events (e.g. 19 June in Frioul, 15 June in Lampedusa), but these differences remain in the minority.

Similar comparisons have been realized for 30 AERONET stations (see their locations in Fig. 1), the results are presented in a Taylor diagram (Fig. 4, adapted to daily time series from Taylor, 2001). This diagram represents three statistics: the correlation coefficient is the azimuth angle, the radial distance from the origin is the standard deviation normalized by observations, and the distance to the "REF" point on the x axis is the root-mean-square error (RMSE). The average temporal correlation coefficient for CNRM-RCSM5 is 0.70, while the ratio between simulated and observed standard deviations is 1.01, revealing the ability of the aerosol scheme to reproduce AOD daily variability. In addition, CNRM-RCSM5 has no station with very low scores and has a low

mean bias both when considering all 30 stations (0.02) and only the stations to the south of 33° N (0.03).

Additionally, the daily values for the satellite products have been added in Figs. 3 and 4 as information for data users. It is indeed important to note that in terms of daily variability, (1) MODIS and AERUS-GEO have a higher temporal correlation with AERONET (0.73 and 0.76 respectively) than MISR (0.15), probably because of a reduced number of available retrievals with this instrument; (2) AERUS-GEO has the best scores among the satellite products; (3) MODIS and AERUS-GEO have, however, respectively 5 and 3 stations with RMSE higher than 1.25; and (4) all these products have a higher mean bias than CNRM-RCSM5.

3.3 Contribution of aerosol species to AOD

Satellites and ground-based measurements do not provide the contribution of the different aerosol types to AOD (the distinction between coarse and fine modes is not sufficient), which is the reason why a comparison to the MACC reanalysis (Morcrette et al., 2009; Benedetti et al., 2011) and the AOD climatology from Nabat et al. (2013), named NAB13 thereafter, is presented in this section. Note that total AOD

Table 3. Total AOD and components for the five aerosol types simulated by CNRM-RCSM5 and the MACC reanalysis in summer 2012 over Europe (continental area up to 30° E), the Mediterranean Sea and northern Africa (continental area up to 25° N). Averages in summer from NAB13, the climatology of Nabat et al. (2013), have also been indicated with the minimum and maximum summer values (period 2003–2009). Total AOD from satellite data (MODIS, MISR, AERUS-GEO) is also given.

Europe	CNRM-RCSM5	MACC	NAB13	MODIS	MISR	AERUS-GEO
Sea salt	0.01	0.02	0.00 [0.00–0.00]	–	–	–
Desert dust	0.04	0.06	0.05 [0.04–0.05]	–	–	–
Organic matter	0.04	0.02	0.02 [0.02–0.03]	–	–	–
Black carbon	0.01	0.01	0.01 [0.01–0.01]	–	–	–
Sulphate	0.08	0.10	0.10 [0.08–0.12]	–	–	–
Total	0.18	0.21	0.18 [0.16–0.20]	0.16	0.15	0.15
Mediterranean						
Sea salt	0.01	0.02	0.01 [0.00–0.01]	–	–	–
Desert dust	0.11	0.10	0.12 [0.10–0.13]	–	–	–
Organic matter	0.03	0.02	0.01 [0.01–0.02]	–	–	–
Black carbon	0.01	0.01	0.01 [0.00–0.01]	–	–	–
Sulphate	0.07	0.09	0.08 [0.07–0.10]	–	–	–
Total	0.23	0.24	0.23 [0.19–0.25]	0.20	0.22	0.18
Africa						
Sea salt	0.00	0.01	0.00 [0.00–0.00]	–	–	–
Desert dust	0.37	0.18	0.31 [0.25–0.33]	–	–	–
Organic matter	0.02	0.02	0.01 [0.01–0.02]	–	–	–
Black carbon	0.01	0.01	0.01 [0.01–0.01]	–	–	–
Sulphate	0.05	0.07	0.08 [0.06–0.09]	–	–	–
Total	0.45	0.29	0.41 [0.33–0.44]	0.33	0.32	0.21

of NAB13 corresponds to MODIS AOD by definition of this product and that the total AOD of MACC has been added in Figs. 2, 3, 4 and Table 2 as information for data users.

Figure 5 presents the mean AOD for summer 2012 for the five simulated aerosol types. Dust aerosols prevail in the southern part of the domain because of sources in Sahara and in the Arabian peninsula, while anthropogenic particles, especially sulphate and organic matter, are responsible for local maxima in AOD in Europe. Sea-salt particles are essentially simulated over the Atlantic Ocean, as well as the western Mediterranean Sea in lower quantities.

The different contributions to AOD for each aerosol type are given in Table 3 for CNRM-RCSM5, MACC and NAB13. NAB13 is based on both model and satellite data, and MACC is based on model and data assimilation. NAB13, which gives reliable estimations of the different AOD components, is only available on the 2003–2009 period, so that the average over this period with the minimum and maximum values have been indicated. Averages have been calculated on the three domains defined in Nabat et al. (2013): Europe, the Mediterranean Sea and northern Africa.

Over Europe, CNRM-RCSM5 is very close to NAB13 for total AOD (0.18 on average) and the five aerosol types, even if the sharing between organic matter and sulphate aerosols is slightly different. MACC simulate more dust and sulphate particles, but the three satellites' data have lower AOD (be-

tween 0.15 and 0.16) so that CNRM-RCSM5 AOD is median. Over the Mediterranean Sea a good agreement is shown among CNRM-RCSM5 (0.23 for total AOD), MACC (0.24) and NAB13 (0.23). In addition, the proportion among the different aerosol types is similar in the three data sets. However, as in Europe, satellite data have lower AOD (between 0.18 and 0.22).

More variability is noted with regards to AOD over northern Africa, notably because of the dust component. CNRM-RCSM5 shows higher AOD (0.45) than NAB13 (0.41), MACC (0.32) and the satellite data (between 0.21 and 0.33). However, interannual variability is stronger in this region as shown by the larger amplitude in NAB13 (0.33–0.44). Moreover, MACC does not assimilate AOD over the Sahara because the standard algorithm of MODIS cannot retrieve AOD on bright surface, so that an underestimation of dust aerosols in MACC had been identified (Nabat et al., 2013).

In summary, the evaluation of AOD for each aerosol type is complicated because of the heterogeneity among the different data sets, but the contribution of aerosol types to AOD in CNRM-RCSM5 is close to that in MACC and NAB13. It is worth mentioning that CNRM-RCSM5 does not include the nitrate component. However, dust aerosols constitute the main focus of the following paragraphs.

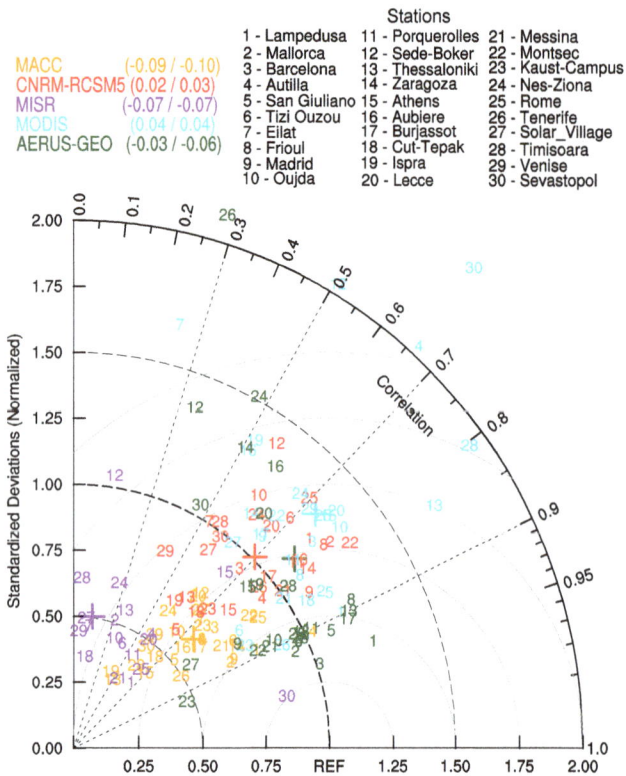

Figure 4. Taylor diagram evaluating CNRM-RCSM5 (red), MACC (orange) and satellite (MODIS in blue, MISR in purple and AERUS-GEO in green) data against 30 AERONET ground-based observations in terms of daily AOD in summer 2012. Averages over the 30 stations for each data set are indicated with crosses. The mean bias against AERONET is indicated in the caption between brackets (all 30 stations/9 stations located to the north of 33° N).

3.4 Dust extinction vertical profile

CNRM-RCSM5 has shown its ability to reproduce AOD daily evolution correctly, which is a parameter often evaluated in climate models. However, aerosol direct and semi-direct forcing also depend on the profile and size distribution of particles, rarely evaluated given the scarcity of observations and affected by large uncertainties (Textor et al., 2006). Even if total AOD is necessary to evaluate AOD against in situ or satellite measurements that cannot separate the different aerosol types, more attention is now given to the dust component which is the focus of this study. The TRAQA campaign has well documented a dust outbreak over the Mediterranean Sea, which is useful for this evaluation. However, a deeper evaluation of the other aerosol components is outside of the scope of this paper.

The dust plume observed in the TRAQA campaign comes from the uplift of dust particles in western Africa between 21 and 23 June. These particles have been transported along the African coast up to southern Spain, driven by the presence of a low pressure system over Morocco and a high pres-

sure area over the Azores. From 26 June, a low formed in the bay of Biscay generated a south-westerly flow, bringing the dust plume over northern Spain. Successively moving to the southeast, dust particles have also been transported over the Mediterranean Sea. Figure 6 presents the vertical distribution of aerosols during the dust outbreak observed by lidars in the TRAQA campaign in terms of extinction coefficient in Barcelona at 532 nm and in San Giuliano at 355 nm. Dust aerosols first reach Spain on 27 June, transported in the mid-troposphere, as noted in the profile between 2000 and 5000 m with a maximum extinction ($0.18\,\text{km}^{-1}$) at 3500 m. The two-component elastic lidar inversion algorithm constrained with an AERONET AOD of 0.32 gave a column-equivalent lidar ratio of 54 sr. This value is in the range of 50–70 sr established by Tesche et al. (2009) of desert dust lidar ratio observations by Raman lidar, which makes us confident of the result of the lidar inversion. The altitude of these dust particles is quite similar in CNRM-RCSM5 despite an underestimation of the intensity of the dust outbreak and a slight overestimation in the higher layers. Under this dust layer, the presence of sulphate aerosols is noted in the model, with an extinction coefficient close to observations ($0.03\,\text{km}^{-1}$). In San Giuliano, where the dust plume has arrived 3 days later, its altitude is also similar in CNRM-RCSM5 and observations: between 2000 and 5000 m. As in Barcelona, extinction is slightly overestimated in the high troposphere (above 6500 m).

In summary, the dust extinction simulated profiles have been evaluated against these lidar profiles, showing the variability in the altitudes of dust aerosols. It should also be mentioned that two profiles are not sufficient to conclude on the ability of the model to estimate the dust vertical distribution. This kind of comparison would need to be done for other places and situations; however, it is a difficult exercise because evaluating only the aerosol vertical distribution requires finding cases where adequate observations are available and where the model correctly simulates the transport of dust aerosols.

3.5 Dust vertical size distribution

Size distribution is also an essential physical parameter for aerosol–climate studies, as optical properties depend on the particle size. Figure 7 presents the size distribution observed during a sounding realized by the ATR42 during the TRAQA campaign as well as the simulated distribution. Note that the bin scheme used in CNRM-RCSM5 does not enable the model to reproduce exactly the observed distribution, but the division in three bins for dust particles notably can still be evaluated. This sounding took place in the Mediterranean Sea (43.05° N, 9.55° E) on 29 June, when the dust plume has been transported over this area. In the lower layers, a first maximum is observed in the smallest particles (around $0.1\,\mu\text{m}$), probably due to sulphate aerosols, as represented by CNRM-RCSM5. The observed distribution shows that

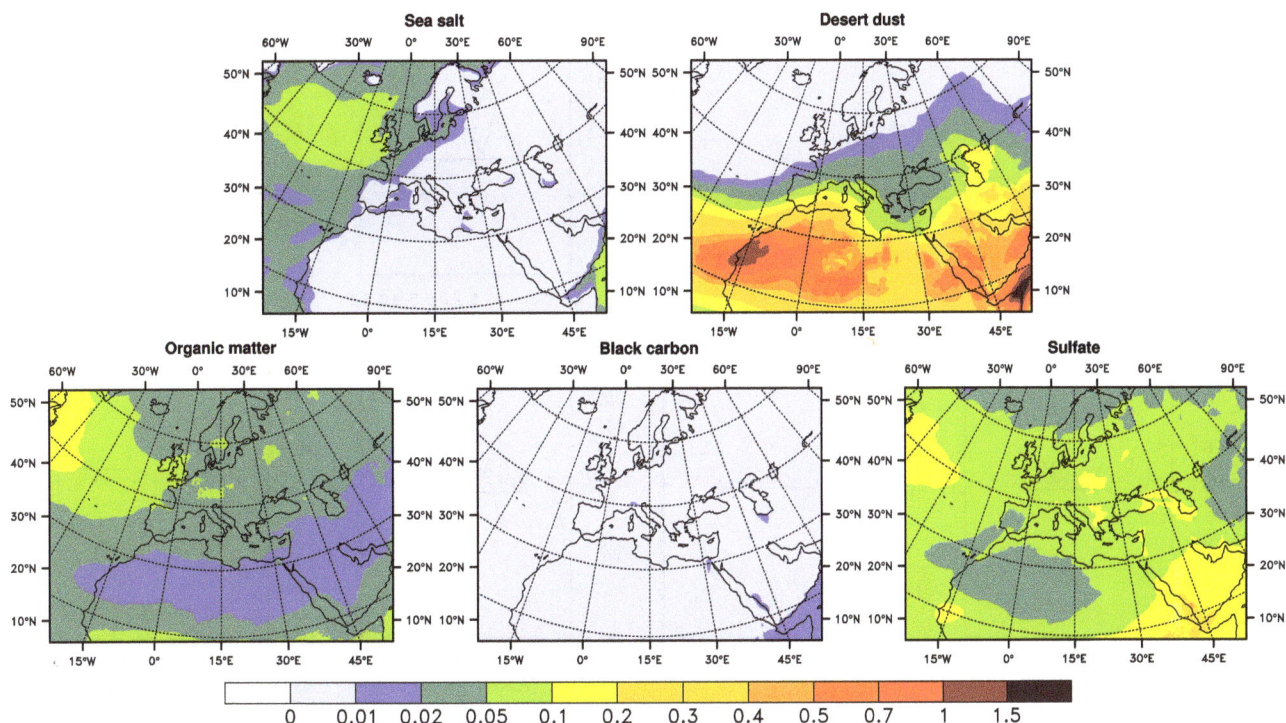

Figure 5. Mean aerosol optical depth at 550 nm in summer 2012 (JJA) simulated by CNRM-RCSM5 for the five aerosol types (sea salt, desert dust, organic matter, black carbon and sulphate).

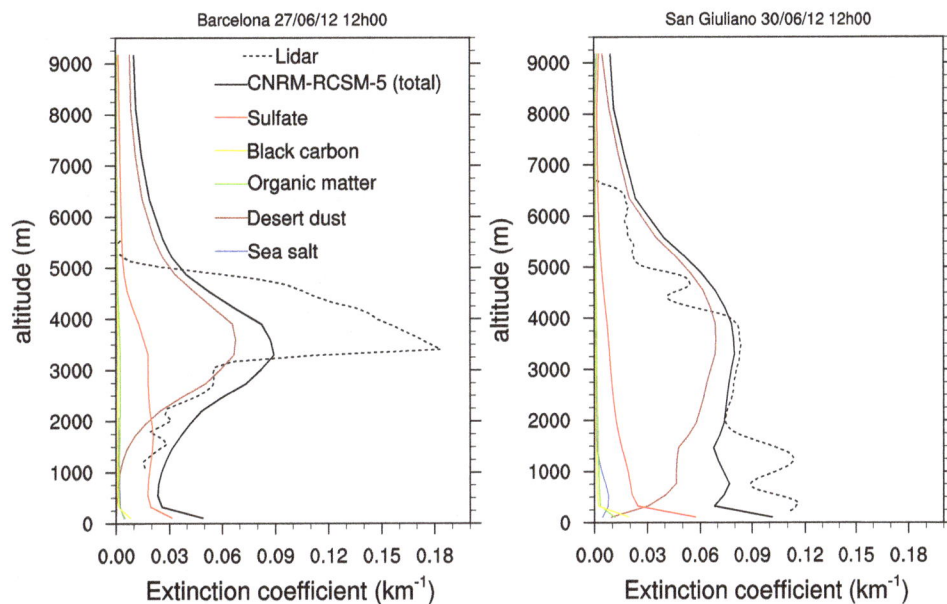

Figure 6. Aerosol extinction coefficient simulated by CNRM-RCSM5 (full black lines) and observed by a ground-based lidar (dotted black lines) in Barcelona on 27 June at 12:00 UTC (left) and in San Giuliano (Corsica) on 30 June 2012 at 12:00 UTC (right). The different coloured lines represent the contribution of each aerosol type to the extinction coefficient.

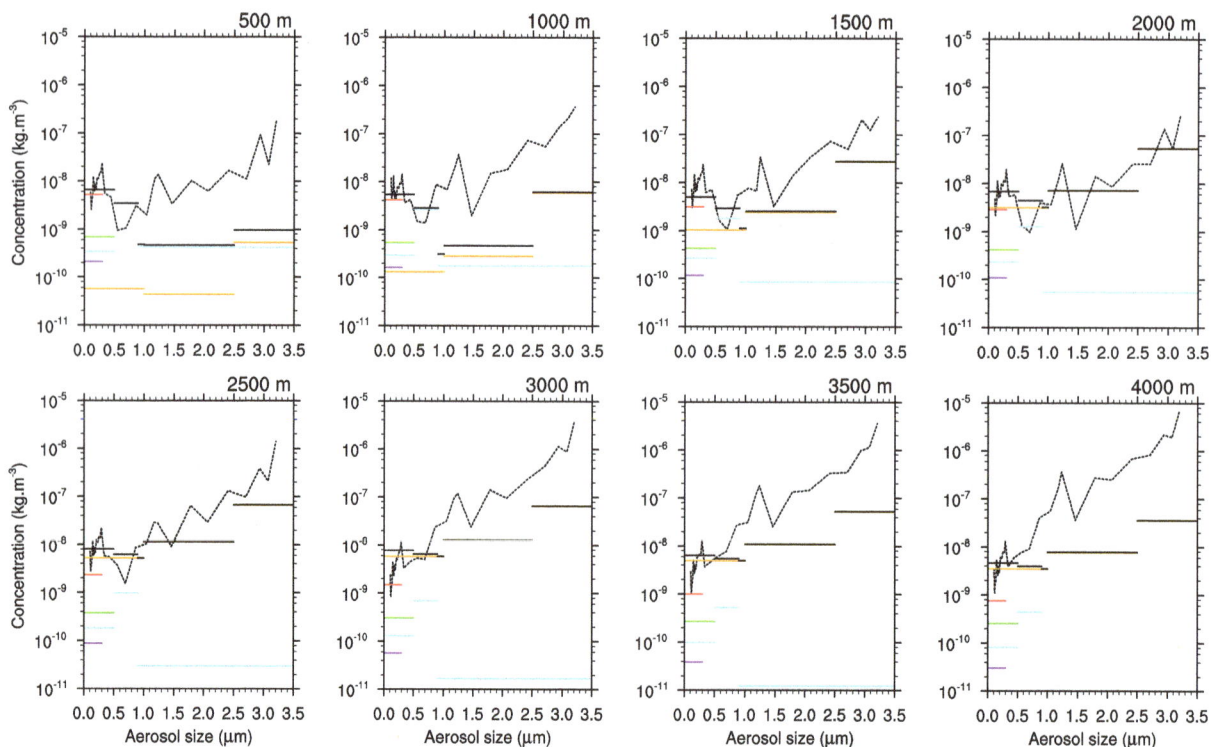

Figure 7. Dust particle size distribution observed by the PCASP instrument onboard ATR42 (flight 22) on 29 June at 08:00 UTC (dashed black lines), the dust refractive index has been adjusted (1.53–0.002i). Full coloured lines indicate the aerosol concentration for each aerosol bin of CNRM-RCSM5 (red is sulphate, blue is sea salt, orange is dust, green is organic matter and purple is black carbon), while full black lines indicate the total concentration ($kg\,m^{-3}$).

mass concentration is higher for larger particles, especially between 2000 and 4000m, where dust aerosols are located. This distribution is simulated by CNRM-RCSM5, notably between 2000 and 3000 m. Above 3000 m, coarse particles (larger than 2.0 μm) are underestimated. However, these particles have less impact on extinction in SW radiation than submicronic particles, but they could play a role in other processes (e.g. deposition).

These results finally show that the aerosol vertical and size distributions simulated by CNRM-RCSM5 reproduce the main patterns seen in observations from the TRAQA campaign, even if the simulated profile in Barcelona shows an underestimated extinction peak between 3 and 5 km in altitude.

To summarize, we have shown in this section the strengths and the weaknesses of CNRM-RCSM5 to simulate the evolution of aerosols during summer 2012 in terms of spatial pattern and daily variability, as well as the vertical profiles and size distribution of dust particles. This model will be used in the following section to study the impact of dust outbreaks on meteorological parameters (radiation, temperature) in summer 2012. In addition, an intercomparison modelling study about this dust event observed in the TRAQA campaign will be the subject of a parallel study led by Sara Basart.

4 Aerosol radiative effects

As seen previously in the AOD temporal series, the Mediterranean basin has been affected by frequent dust outbreaks in summer 2012. This section aims at assessing their impact on different meteorological parameters.

4.1 Direct radiative forcing (DRF)

Figure 8 first shows the daily direct SW DRF of aerosols in PROG. DRF is calculated online during the simulation, calling twice the radiation code: with and without aerosols. A negative forcing of aerosols at the surface is noted. It is stronger over regions under dust influence – northern Africa, Arabian peninsula and the tropical Atlantic Ocean – reaching -20 to $-50\,Wm^{-2}$, in line with Nabat et al. (2015). Over Europe and the northern Atlantic, aerosol DRF ranges from -10 to $-15\,Wm^{-2}$ notably because of sulphate aerosols. Compared to estimations from literature such as the studies of di Sarra et al. (2008) and Di Biagio et al. (2010), who have found an average DRF of -30 and $-26\,Wm^{-2}$ respectively in Lampedusa, the values given by CNRM-RCSM5 have the same order of magnitude even if they can reach larger forcings. Also note that the Atlantic Ocean off Africa, under the influence of dust export, shows the highest variability.

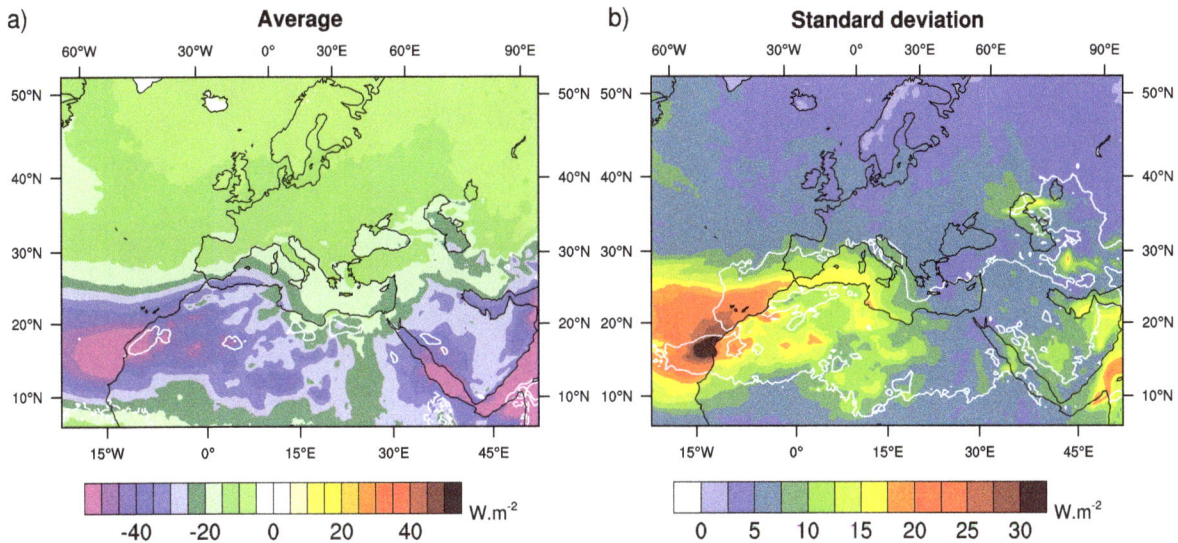

Figure 8. Aerosol SW direct radiative forcing (DRF): (**a**) Average in summer 2012 for PROG (colours) and the PROG-PROG-M difference (white lines, interval is 5 Wm^{-2}). (**b**) Standard deviation of daily DRF for PROG (colours). The white line indicated the region where the ratio between the standard deviations of PROG and PROG-M is higher than 2.

4.2 At the daily scale

As dust aerosols can interact with solar and thermal radiation, consequences on meteorological parameters such as surface radiation and temperature might be expected. In the present work, an effort has been made to gather colocalized measurements of AOD, SW radiation and 2 m temperature or sea surface temperature. The list of the 14 corresponding stations in the Mediterranean basin used in this study is presented in Table 1.

Daily series of solar surface radiation (SSR), cloud cover and surface temperature are presented in details for two stations representative of the Mediterranean basin, namely Lampedusa (LAM) and the buoy in the Gulf of Lions (LIO). Lampedusa is located in an island close to dust-emitting regions where clear-sky conditions are frequent in summer, while LIO is in the northwestern Mediterranean, where more clouds are observed. Figures 9 and 10 present respectively in LAM and in LIO the daily series of AOD, downward SSR, cloud cover and surface temperature (2 m temperature and SST respectively), observed and simulated by PROG, PROG-M and NO.

First of all, NO is the only CNRM-RCSM5 simulation to have a high bias against observed SSR ($+18.0$ Wm^{-2} in LAM, 31.2 Wm^{-2} in LIO) compared to PROG-M (-6.0 Wm^{-2} in LAM, 13.6 Wm^{-2} in LIO) and PROG (-3.5 Wm^{-2}, 15.9 Wm^{-2} in LIO) due to the absence of aerosols in NO. While the aerosol climatology is enough to reduce the bias in PROG-M, PROG has the highest temporal correlation (0.87 against 0.81 for NO and 0.85 for PROG-M in LAM), and its standard deviation is the closest to observations (a ratio of 0.88 against 0.74 both for NO and PROG-

M in LAM). Indeed, PROG-M and NO clearly miss some variations of SSR. When AOD is high (e.g. 21/06, 3–12/07, 29/07, 7/08 in LAM, 19/06, 27/07, 20/08 in LIO), PROG-M and NO overestimate SSR, especially in case of low cloud cover. Inversely when AOD is low (e.g. 24/06, 20/07, 10/08 in LAM, 5/06, 27/08 in LIO), PROG-M underestimates SSR while NO benefits in this case from the absence of aerosols. ERA-Interim has a monthly aerosol climatology similar to PROG-M except that the aerosol climatology used in ERA-Interim (Tegen et al., 1997) is probably less realistic and simulates radiation variations lower than observed. As a result, the effect of aerosols on surface radiation has been identified in both stations.

With regards to land surface temperature in LAM and SST in LIO, the three CNRM-RCSM5 simulations have similar temporal correlations (between 0.72 and 0.73 for LAM, 0.98 for SST in LIO), while PROG-M and PROG are on average cooler than NO because of the aerosol forcing. Even during dust outbreaks, it is not possible to state that average temperature in PROG is closer to observations. With regards to standard deviations, the daily variability is reduced in PROG (0.89 in LAM against 0.92 for PROG-M and 0.95 for NO). The aerosol forcing during dust events could indeed decrease the maximum daily temperature, while the effect of dust particles on thermal surface radiation (TSR) could increase night-time temperature and thus reduce T2m diurnal variability.

In order to confirm these results in the other stations, the evaluation of surface radiation and 2 m temperature for the three simulations and the ERA-Interim reanalysis in the 14 stations is presented respectively in Tables 4 and 5. As far as radiation is concerned, the bias is reduced both in PROG and

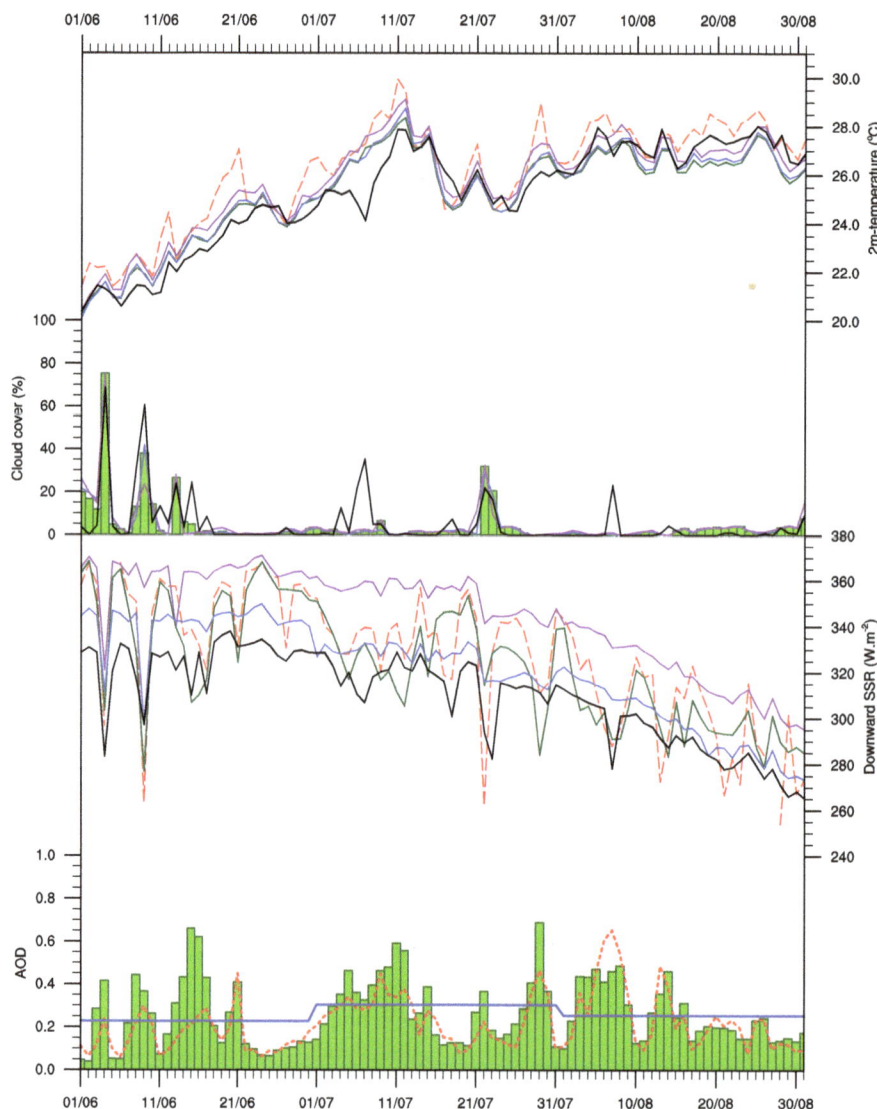

Figure 9. Cloud cover (%, green bars for PROG, curves for the other simulations), 2 m temperature (°C, curves), cloud cover (%, green bars for PROG, curves for the other simulations), downward SSR (Wm^{-2}, curves) and AOD (green bars for PROG, blue line for PROG-M), from top to bottom, in Lampedusa (Italy) for PROG (green), PROG-M (blue), NO (purple), ERA-Interim (black) and observations (dashed red).

PROG-M, reaching a level close to ERA-Interim (between 11 and 13 Wm^{-2}). A net improvement is noted in temporal correlation, since it is higher in PROG than in PROG-M and NO in every station. Daily variability in SSR is also higher in PROG for most stations, representing an improvement compared to observations except where this variability was already overestimated (e.g. Ajaccio). It is worth mentioning that in Sde Boker PROG gets closer to observations by reducing SSR variability. A misrepresentation of cloud processes could also explain some of the discrepancies with observations. The lack of cloud cover in CNRM-RCSM shown in Nabat et al. (2015) could explain the remaining bias. ERA-Interim, which does not have the daily aerosol variations and consequently misses some peaks in surface radiation, suc-

ceeds in getting a high average correlation coefficient (0.79) probably because of a better representation of clouds. Moreover, changes in water vapour column amount may also affect the SSR to a lesser extent.

As far as surface temperature is concerned, no change in correlation coefficient is noted. The PROG simulation is cooler than NO and PROG-M, increasing the negative bias. Nevertheless the daily variability is slightly reduced, getting closer to observed variability. In addition, it is worth mentioning that ERA-Interim has the highest scores in terms of correlation and variability (standard deviation), probably benefiting from the assimilation of surface temperature (Dee et al., 2011).

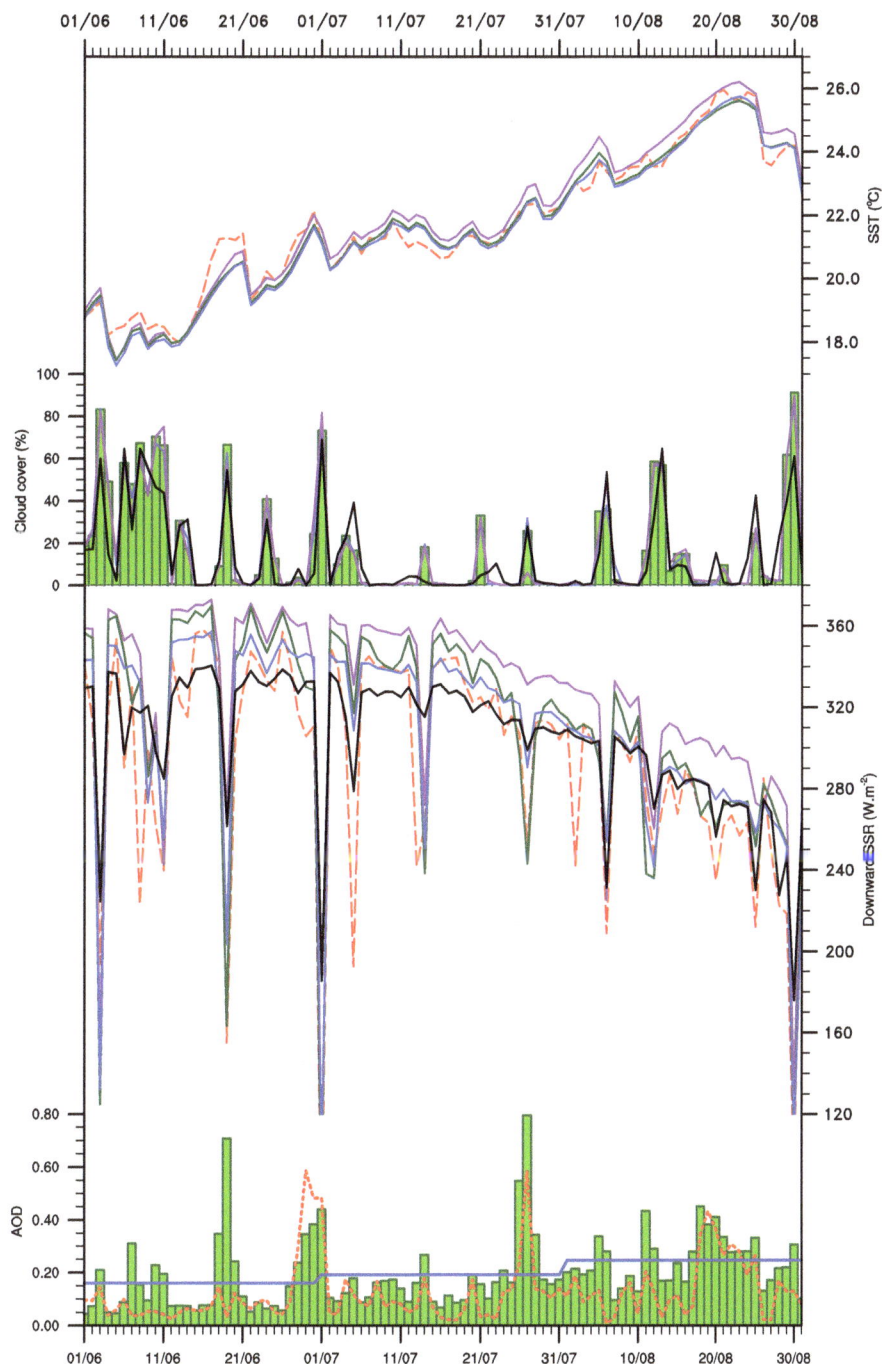

Figure 10. Same as Fig. 9 but for the buoy in the Gulf of Lions (LIO) and SST instead of 2 m temperature.

As a result, these comparisons show that the prognostic aerosol scheme used in PROG enables the model to better reproduce the evolution of surface radiation, which cannot be done properly with an aerosol climatology. Besides, no improvement has been shown in the scores of land and sea surface temperature. However, aerosol maxima over the Mediterranean could be associated to particular weather conditions which are responsible for effects on radiation and temperature that are not due to aerosols. That is the reason why a composite study to isolate the effect of dust aerosols is carried out in the following section.

Table 4. Evaluation of daily SSR simulated by NO, PROG-M, PROG and ERA-Interim against 14 ground-based measurements located around the Mediterranean basin, in terms of bias (Wm^{-2}), temporal correlation coefficient and standard deviation (SD) ratio.

Bias	MUR	BAR	MAL	ALI	AJA	CAR	MON	NIC	PER	FES	LIO	AZU	LAM	SED	MOY
NO	31.0	2.8	54.3	39.0	18.0	22.3	35.9	37.6	34.7	48.2	31.2	35.1	18.0	5.6	29.6
PROG-M	7.6	−8.5	35.1	18.1	2.4	10.1	20.8	19.6	19.3	13.6	13.6	16.6	−6.0	−13.4	10.6
PROG	9.7	−7.5	36.0	21.2	5.1	11.5	24.0	22.9	21.0	16.5	15.9	19.1	−3.5	−11.7	12.9
ERA-Interim	12.8	4.6	53.7	25.4	−1.0	−4.3	17.0	10.1	27.7	34.7	10.2	7.2	−16.8	−12.9	12.0

Corr.															
NO	0.72	0.76	0.66	0.62	0.87	0.89	0.71	0.67	0.76	0.39	0.87	0.86	0.81	0.84	0.75
PROG-M	0.76	0.77	0.65	0.67	0.89	0.87	0.70	0.69	0.77	0.49	0.88	0.87	0.85	0.89	0.76
PROG	0.77	0.79	0.69	0.74	0.89	0.91	0.75	0.69	0.78	0.53	0.89	0.90	0.87	0.90	0.79
ERA-Interim	0.79	0.81	0.88	0.81	0.88	0.88	0.77	0.68	0.75	0.37	0.90	0.76	0.87	0.88	0.79

SD															
NO	0.79	1.20	0.84	1.16	1.11	0.96	0.97	0.81	0.96	0.93	0.92	1.00	0.74	1.15	0.97
PROG-M	0.79	1.10	0.82	1.11	1.10	0.86	0.92	0.81	0.93	0.94	0.91	1.04	0.74	0.99	0.93
PROG	0.95	1.16	1.01	1.20	1.17	0.94	0.98	0.88	0.99	1.01	1.01	1.12	0.88	1.07	1.03
ERA-Interim	0.58	0.72	0.69	0.78	0.78	0.77	0.69	0.72	0.63	0.53	0.61	0.90	0.67	0.92	0.71

Table 5. Evaluation of daily 2 m temperature simulated by NO, PROG-M, PROG and ERA-Interim against 13 ground-based measurements located around the Mediterranean basin, in terms of bias ($°$C), temporal correlation coefficient and standard deviation (SD) ratio.

Bias	MUR	BAR	MAL	ALI	AJA	CAR	MON	NIC	PER	FES	LIO	AZU	LAM	MOY
NO	0.3	−1.6	1.2	−0.5	−1.5	0.9	−1.5	−0.0	−2.0	0.0	0.6	1.6	−0.4	−0.3
PROG-M	−0.6	−1.7	0.8	−0.7	−1.7	0.8	−1.7	−0.3	−2.2	−0.4	0.4	1.4	−0.8	−0.5
PROG	−0.8	−1.9	0.7	−0.8	−1.8	0.8	−1.7	−0.3	−2.2	−0.4	0.4	1.4	−0.8	−0.6
ERA-Interim	−2.7	−2.8	−1.2	−0.1	0.1	−2.8	−1.3	−1.4	−1.6	−0.9	0.4	0.6	−0.5	−1.1

Corr.														
NO	0.76	0.87	0.91	0.76	0.88	0.92	0.77	0.79	0.87	0.91	0.97	0.82	0.96	0.86
PROG-M	0.77	0.89	0.92	0.77	0.88	0.92	0.77	0.81	0.88	0.92	0.97	0.81	0.96	0.86
PROG	0.76	0.88	0.92	0.75	0.88	0.92	0.77	0.80	0.89	0.92	0.97	0.81	0.96	0.86
ERA-Interim	0.88	0.98	0.88	0.75	0.86	0.92	0.89	0.90	0.89	0.96	0.93	0.81	0.90	0.89

SD														
NO	1.36	1.09	1.25	1.44	1.45	1.16	0.90	1.42	1.37	0.96	1.14	1.08	0.97	1.20
PROG-M	1.31	1.10	1.26	1.38	1.45	1.15	0.87	1.41	1.37	0.96	1.10	1.05	0.95	1.18
PROG	1.27	1.04	1.20	1.34	1.42	1.12	0.87	1.36	1.35	0.97	1.08	1.03	0.93	1.15
ERA-Interim	1.04	0.76	0.92	1.36	1.05	1.03	0.93	0.98	0.82	0.88	0.95	0.99	1.00	0.98

4.3 Composite analysis

4.3.1 Methodology

This section aims at highlighting the simulated and observed differences between days of high aerosol load and the set of all the days in terms of several meteorological parameters (radiation, temperature, cloud cover, etc.). For the 14 stations defined previously, the days of high AOD, called thereafter "dusty" days as dust aerosols are mostly responsible for these AOD maxima, have been selected over the 92 days of summer 2012 (June–July–August). A day is considered as a dusty day provided that observed AOD is higher than 0.2 and that simulated dust AOD in PROG is higher than 0.2. Days when observations were not available have been removed.

Average differences for several parameters have then been calculated between the dusty days and the set of all the days for the three simulations (NO, PROG-M and PROG) and observations. The differences obtained for NO will enable us to estimate the meteorological effect due only to changes in weather parameters (cloud cover, wind, etc.) without considering the aerosols; those for PROG-M can estimate the average effect of having an aerosol climatology; and those for PROG can estimate the added value of prognostic aerosols. The objective is to isolate the effects of aerosols from weather changes that are systematically observed during dust outbreaks. This method is first presented for the station of Murcia, whose results are representative of the whole Mediterranean basin, and then generalized to the 14 stations.

4.3.2 Case of Murcia

In Murcia, 23 days have been identified as dusty days over the 83 days when observations are available; results are presented in Table 6. First of all, as expected, the difference in AOD between dusty days and the set of all the days is clearly positive in the PROG simulation (0.19), very low in PROG-M (−0.01) but not necessarily zero, because the number of dusty days varies from 1 month to another (AOD is monthly constant in PROG-M), and equal to zero in NO (no aerosols). This difference in AOD is similar in the observations and PROG, confirming the ability of CNRM-RCSM5 to reproduce aerosol daily variability and making the comparison for other parameters possible. The higher AOD during dusty days leads to a decrease in downward SSR. The difference with the set of all the days reaches $-22\,\text{Wm}^{-2}$ against only −6 and $-7\,\text{Wm}^{-2}$ for NO and PROG-M respectively, while measurements in the station show a difference of $-19\,\text{Wm}^{-2}$. The difference in NO ($-6\,\text{Wm}^{-2}$) can be considered as the "weather effect" that is due to the choice of the days (meteorological and astronomical variations). The duration of sunshine indeed varies during summer and reaches its maximum at the solstice (21 June), which can explain a part of the radiation differences in NO, in addition to changes in cloud cover. PROG-M, which has a monthly climatology of aerosols, is useful to identify changes in atmospheric circulation and cloud cover due to a monthly climatology of aerosols ($-1\,\text{Wm}^{-2}$). The difference between PROG-M and PROG gives the contribution of the daily variability of aerosols that is necessary to reproduce observed radiation measurements. Few changes among the three simulations are observed in cloud cover and TSR.

Temperature is also affected by weather changes, as dusty days are 1.6 °C higher in NO than the set of all the days. This is probably explained by the predominance of stronger southern fluxes during dusty days that can transport aerosols from Sahara to the Mediterranean basin. Figure 11 indeed shows the average circulation at 850 hPa during dusty days and the set of all the days, indicating a reinforcement of south-westerly winds in southern Spain advecting warm air. However, this increase in temperature during dusty days is lower in PROG than in PROG-M and NO, which is closer to observed variations of temperature. This decrease of −0.2 °C between PROG-M and PROG is caused by dust aerosols that have reduced incoming solar radiation. In other words, without prognostic aerosols the warming simulated by CNRM-RCSM during dusty days is too strong compared to observations, which is corrected in PROG. A similar impact is observed in soil temperature.

As a result, radiation and temperature in Murcia have been shown to be better reproduced in the PROG simulation,

showing the added value of a prognostic scheme compared to monthly climatologies to reproduce local meteorological variations.

4.3.3 Generalization

A similar composite study has been carried out for other stations (defined in Table 1) where daily radiation and temperature data were available. Figure 12 presents the results per station for six parameters (AOD, solar and thermal surface radiation, cloud cover, 2 m and soil temperature) for the NO, PROG-M and PROG simulations, as well as for observations when available, while the average composites are given in Table 7.

As in Murcia, the difference in AOD between dusty days and the set of all the days is for every station similar in observations (0.22 on average) and the PROG simulation (0.21). The difference in PROG-M comes only from the number of dusty days varying from 1 month to another. As a consequence, measurements reveal that downward SSR is on average 23 Wm^{-2} lower during dusty days, which is correctly reproduced by PROG ($-23\,\text{Wm}^{-2}$). A part of this decrease ($-2\,\text{Wm}^{-2}$) is explained by weather changes as simulated by NO, while added an aerosol climatology does not bring significant differences ($-3\,\text{Wm}^{-2}$). Additionally, the decrease of SSR in dusty days varies from one station to another (ranging from −2 to $-53\,\text{Wm}^{-2}$). The amplitude of the increase in AOD on dusty days and changes in weather conditions explain this variability. For example in Mallorca, an increase of 6 % in cloud cover on dusty days amplifies the dimming due to aerosol loads.

With regards to downward TSR, an average increase of $14\,\text{Wm}^{-2}$ is simulated by PROG on dusty days, but it is mainly due to weather conditions as NO and PROG-M also show an increase of 12 Wm^{-2}. Dust aerosols would consequently only represent an increase of 2 Wm^{-2}. Unfortunately, few LW observations are available. The measurements in the Gulf of Lions and in Lampedusa show a lower increase than the simulations.

More observations are available for T2m, revealing a general increase of temperature on dusty days (on average 1.4 °C). As in Murcia, this increase is probably due to warm advection caused by southerly to south-westerly winds responsible of these dust outbreaks. NO indeed simulates an average increase of 1.7 °C but reduced to 1.5 °C in PROG, indicating the cooling due to dust aerosols, which makes the simulation closer to observations. This improvement is noted in 10 out of the 13 stations considered in the study (Fig. 12) – these 10 stations being the 9 continental stations and the buoy Azur. The other stations either do not show a cooling (Ajaccio) or this cooling is not in line with observations (buoy of the Gulf of Lions, Lampedusa). For these two latter stations, sea surface temperature also increases on dusty days (up to 2.0 °C in the Gulf of Lions in NO), while PROG-M and PROG both alleviate this increase by 0.1 °C. However,

Table 6. Composite study for Murcia: differences between dusty days and the set of all the days in observations (OBS), NO, PROG-M and PROG for AOD, downward SSR (Wm^{-2}), cloud cover (%), downward TSR (Wm^{-2}), 2 m temperature (°C) and soil temperature (Ts, °C). The contribution of the different effects, namely weather, aerosol (mean) and aerosol (variability), have been added.

Parameter	OBS	NO	PROG-M	PROG	Weather	Aerosol (mean)	Aerosol (var)
AOD	0.15	0.00	−0.01	0.19	0.00	−0.01	0.20
SSR	−19	−6	−7	−22	−6	−1	−15
Cloud cover	–	1	2	1	1	1	−1
TSR	–	10	9	11	10	−1	2
T2m	1.2	1.6	1.5	1.3	1.6	−0.1	−0.2
Ts	–	1.5	1.4	1.2	1.5	−0.1	−0.2

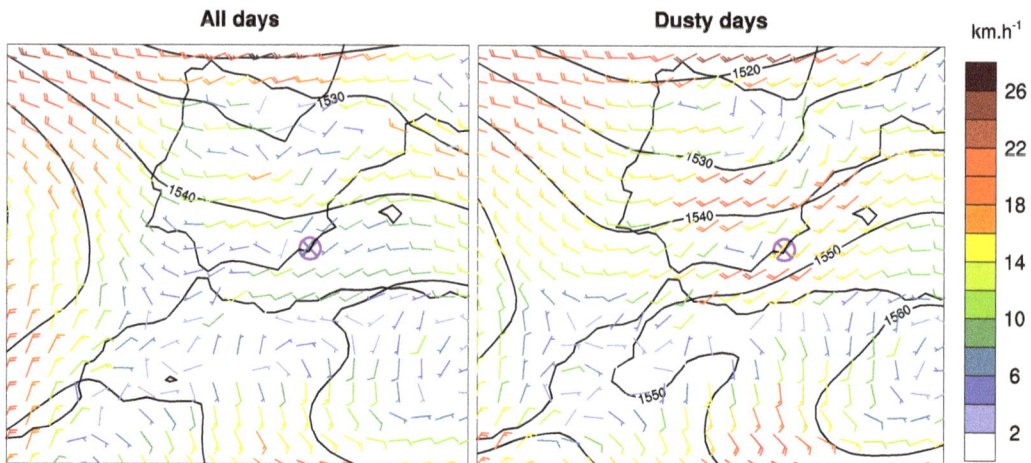

Figure 11. Average wind ($km\,h^{-1}$, coloured barbs) and geopotential (mgp, black lines) at 850 hPa for the set of all the days (left) and the dusty days (right) defined in Murcia (purple cross).

this reduction cannot be confirmed by observations. Maybe the 3-month period is not long enough to identify the daily effects of aerosols on SST. With regards to land soil temperature, a cooling of −0.3 °C due to dust aerosols is simulated by PROG, in relationship with the cooling in T2m.

In fact, this composite analysis has shown that significant differences are observed between dusty days and the set of all the days, which come both from weather changes (notably due to south-westerly winds bringing warm air) and from the presence of dust aerosols that alleviate this warming by reducing incoming solar radiation. These results underline the importance of the use of prognostic aerosols to represent daily variations in weather parameters such as temperature and radiation.

4.4   Impact of daily aerosol variability on the summer average

The question that arises from the impact of aerosols shown on surface radiation and temperature during dusty days is whether using an aerosol prognostic scheme instead of a monthly climatology also has an impact on the summer average.

As far as DRF is concerned, average differences in summer 2012 between PROG and PROG-M are presented in Fig. 13 both for SW (a) and LW (b) radiation. The intensity of the average aerosol forcing is slightly lower ($3\,Wm^{-2}$) in PROG-M than in PROG for the SW component, while very few differences are observed for LW radiation. Moreover, the daily standard deviation of SW DRF is higher in PROG than in PROG-M, particularly over northern Africa and the Mediterranean Sea, where it is more than twice as high (Fig. 8b). Indeed, dust emission is not a continuous phenomenon, because it is associated with episodes of strong wind over northern Africa. Consequently, dust particles show high variability over the Mediterranean basin that PROG-M cannot take into account, contrary to PROG. The only daily variations of DRF in PROG-M are due to cloud cover variations, as the aerosol effect can be partially masked by the presence of clouds.

As a consequence, the aerosol effect on surface temperature is on average slightly different in PROG-M compared to PROG (Fig. 13c). The general cooling, due to the presence of aerosols that scatter and absorb incident solar radiation, preventing it from reaching the surface, is either reinforced (e.g. in the south-western Mediterranean) or alleviated (e.g.

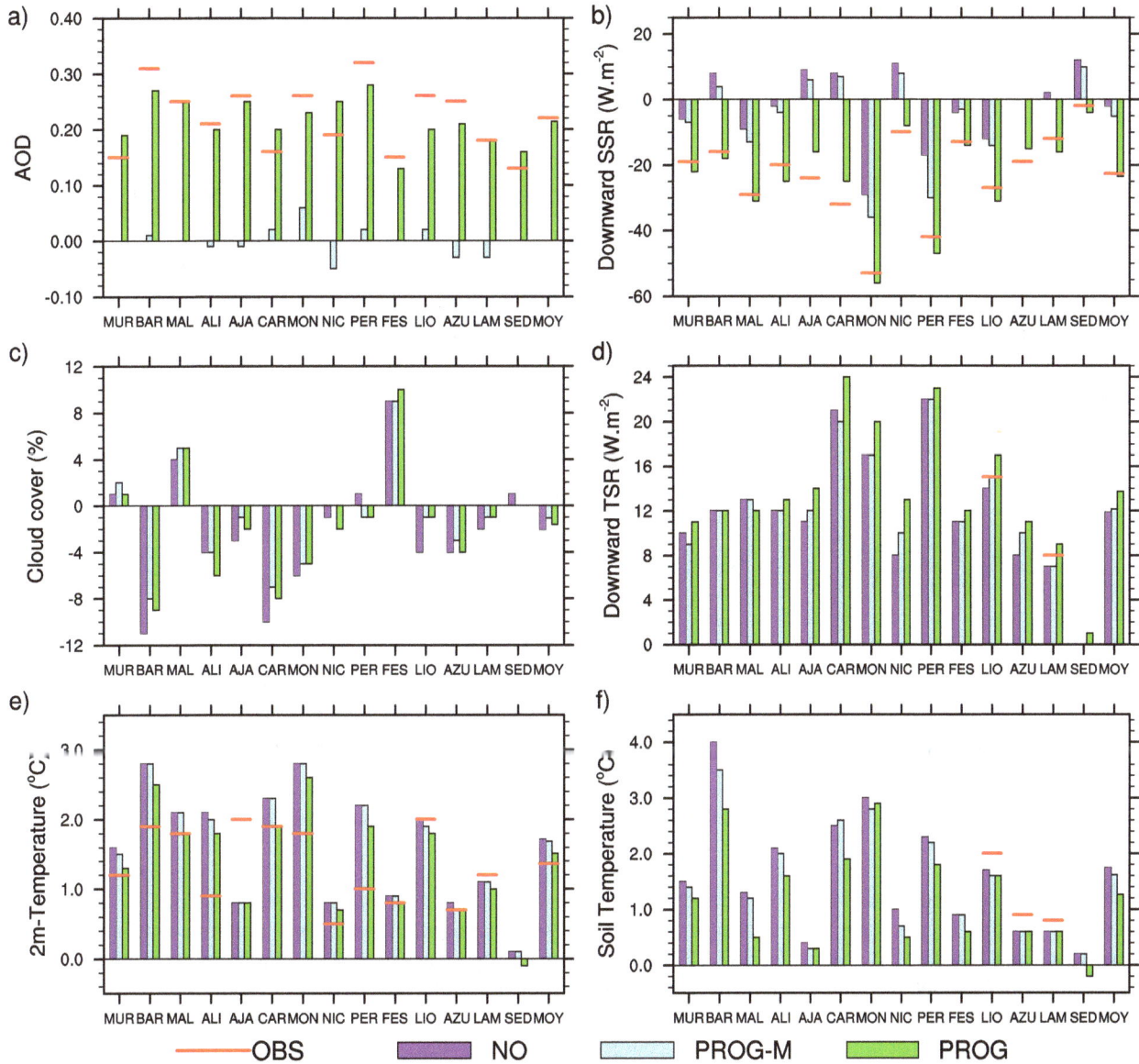

Figure 12. Average AOD (**a**), downward SSR (**b**), cloud cover (**c**), downward TSR (**d**), 2 m temperature (**e**) and soil temperature (**f**) differences between the dusty and the set of all the days in 14 stations (presented in Table 1) in summer 2012 for the NO, PROG-M and PROG simulations, as well as observations (AERUS-GEO for AOD, ground-based measurements for the other parameters). For Lampedusa and the buoys in the Gulf of Lions and Azur, 2 m temperature has been replaced by SST.

Table 7. Same as Table 6 but for the average over the 14 stations defined in Table 1.

Parameter	OBS	NO	PROG-M	PROG	Weather	Aerosol (mean)	Aerosol (var)
AOD	0.22	0.00	0.00	0.21	0.00	0.00	0.21
SSR	−23	−2	−5	−23	−2	−3	−18
Cloud cover	−	−2	−1	−2	−2	1	−1
TSR	−	12	12	14	12	0	2
T2m	1.4	1.7	1.7	1.5	2.0	0.0	−0.2
Land soil temperature	−	1.7	1.6	1.3	1.7	−0.1	−0.3
SST	1.3	0.9	0.9	0.9	1.3	0.0	0.0

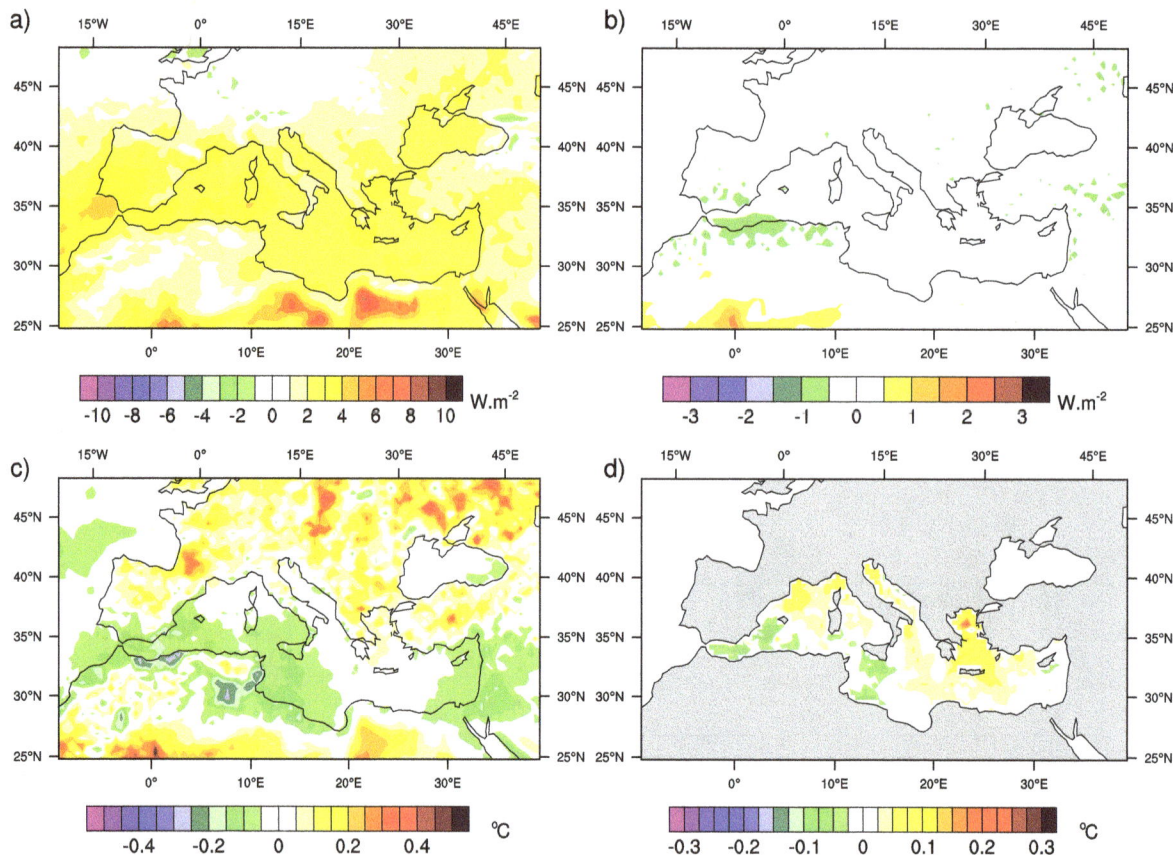

Figure 13. Average difference in summer 2012 between the PROG and PROG-M simulations in terms of (**a**) SW surface direct radiative forcing (Wm^{-2}), (**b**) LW surface direct radiative forcing (Wm^{-2}), (**c**) 2 m temperature (°C) and (**d**) sea surface temperature (°C).

in eastern Europe) when using an aerosol interactive scheme instead of a monthly climatology. A similar difference between PROG and PROG-M is found for SST (Fig. 13d). These changes are probably due to the interactions between aerosols and weather conditions. As seen previously in the composite study, the fact that high dust loads often occur in southern fluxes could modify their impact on weather and climate. Moreover, when using an aerosol climatology, the variability of the atmospheric aerosol content is weaker, and the extreme values of AOD are not represented in the model.

Over the Mediterranean, while frequent AOD peaks are observed in the south-west due to frequent dust outbreaks, the latter less often reach the Gulf of Lions and hence there are less frequently AOD peaks there. The AOD standard deviation in PROG is, for example, 0.22 for the Strait of Gibraltar and only 0.14 for the Gulf of Lions. In addition, there are more days in the Strait of Gibraltar (32) where AOD is much higher (difference higher than 0.1) in PROG than in PROG-M and in the Gulf of Lions (15), despite common averages. Consequently, the aerosol effect can be more important in the Strait of Gibraltar than in the Gulf of Lions, which must explain a cooler SST in the Strait of Gibraltar. In addition, the days when AOD is high in the Gulf of Lions are often cloudy,

which alleviate the effect of aerosols. Indeed, dust outbreaks over the northern basins are more frequent under southerly winds (Gkikas et al., 2012) that also favour humidity advection and cloud cover.

In summary, the choice of using an aerosol prognostic scheme instead of a monthly climatology has not only an impact on daily weather and climate variability but also on the summer average. This second impact has never been shown before over the Mediterranean to our knowledge.

4.5 Discussion

This study has shown the radiative effects of dust aerosols in summer 2012 over the Mediterranean, but some points need to be discussed.

First, the choice to focus on a particular summer has been motivated by the fact that summer 2012 was particularly affected by dust outbreaks. Thus, a high number of dusty days were noted, providing an interesting case to estimate the radiative effects of dust aerosols. However, one can wonder if the results would change during a summer with few dust outbreaks, notably with regards to the impact of the choice of prognostic aerosols. As a matter of fact, the composite study and the analysis of the utility of prognostic aerosols should be redone for a longer period to better understand the interactions between dust aerosols and regional climate, even if finding adequate observations may represent an obstacle. It would be also interesting to consider the effects of dust aerosols during the other seasons.

In addition, the choice of using the spectral nudging method may have influenced the results, as it can be seen as a limitation of the effect of aerosols on the atmosphere. Indeed, this relaxation towards the ERA-Interim inside the regional domain could, for example, prevent aerosols from modifying temperature and humidity profiles above 700 hPa and thus have stronger semi-direct effects. This point is particularly interesting with regards to the impact of the choice of prognostic aerosols instead of monthly AOD means. Nevertheless, the spectral nudging method is essential to represent the real chronology of dust events, making the comparison to observations possible. With regards to the uncertainties of the model outputs, they will be more deeply evaluated in a multi-model exercise currently carried out in the framework of the TRAQA/ChArMEx campaign.

Finally, the low complexity of the aerosol scheme used in the present work could constitute another limitation. In particular, the low number of bins for dust aerosols (only three), the absence of detailed processes representing the formation of secondary aerosols, the choice of a bulk approach for aerosol modelling and the absence of internal mixing are limitations to the present work. Future developments on this aerosol scheme will be carried out to improve the representation of aerosols in the model. For example, the implementation of the Ångstrom exponent will make the definition of dusty days for the composite study more robust. However, some of the simplifications remain necessary to keep a low numerical cost in order to be able to carry out easily multi-annual climate simulations with a coupling between the different components of the regional climate system (atmosphere, aerosols, land surface and ocean). Moreover, this scheme does not take into account the second indirect effect of aerosols because of the huge uncertainties in their parameterizations (Quaas et al., 2009).

5 Conclusions

A prognostic aerosol scheme has recently been added in the regional climate model ALADIN-Climate, enabling for the first time a regional coupled system model (CNRM-RCSM5) including the atmosphere, prognostic aerosols, land surface and the ocean components over the Mediterranean region. Simulations have been carried out in summer 2012, first to evaluate the aerosols produced by the model and then to estimate the radiative effects of dust outbreaks over the Mediterranean region.

CNRM-RCSM5 has shown its ability to reproduce the spatial and temporal variability of AOD over the Mediterranean region in summer 2012. The general spatial patterns, notably the locations of regions with high AOD, are in agreement with satellite data, while the distribution in the main different aerosol types is close to the MACC reanalysis and the independent climatology from Nabat et al. (2013). Daily variability is also correctly simulated by the model, since the evaluation against 30 stations from the AERONET network shows a mean bias of 0.02, an average correlation coefficient of 0.70 and an average ratio of standard deviations of 1.01 as good as satellite data. In addition, the TRAQA campaign has provided lidar and airborne measurements of a strong dust outbreak that occurred at the end of June 2012. The aerosol vertical distributions observed in Barcelona and in Corsica show that the model is able to reproduce the altitude of maximum extinction, even when a slight overestimation has been noted in the upper troposphere. With regards to dust size distribution, the three-bin scheme used in ALADIN-Climate simulates higher mass concentrations for the largest particles, as well as a second maxima for submicronic particles, as observed during the TRAQA campaign.

The simulated aerosol surface SW DRF is negative, ranging from $-10\,\mathrm{Wm^{-2}}$ in Europe to $-50\,\mathrm{Wm^{-2}}$ in Africa, in line with previous studies. However, here the aerosol DRF is shown to have much variability when using a prognostic aerosol scheme instead of a monthly climatology. As a consequence, thanks to the prognostic aerosol scheme, downward SSR is better reproduced compared to ground-based measurements from several stations across the Mediterranean, both on days of high AOD (lower SSR) and low AOD (higher SSR), as correlation and standard deviation are improved. The forcing due to the dust outbreaks also causes extra cooling in surface temperature, but it is insufficient to improve significantly the correlation. However, the average difference between a simulation using a prognostic aerosol scheme and an aerosol climatology shows a cooling of 0.1 to 0.2 °C both in T2m and SST close to the dust sources, notably in the south-western Mediterranean. Dynamics can also change in the two simulations and thus modify surface temperature.

A composite study has been realized in 14 stations across the Mediterranean to identify more precisely the differences between dusty days and the set of all the days. During dusty days, SSR is shown to be reduced on average by $28\,\mathrm{Wm^{-2}}$ mostly because of the dimming of aerosols ($-17\,\mathrm{Wm^{-2}}$) but also because of weather conditions ($-10\,\mathrm{Wm^{-2}}$). In parallel, dust outbreaks that are responsible for dusty days also bring warm air, which explains why T2m is observed 1.6 °C higher on dusty days. This warming is too strong (2.0 °C) when considering only an aerosol climatology. The prognostic scheme reduces this average warming of 0.2 °C, getting closer to observations.

Finally, this study has shown the improvement brought by a prognostic aerosol scheme compared to a monthly climatology in terms of radiation and temperature during a summer. This methodology could be applied on multi-annual simulations to evaluate the impact of prognostic aerosols at the climate scale. Differences could be expected not only in terms of variability but also in average climate as suggested by the differences shown in average SST in summer 2012 in the present work.

Acknowledgements. We would like to thank Météo-France for the financial support of the first author and the surface radiation and temperature in French stations. This work is part of the Med-CORDEX initiative (www.medcordex.eu) and a contribution to the HyMeX and ChArMEx programmes. ChArMEx is the atmospheric component of the French multidisciplinary program MISTRALS (Mediterranean Integrated Studies aT Regional And Local Scales). ChArMEx-France was principally funded by INSU, ADEME, ANR, CNES, CTC (Corsica region), EU/FEDER, Météo-France and CEA. The aircraft was operated by SAFIRE. TRAQA was funded by ADEME/PRIMEQUAL and MISTRALS/ChArMEx programmes and Observatoire Midi-Pyrénées. This research has received funding from the French National Research Agency (ANR) projects ADRIMED (contract ANR-11-BS56-0006) and REMEMBER (contract ANR-12-SENV-0001) and MORDICUS (contract ANR-13-SENV-0002), as well as from the FP7 European Commission project CLIMRUN (contract FP7-ENV-2010-265192). The authors acknowledge AEMET for supplying the data and the HyMeX database teams (ESPRI/IPSL and SEDOO/Observatoire Midi-Pyrénées) for their help in accessing the data. We also thank the PI investigators of the different AERONET stations and their staff for establishing and maintaining all the sites used in the present work. The MACC data come from the ECMWF website; MACC was funded between 2009 and 2011 as part of the 7th Framework Programme, pilot core GMES Atmospheric Service, under contract number 218793. Measurements at Lampedusa were supported by the Italian Ministry for University and Research through projects NextData and Ritmare. Lidar measurements in Barcelona were supported by the 7th Framework Programme project Aerosols, Clouds, and Trace Gases Research Infrastructure Network (ACTRIS) (grant agreement no. 262254) and by the Spanish Ministry of Science and Innovation and FEDER funds under the projects TEC2012-34575, TEC2009-09106/TEC, CGL2011-13580-E/CLI and CGL2011-16124-E/CLI. We also thank Dominique Carrer, Xavier Ceamanos and the ICARE centre for the AERUS-GEO data set. The data of the two Mediterranean buoys were obtained from the HyMeX program, sponsored by grants from MISTRALS/HyMeX and Météo-France. We also thank the Direction de la météorologie nationale in Morocco for providing us with the data in Fès. We thank Anaïs Culot and Emilie Bruhier for their help with the analysis of observation data and on the composite study.

Edited by: O. Dubovik

References

Artale, V., Calmanti, S., Carillo, A., Dell'Aquila, A., Herrmann, M., Pisacane, G., Ruti, P. M., Sannino, G., Struglia, M. V., Giorgi, F., Bi, X., Pal, J. S., Rauscher, S., and the PROTHEUS Group: An atmosphere-ocean regional climate model for the Mediterranean area: assessment of a present climate simulation, Clim. Dynam., 35, 721–740, doi:10.1007/s00382-009-0691-8, 2010.

Barnaba, F. and Gobbi, G. P.: Aerosol seasonal variability over the Mediterranean region and relative impact of maritime, continental and Saharan dust particles over the basin from MODIS data in the year 2001, Atmos. Chem. Phys., 4, 2367–2391, doi:10.5194/acp-4-2367-2004, 2004.

Benedetti, A., Kaiser, J. W., and Morcrette, J.-J.: [global climate] aerosols [in "state of the climate in 2010"], B. Am. Meteorol. Soc., 92, S65–S67, 2011.

Benkovitz, C. M., Scholz, M. T., Pacyna, J., Tarrason, L., Dignon, J., Voldner, E. C., Spiro, P. A., Logan, J. A., and Graedel, T. E.: Global gridded inventories of anthropogenic emissions of sulfur and nitrogen, J. Geophys. Res., 101, 29239–29253, 1996.

Bergamo, A., Tafuro, A. M., Kinne, S., De Tomasi, F., and Perrone, M. R.: Monthly-averaged anthropogenic aerosol direct radiative forcing over the Mediterranean based on AERONET aerosol properties, Atmos. Chem. Phys., 8, 6995–7014, doi:10.5194/acp-8-6995-2008, 2008.

Beuvier, J., Sevault, F., Herrmann, M., Kontoyiannis, H., Ludwig, W., Rixen, M., Stanev, E., Béranger, K., and Somot, S.: Modeling the Mediterranean Sea interannual variability during 1961–2000: Focus on the Eastern Mediterranean Transient, J. Geophys. Res., 115, C08017, doi:10.1029/2009JC005950, 2010.

Carrer, D., Roujean, J.-L., Hautecoeur, O., and Elias, T.: Daily estimates of aerosol optical thickness over land surface based on a directional and temporal analysis of SEVIRI MSG visible observations, J. Geophys. Res., 115, D10208, doi:10.1029/2009JD012272, 2010.

Carrer, D., Ceamanos, X., Six, B., and Roujean, J.-L.: AERUS-GEO: A newly available satellite- derived aerosol optical depth product over Europe and Africa, Geophys. Res. Lett., 41, 7731–7738, doi:10.1002/2014GL061707, 2014.

Colin, J., Déqué, M., Radu, R., and Somot, S.: Sensitivity study of heavy precipitation in Limited Area Model climate simulations: influence of the size of the domain and the use of the spectral nudging technique, Tellus, 62A, 591–604, 2010.

Dee, D. P., Uppala, S. M., Simmons, A. J., Berrisford, P., Poli, P., Kobayashi, S., Andrae, U., Balmaseda, M. A., Balsamo, G., Bauer, P., Bechtold, P., Beljaars, A. C. M., van de Berg, L., Bidlot, J., Bormann, N., Delsol, C., Dragani, R., Fuentes, M., Geer, A. J., Haimbergere, L., Healy, S. B., Hersbach, H., Hólm, E. V., Isaksen, L., Kallberg, P., Köhler, M., Matricardi, M., McNally, A. P., Monge-Sanzf, B. M., Morcrette, J.-J., Park, B.-K., Peubey, C., de Rosnaya, P., Tavolato, C., Thépaut, J.-N., and Vitart, F.: The ERA-Interim reanalysis: configuration and performance of the data assimilation system, Q. J. Roy. Meteor. Soc., 137, 553–597, doi:10.1002/qj.828, 2011.

Déqué, M. and Somot, S.: Extreme precipitation and high resolution with Aladin, Idöjaras Quaterly Journal of the Hungarian Meteorological Service, 112, 179–190, 2008.

Di Biagio, C., di Sarra, A., and Meloni, D.: Large atmospheric shortwave radiative forcing by Mediterranean aerosols derived from simultaneous ground-based and spaceborne observations and dependence on the aerosol type and single scattering albedo, J. Geophys. Res., 115, D10209, doi:10.1029/2009JD012697, 2010.

di Sarra, A., Pace, G., Meloni, D., De Silvestri, L., Piacentino, S., and Monteleone, F.: Surface shortwave radiative forcing of different aerosol types in the central Mediterranean, Geophys. Res. Lett., 35, L02714, doi:10.1029/2007GL032395, 2008.

Gao, X., Pal, J. S., and Giorgi, F.: Projected changes in mean and extreme precipitation over the Mediterranean region from a high resolution double nested RCM simulation, Geophys. Res. Lett., 33, L03706, doi:10.1029/2005GL024954, 2006.

Gibelin, A.-L. and Déqué, M.: Anthropogenic climate change over the Mediterranean region simulated by a global variable resolution model, Clim. Dynam., 20, 327–339, doi:10.1007/s00382-002-0277-1, 2003.

Ginoux, P., Chin, M., Tegen, I., Prospero, J., Holben, B. N., Dubovik, O., and Lin, S.-J.: Sources and distributions of dust aerosols simulated with the GOCART model, J. Geophys. Res., 106, 20255–20274, 2001.

Giorgi, F. and Lionello, P.: Climate change projections for the Mediterranean region, Global and Planetary Change, 63, 90–104, doi:10.1016/j.gloplacha.2007.09.005, 2008.

Giorgi, F., Coppola, E., Solmon, F., Mariotti, L., Sylla, M. B., Bi, X., Elguindi, N., Diro, G. T., Nair, V., Giuliani, G., Cozzini, S., Guettler, I., O'Brien, T. A., Tawfik, A. B., Shalaby, A., Zakey, A. S., Steiner, A. L., Stordal, F., Sloan, L. C., and Brankovic, C.: RegCM4: model description and preliminary tests over multiple CORDEX domains., Clim. Res., 52, 7–29, doi:10.3354/cr01018, 2012.

Gkikas, A., Houssos, E., Hatzianastassiou, N., Papadimas, C., and Bartzokas, A.: Synoptic conditions favouring the occurrence of aerosol episodes over the broader Mediterranean basin, Q. J. Roy. Meteor. Soc., 138, 932–949, doi:10.1002/qj.978, 2012.

Gkikas, A., Hatzianastassiou, N., Mihalopoulos, N., Katsoulis, V., Kazadzis, S., Pey, J., Querol, X., and Torres, O.: The regime of intense desert dust episodes in the Mediterranean based on contemporary satellite observations and ground measurements, Atmos. Chem. Phys., 13, 12135–12154, doi:10.5194/acp-13-12135-2013, 2013.

Guelle, W., Schulz, M., Balkanski, Y., and Dentener, F.: Influence of the source formulation on modeling the atmospheric global distribution of the sea salt aerosol, J. Geophys. Res., 106, 27509–27524, 2001.

Guieu, C., Dulac, F., Desboeufs, K., Wagener, T., Pulido-Villena, E., Grisoni, J.-M., Louis, F., Ridame, C., Blain, S., Brunet, C., Bon Nguyen, E., Tran, S., Labiadh, M., and Dominici, J.-M.: Large clean mesocosms and simulated dust deposition: a new methodology to investigate responses of marine oligotrophic ecosystems to atmospheric inputs, Biogeosciences, 7, 2765–2784, doi:10.5194/bg-7-2765-2010, 2010.

Herrmann, M., Somot, S., Calmanti, S., Dubois, C., and Sevault, F.: Representation of spatial and temporal variability of daily wind speed and of intense wind events over the Mediterranean Sea using dynamical downscaling: impact of the regional climate model configuration, Nat. Hazards Earth Syst. Sci., 11, 1983–2001, doi:10.5194/nhess-11-1983-2011, 2011.

Holben, B. N., Eck, T. F., Slutsker, I., Tanré, D., Buis, J. P., Setzer, A., Vermote, E., Reagan, J. A., Kaufman, Y., Nakajima, T., Lavenu, F., Jankowiak, I., and Smirnov, A.: AERONET-A Federated Instrument Network and Data Archive for Aerosol Characterization, Remote Sens. Environ., 66, 1–16, doi:10.1016/S0034-4257(98)00031-5, 1998.

Holben, B. N., Tanré, D., Smirnov, A., Eck, T. F., Slutsker, I., Abuhassan, N., Newcomb, W. W., Schafer, J. S., Chatenet, B., Lavenu, F., Kaufman, Y. J., Castle, J. V., Setzer, A., Markham, B., Clark, D., Frouin, R., Halthore, R., Karneli, A., O'Neill, N. T., Pietras, C., Pinker, R. T., Voss, K., and Zibordi, G.: An emerging ground-based aerosol climatology: Aerosol optical depth from AERONET, J. Geophys. Res., 106, 12067–12097, doi:10.1029/2001JD900014, 2001.

Huneeus, N., Schulz, M., Balkanski, Y., Griesfeller, J., Prospero, J., Kinne, S., Bauer, S., Boucher, O., Chin, M., Dentener, F., Diehl, T., Easter, R., Fillmore, D., Ghan, S., Ginoux, P., Grini, A., Horowitz, L., Koch, D., Krol, M. C., Landing, W., Liu, X., Mahowald, N., Miller, R., Morcrette, J.-J., Myhre, G., Penner, J., Perlwitz, J., Stier, P., Takemura, T., and Zender, C. S.: Global dust model intercomparison in AeroCom phase I, Atmos. Chem. Phys., 11, 7781–7816, doi:10.5194/acp-11-7781-2011, 2011.

Israelevich, P., Ganor, E., Alpert, P., Kishcha, P., and Stupp, A.: Predominant transport paths of Saharan dust over the Mediterranean Sea to Europe, J. Geophys. Res., 117, D02205, doi:10.1029/2011JD016482, 2012.

Jaeglé, L., Quinn, P. K., Bates, T. S., Alexander, B., and Lin, J.-T.: Global distribution of sea salt aerosols: new constraints from in situ and remote sensing observations, Atmos. Chem. Phys., 11, 3137–3157, doi:10.5194/acp-11-3137-2011, 2011.

Kahn, R. A., Gaitley, B. J., Martonchik, J. V., Diner, D. J., Crean, K. A., and Holben, B.: Multiangle imaging spectroradiometer (misr) global aerosol optical depth validation based on 2 years of coincident aerosol robotic network (aeronet) observations, J. Geophys. Res., 110, D10S04, doi:10.1029/2004JD004706, 2005.

Kahn, R. A., Gaitley, B. J., Garay, M. J., Diner, D. J., Eck, T. F., Smirnov, A., and Holben, B. N.: Multiangle Imaging SpectroRadiometer global aerosol product assessment by comparison with the Aerosol Robotic Network, J. Geophys. Res., 115, D23209, doi:10.1029/2010JD014601, 2010.

Knippertz, P. and Todd, M. C.: Mineral dust aerosols over the Sahara: meteorological controls on emission and transport and implications for modeling, Rev. Geophys., 50, RG1007, doi:10.1029/2011RG000362, 2012.

Kok, J. F.: A scaling theory for the size distribution of emitted dust aerosols suggests climate models underestimate the size of the global dust cycle, P. Natl. Acad. Sci. USA, 108, 1016–1021, doi:10.1073/pnas.1014798108, 2011.

Krzic, A., Tosic, I., Djurdjevic, V., Veljovic, K., and Rajkovic, B.: Changes in climate indices for Serbia according to the SRES-A1B and SRES-A2 scenarios, Clim. Res., 49, 73–86, doi:10.3354/cr01008, 2011.

Lamarque, J.-F., Bond, T. C., Eyring, V., Granier, C., Heil, A., Klimont, Z., Lee, D., Liousse, C., Mieville, A., Owen, B., Schultz, M. G., Shindell, D., Smith, S. J., Stehfest, E., Van Aardenne, J., Cooper, O. R., Kainuma, M., Mahowald, N., McConnell, J. R., Naik, V., Riahi, K., and van Vuuren, D. P.: Historical (1850–2000) gridded anthropogenic and biomass burning emissions of reactive gases and aerosols: methodology and application, Atmos. Chem. Phys., 10, 7017–7039, doi:10.5194/acp-10-7017-2010, 2010.

Lamarque, J.-F., Shindell, D. T., Josse, B., Young, P. J., Cionni, I., Eyring, V., Bergmann, D., Cameron-Smith, P., Collins, W. J., Doherty, R., Dalsoren, S., Faluvegi, G., Folberth, G., Ghan, S. J., Horowitz, L. W., Lee, Y. H., MacKenzie, I. A., Nagashima, T., Naik, V., Plummer, D., Righi, M., Rumbold, S. T., Schulz, M., Skeie, R. B., Stevenson, D. S., Strode, S., Sudo, K., Szopa, S., Voulgarakis, A., and Zeng, G.: The Atmospheric Chemistry and Climate Model Intercomparison Project (ACCMIP): overview and description of models, simulations and climate diagnostics, Geosci. Model Dev., 6, 179–206, doi:10.5194/gmd-6-179-2013, 2013.

Lelieveld, J., Berresheim, H., Borrmann, S., Crutzen, P. J., Dentener, F. J., Fischer, H., Feichter, J., Flatau, P. J., Heland, J., Holzinger, R., Korrmann, R., Lawrence, M. G., Levin, Z., Markowicz, K. M., Mihalopoulos, N., Minikin, A., Ramanathan, V., de Reus, M., Roelofs, G. J., Scheeren, H. A., Sciare, J., Schlager, H., Schultz, M., Siegmund, P., Steil, B., Stephanou, E. G., Stier, P., Traub, M., Warneke, C., Williams, J., and Ziereis, H.: Global Air Pollution Crossroads over the Mediterranean, Science, 298, 794–799, doi:10.1126/science.1075457, 2002.

Levy, R. C., Remer, L. A., Mattoo, S., Vermote, E. F., and Kaufman, Y. J.: Second-generation operational algorithm: Retrieval of aerosol properties over land from inversion of Moderate Resolution Imaging Spectroradiometer spectral reflectance, J. Geophys. Res., 112, D13211, doi:10.1029/2006JD007811, 2007.

Levy, R. C., Remer, L. A., Kleidman, R. G., Mattoo, S., Ichoku, C., Kahn, R., and Eck, T. F.: Global evaluation of the Collection 5 MODIS dark-target aerosol products over land, Atmos. Chem. Phys., 10, 10399–10420, doi:10.5194/acp-10-10399-2010, 2010.

Lucas-Picher, P., Somot, S., Déqué, M., Decharme, B., and Alias, A.: Evaluation of the regional climate model ALADIN to simulate the climate over North America in the CORDEX framework, Clim. Dynam., 41, 1117–1137, doi:10.1007/s00382-012-1613-8, 2013.

Léon, J.-F., Augustin, P., Mallet, M., Pont, V., Dulac, F., Fourmentin, M., and Lambert, D.: Aerosol vertical distribution, optical properties and transport over Corsica (Western Mediterranean), Atmos. Chem. Phys. Discuss., submitted, 2015.

L'Hévéder, B., Li, L., Sevault, F., and Somot, S.: Interannual variability of deep convection in the Northwestern Mediterranean simulated with a coupled AORCM, Clim. Dynam., Clim. Dynam., 41, 937–960, doi:10.1007/s00382-012-1527-5, 2012.

Mahowald, N., Albani, S., Kok, J. F., Engelstaeder, S., Scanza, R., Ward, D. S., and Flanner, M. G.: The size distribution of desert dust aerosols and its impact on the Earth system, Aeolian Research, 15, 53–71, doi:10.1016/j.aeolia.2013.09.002, 2013.

Mariotti, A. and Dell'Aquila, A.: Decadal climate variability in the Mediterranean region: roles of large-scale forcings and regional processes, Clim. Dynam., 38, 1129–1145, doi:10.1007/s00382-011-1056-7, 2012.

Marticorena, B. and Bergametti, G.: Modeling the atmosphere dust cycle: 1. Design of a soil-derived dust emission scheme, J. Geophys. Res., 100, 16415–16430, 1995.

Masson, V., Champeaux, J., Chauvin, F., Meriguet, C., and Lacaze, R.: A global database of land surface parameters at 1-km resolution in meteorological and climate models, J. Climate, 16, 1261–1282, 2003.

Meier, J., Tegen, I., Heinold, B., and Wolke, R.: Direct and semi-direct radiative effects of absorbing aerosols in Europe: Results from a regional model, Geophys. Res. Lett., 39, L09802, doi:10.1029/2012GL050994, 2012.

Michou, M., Nabat, P., and Saint-Martin, D.: Development and basic evaluation of a prognostic aerosol scheme (v1) in the CNRM Climate Model CNRM-CM6, Geosci. Model Dev., 8, 501–531, doi:10.5194/gmd-8-501-2015, 2015.

Middleton, N. J. and Goudie, A. S.: Saharan dust: sources and trajectories, T. I. Brit. Geogr., 26, 165–181, doi:10.1111/1475-5661.00013, 2001.

Mlawer, E. J., Taubman, S. J., Brown, P. D., Iacono, M. J., and Clough, S. A.: Radiative transfer for inhomogeneous atmospheres: RRTM, a validated correlated-k model for the longwave, J. Geophys. Res., 102, 16663–16682, 1997.

Morcrette, J.-J.: Description of the Radiation Scheme in the ECMWF Model, Tech. rep., ECMWF, Reading, UK, 165 pp., 1989.

Morcrette, J.-J., Boucher, O., Jones, L., Salmond, D., Bechtold, P., Beljaars, A., Benedetti, A., Bonet, A., Kaiser, J. W., Razinger, M., Schulz, M., Serrar, S., Simmons, J., Sofiev, M., Suttie, M., Tompkins, A. M., and Untch, A.: Aerosol analysis and forecast in the european centre for medium-range weather forecasts integrated forecast system: Forward modeling, J. Geophys. Res., 114, D06206, doi:10.1029/2008JD011235, 2009.

Moulin, C., Guillard, F., Dulac, F., and Lambert, C. E.: Long-term daily monitoring of Saharan dust load over ocean using Meteosat ISCCP-B2 data 1. Methodology and preliminary results for 1983-1994 in the Mediterranean, J. Geophys. Res., 102, 16947–16958, 1997.

Moulin, C., Lambert, C. E., Dayan, U., Masson, V., Ramonet, M., Bousquet, P., Legrand, M., Balkanski, Y. J., Guelle, W., Marticorena, B., Bergametti, G., and Dulac, F.: Satellite climatology of African dust transport in the Mediterranean atmosphere, J. Geophys. Res., 103, 13137–13144, doi:10.1029/98JD00171, 1998.

Nabat, P., Solmon, F., Mallet, M., Kok, J. F., and Somot, S.: Dust emission size distribution impact on aerosol budget and radiative forcing over the Mediterranean region: a regional climate model approach, Atmos. Chem. Phys., 12, 10545–10567, doi:10.5194/acp-12-10545-2012, 2012.

Nabat, P., Somot, S., Mallet, M., Chiapello, I., Morcrette, J. J., Solmon, F., Szopa, S., Dulac, F., Collins, W., Ghan, S., Horowitz, L. W., Lamarque, J. F., Lee, Y. H., Naik, V., Nagashima, T., Shindell, D., and Skeie, R.: A 4-D climatology (1979–2009) of the monthly tropospheric aerosol optical depth distribution over the Mediterranean region from a comparative evaluation and blending of remote sensing and model products, Atmos. Meas. Tech., 6, 1287–1314, doi:10.5194/amt-6-1287-2013, 2013.

Nabat, P., Somot, S., Mallet, M., Sevault, F., Chiacchio, M., and Wild, M.: Direct and semi-direct aerosol radiative effect on the Mediterranean climate variability using a Regional Climate System Model, Clim. Dynam., 44, 1127–1155, doi:10.1007/s00382-014-2205-6, 2015.

Noilhan, J. and Mahfouf, J.-F.: The ISBA land surface parameterisation scheme, Global Planet. Change, 13, 145–159, doi:10.1016/0921-8181(95)00043-7, 1996.

Papadimas, C. D., Hatzianastassiou, N., Mihalopoulos, N., Querol, X., and Vardavas, I.: Spatial and temporal variability in aerosol properties over the Mediterranean basin based on 6-year (2000–2006) MODIS data, J. Geophys. Res., 113, D11205, doi:10.1029/2007JD009189, 2008.

Pappalardo, G., Amodeo, A., Apituley, A., Comeron, A., Freudenthaler, V., Linné, H., Ansmann, A., Bösenberg, J., D'Amico, G., Mattis, I., Mona, L., Wandinger, U., Amiridis, V., Alados-Arboledas, L., Nicolae, D., and Wiegner, M.: EARLINET: towards an advanced sustainable European aerosol lidar network,

Atmos. Meas. Tech. Discuss., 7, 2929–2980, doi:10.5194/amtd-7-2929-2014, 2014.

Quaas, J., Ming, Y., Menon, S., Takemura, T., Wang, M., Penner, J. E., Gettelman, A., Lohmann, U., Bellouin, N., Boucher, O., Sayer, A. M., Thomas, G. E., McComiskey, A., Feingold, G., Hoose, C., Kristjánsson, J. E., Liu, X., Balkanski, Y., Donner, L. J., Ginoux, P. A., Stier, P., Grandey, B., Feichter, J., Sednev, I., Bauer, S. E., Koch, D., Grainger, R. G., Kirkevåg, A., Iversen, T., Seland, Ø., Easter, R., Ghan, S. J., Rasch, P. J., Morrison, H., Lamarque, J.-F., Iacono, M. J., Kinne, S., and Schulz, M.: Aerosol indirect effects – general circulation model intercomparison and evaluation with satellite data, Atmos. Chem. Phys., 9, 8697–8717, doi:10.5194/acp-9-8697-2009, 2009.

Radu, R., Déqué, M., and Somot, S.: Spectral nudging in a spectral regional climate model, Tellus, 60A, 898–910, doi:10.1111/j.1600-0870.2008.00341.x, 2008.

Reba, M. N. M., Rocadenbosch, F., Sicard, M., Kumar, D., and Tomás, S.: On the lidar ratio estimation from the synergy between AERONET sun-photometer data and elastic lidar inversion, Proc. of the 25th International Laser Radar Conference, Saint-Petersburg (Rusia), 5–9 July 2010, 1102–1105, ISBN 978-5-94458-109-9, 2010.

Santese, M., Perrone, M. R., Zakey, A. S., De Tomasi, F., and Giorgi, F.: Modeling of Saharan dust outbreaks over the Mediterranean by RegCM3: case studies, Atmos. Chem. Phys., 10, 133–156, doi:10.5194/acp-10-133-2010, 2010.

Schulz, M., de Leeuw, G., and Balkanski, Y.: Sea-salt aerosol source functions and emissions, in: Emission of Atmospheric Trace Compounds, edited by: Granier, C., Artaxo, P., and Reeves, C. E., Kluwer Acad., Norwell, Mass., Springer Netherlands, 333–359, 2004.

Schulz, M., Textor, C., Kinne, S., Balkanski, Y., Bauer, S., Berntsen, T., Berglen, T., Boucher, O., Dentener, F., Guibert, S., Isaksen, I. S. A., Iversen, T., Koch, D., Kirkevåg, A., Liu, X., Montanaro, V., Myhre, G., Penner, J. E., Pitari, G., Reddy, S., Seland, Ø., Stier, P., and Takemura, T.: Radiative forcing by aerosols as derived from the AeroCom present-day and pre-industrial simulations, Atmos. Chem. Phys., 6, 5225–5246, doi:10.5194/acp-6-5225-2006, 2006.

Somot, S., Sevault, F., Déqué, M., and Crépon, M.: 21st century climate change scenario for the Mediterranean using a coupled atmosphere–ocean regional climate model, Global Planet. Change, 63, 112–126, doi:10.1016/j.gloplacha.2007.10.003, 2008.

Spyrou, C., Kallos, G., Mitsakou, C., Athanasiadis, P., Kalogeri, C., and Iacono, M. J.: Modeling the radiative effects of desert dust on weather and regional climate, Atmos. Chem. Phys., 13, 5489–5504, doi:10.5194/acp-13-5489-2013, 2013.

Tanré, D., Geleyn, J., and Slingo, J.: First results of the introduction of an advanced aerosol radiation interaction in ECMWF low resolution global model, in Aerosols and Their Climatic Effects, edited by: Gerber, H. and Deepak, A., pp. 133–177, A. Deepak, Hampton, Va, 1984.

Tanré, D., Kaufman, Y. J., Herman, M., and Mattoo, S.: Remote sensing of aerosol properties over oceans using the MODIS/EOS spectral radiances, J. Geophys. Res., 102, 16971–16988, 1997.

Taylor, K. E.: Summarizing multiple aspects of model performance in a single diagram, J. Geophys. Res., 106, 7183–7192, doi:10.1029/2000JD900719, 2001.

Tegen, I., Hollrig, P., Chin, M., Fung, I., Jacob, D., and Penner, J.: Contribution of different aerosol species to the global aerosol extinction optical thickness: Estimates from model results, J. Geophys. Res., 102, 23895–23915, 1997.

Tesche, M., Ansmann, A., Müller, D., Althausen, D., Mattis, I., Heese, B., Freundenthaler, V., Wiegner, M., Esselborn, M., Pisani, G., and Knippertz, P.: Vertical profiling of Saharan dust with Raman lidars and airborne HSRL in southern Morocco during SAMUM, Tellus B, 61, 144–164, doi:10.1111/j.1600-0889.2008.00390.x, 2009.

Textor, C., Schulz, M., Guibert, S., Kinne, S., Balkanski, Y., Bauer, S., Berntsen, T., Berglen, T., Boucher, O., Chin, M., Dentener, F., Diehl, T., Easter, R., Feichter, H., Fillmore, D., Ghan, S., Ginoux, P., Gong, S., Grini, A., Hendricks, J., Horowitz, L., Huang, P., Isaksen, I., Iversen, I., Kloster, S., Koch, D., Kirkevåg, A., Kristjansson, J. E., Krol, M., Lauer, A., Lamarque, J. F., Liu, X., Montanaro, V., Myhre, G., Penner, J., Pitari, G., Reddy, S., Seland, Ø., Stier, P., Takemura, T., and Tie, X.: Analysis and quantification of the diversities of aerosol life cycles within AeroCom, Atmos. Chem. Phys., 6, 1777–1813, doi:10.5194/acp-6-1777-2006, 2006.

Todd, M. C., Karam, D. B., Cavazos, C., Bouet, C., Heinold, B., Baldasano, J. M., Cautenet, G., Koren, I., Perez, C., Solmon, F., Tegen, I., Tulet, P., Washington, R., and Zakey, A.: Quantifying uncertainty in estimates of mineral dust flux: An intercomparison of model performance over the Bodélé Depression, northern Chad, J. Geophys. Res., 113, D24107, doi:10.1029/2008JD010476, 2008.

Turuncoglu, U. U., Giuliani, G., Elguindi, N., and Giorgi, F.: Modelling the Caspian Sea and its catchment area using a coupled regional atmosphere-ocean model (RegCM4-ROMS): model design and preliminary results, Geosci. Model Dev., 6, 283–299, doi:10.5194/gmd-6-283-2013, 2013.

Valcke, S.: The OASIS3 coupler: a European climate modelling community software, Geosci. Model Dev., 6, 373–388, doi:10.5194/gmd-6-373-2013, 2013.

Vogel, B., Vogel, H., Bäumer, D., Bangert, M., Lundgren, K., Rinke, R., and Stanelle, T.: The comprehensive model system COSMO-ART – Radiative impact of aerosol on the state of the atmosphere on the regional scale, Atmos. Chem. Phys., 9, 8661–8680, doi:10.5194/acp-9-8661-2009, 2009.

Woodage, M. J., Slingo, A., Woodward, S., and Comer, R. E.: Simulations of Desert Dust and Biomass Burning Aerosols with a High-Resolution Atmospheric GCM, J. Climate, 23, 1636–1659, doi:10.1175/2009JCLI2994.1, 2010.

Zanis, P.: A study on the direct effect of anthropogenic aerosols on near surface air temperature over Southeastern Europe during summer 2000 based on regional climate modeling, Ann. Geophys., 27, 3977–3988, doi:10.5194/angeo-27-3977-2009, 2009.

Zanis, P., Ntogras, C., Zakey, A., Pytharoulis, I., and Karacostas, T.: Regional climate feedback of anthropogenic aerosols over Europe using RegCM3, Clim. Res., 52, 267–278, doi:10.3354/cr01070, 2012.

6

Spaceborne observations of the lidar ratio of marine aerosols

K. W. Dawson[1], N. Meskhidze[1], D. Josset[2,*], and S. Gassó[3]

[1]Marine, Earth, and Atmospheric Science, North Carolina State University, Raleigh, NC, USA
[2]Science Systems and Applications, Inc./NASA Langley Research Center, Hampton, VA, USA
[3]GESTAR/Morgan State University, Goddard Space Flight Center, Greenbelt, MD, USA
[*]now at: Naval Research Laboratory, Stennis Space Center, Mississippi, USA

Correspondence to: N. Meskhidze (nmeskhidze@ncsu.edu)

Abstract. Retrievals of aerosol optical depth (AOD) from the Cloud-Aerosol Lidar with Orthogonal Polarization (CALIOP) satellite sensor require the assumption of the extinction-to-backscatter ratio, also known as the lidar ratio. This paper evaluates a new method to calculate the lidar ratio of marine aerosols using two independent sources: the AOD from the Synergized Optical Depth of Aerosols (SODA) project and the integrated attenuated backscatter from CALIOP. With this method, the particulate lidar ratio can be derived for individual CALIOP retrievals in single aerosol layer, cloud-free columns over the ocean. Global analyses are carried out using CALIOP level 2, 5 km marine aerosol layer products and the collocated SODA nighttime data from December 2007 to November 2010. The global mean lidar ratio for marine aerosols was found to be 26 sr, roughly 30 % higher than the current value prescribed by the CALIOP standard retrieval algorithm. Data analysis also showed considerable spatiotemporal variability in the calculated lidar ratio over the remote oceans. The calculated marine aerosol lidar ratio is found to vary with the mean ocean surface wind speed (U_{10}). An increase in U_{10} reduces the mean lidar ratio for marine regions from 32 ± 17 sr (for $0 < U_{10} < 4$ m s^{-1}) to 22 ± 7 sr (for $U_{10} > 15$ m s^{-1}). Such changes in the lidar ratio are expected to have a corresponding effect on the marine AOD from CALIOP. The outcomes of this study are relevant for future improvements of the SODA and CALIOP operational product and could lead to more accurate retrievals of marine AOD.

1 Introduction

Marine aerosols are produced through primary emission of sea spray particles and oxidation of phytoplankton-produced dimethylsulfide and biogenic volatile organic carbon. Radiative forcing by marine aerosol comprises a significant portion of the global energy budget. Studies have shown that marine aerosol optical depth (AOD) is approximately 0.15 and, likewise, the contribution of marine aerosol to cloud condensation nuclei is about 60 cm^{-3} (Kaufman et al., 2002; Lewis and Schwartz, 2004). Thus, marine aerosol is an important natural contributor to global aerosol burden affecting both direct (i.e., extinction of solar radiation via scattering and absorption) and indirect (i.e., cloud lifetime and frequency) radiative forcing of climate. As marine aerosols contribute considerably to the preindustrial, natural background and provide the base line on top of which anthropogenic forcing should be quantified, it is very important to properly characterize marine aerosol burden and its spatiotemporal distribution. The incomplete characterization of background aerosols, of which marine particles are a part, was shown to contribute large uncertainty in anthropogenic aerosol forcing calculations and climate simulations (Ghan et al., 2001; Hoose et al., 2009; Wang and Penner, 2009; Meskhidze et al., 2011; Westervelt et al., 2012; Carslaw et al., 2013).

Aerosols over the remote oceans come from natural continental (e.g., mineral dust and biomass burning) and human-induced pollution (Andreae, 2007) in addition to marine sources. Therefore, knowing horizontal and vertical distribution as well as speciation of aerosols becomes extremely important for the correct quantification of marine aerosol radiative properties. The last decade has produced a large body of information regarding the sources and composition of marine aerosol, resulting in a reassessment of the complex role that marine aerosols play in climate and various geophysical phenomena. Passive satellite instruments like the Sea-viewing Wide Field-of-view Sensor (SeaWiFS), the MODerate resolution Imaging Spectroradiometer (MODIS), the Multi-angle Imaging SpectroRadiometer, and the ground-based AErosol RObotic NETwork (AERONET) have contributed immensely to quantitative characteristics of marine aerosol in terms of AOD (the column integrated aerosol extinction), size distribution information, and spectral optical properties. Although passive instruments have been useful for developing a basic picture of marine aerosol distribution, they supply limited information on aerosol speciation and very little data related to aerosol distribution in the vertical column. The introduction of the Cloud-Aerosol Lidar with Orthogonal Polarization (CALIOP) onboard the Cloud-Aerosol Lidar and Infrared Pathfinder Satellite Observations (CALIPSO) platform has eliminated some of the assumptions made by the passive instruments and has provided a more complete picture of the global aerosol distribution wanted by climate scientists. However, CALIOP is an elastic backscatter lidar with no molecular filtering capability and therefore requires the assumption of an extinction-to-backscatter ratio, also known as the lidar ratio, to infer extinction from attenuated backscatter measurements. Depending on the microphysical properties of the aerosol, the lidar ratio can have a wide range of values and therefore a straightforward a priori solution within some reasonable uncertainty range is generally unobtainable without various assumptions or constraints. Theoretical calculations for the lidar ratio can be performed when the physicochemical properties and the size distribution of the particles at the different heights in the vertical column are known; however, the fulfillment of these requirements would make the lidar measurements unnecessary (Ackermann, 1998). The typical solution to this problem is to assign a vertically independent lidar ratio to aerosol retrievals that fit a specific aerosol model as outlined in Omar et al. (2009).

To date, experimental techniques for directly measuring the lidar ratio include the use of high spectral resolution lidar (HSRL, Eloranta, 2005; Hair et al., 2008) and Raman lidar (RL, Ansmann et al., 1990). These instruments are capable of measuring aerosol backscatter and extinction parameters independently and therefore do not require the lidar ratio to be prescribed (e.g., Shipley et al., 1983; Grund and Eloranta, 1991; Piironen and Eloranta, 1994; Müller et al., 2007; Amiridis et al., 2009; Tesche et al., 2009a, b; Burton et al., 2012). On the other hand, Cattrall et al. (2005) use AERONET size distributions inverted from sun photometer data (Holben et al., 1998) to calculate the lidar ratio and then compare their indirect to literature reported direct measurements. They determined that their indirect method (285) compared well to the literature average of direct retrievals (295) (see Tables 3 and 4 in Cattrall et al., 2005). Direct measurements do not suffer the same limitations as indirect ones which require assumptions on size distribution and chemical composition or a molecular extinction profile. The Supplement Table S1 summarizes available retrieval methods and values of some experimentally determined lidar ratios over marine regions. Currently, most lidars do not yet have Raman or high spectral resolution capability and CALIPSO is the only lidar that provides aerosol data at the vast spatiotemporal resolution required for global climate model comparison.

Since the uncertainty in the lidar ratio can significantly affect the accuracy of the aerosol extinction retrieval (see a detailed discussion below), lidar ratios have been constrained by numerous approaches. However, marine aerosol size distribution, chemical composition and refractive index can change significantly with ocean surface wind speed (U_{10}), relative humidity (RH), temperature, salinity and chemical/biological composition of surface sea water (de Leeuw et al., 2011; Lewis and Schwartz, 2004). For this reason, large disagreement exists in the literature regarding the value of maritime aerosol lidar ratio (S_p; subscript "p" indicates particulate). For example, lidar measurements of Ansmann et al. (2001) over the North Atlantic showed $S_p = 24 \pm 5$ sr, whereas measurements using a nighttime lidar at a horizontal orientation off the northern coast of Queensland, Australia, showed maritime aerosol lidar ratios as high as $S_p = 39 \pm 5$ (Young et al., 1993). Using the data from AERONET oceanic sites, Cattrall et al. (2005) derived a lidar ratio of 28 ± 5 sr, a value that compared well with a literature averaged value of $S_p = 29 \pm 5$ sr (for $490 \leq \lambda \leq 550$ nm) for maritime aerosols. Passive techniques have also been used to derive the lidar ratio using an alternative definition of S_p as a function of single scattering albedo and the scattering phase function near 180° (Bréon, 2013). Using the multi-directional measurements of solar radiation from the polarization sensitive passive radiometer POLDER, typical values for clean marine aerosol S_p were derived to be 25 sr at 532 nm (Bréon, 2013). The lidar ratio of 20 ± 6 sr (at 532 nm) was selected for the CALIOP retrieval algorithm based on parameters measured during the Shoreline Environmental Aerosol Study (SEAS) experiment (Masonis et al., 2003; Omar et al., 2009). The SEAS measurements conducted on the beach (downwind of an offshore reef) report a particulate lidar ratio of $S_p = 25.4 \pm 3.5$ sr at 532 nm based on the optical size measurements of marine aerosol, and an average modeled value of $S_p = 20.3$ sr (Masonis et al., 2003). However, it was also shown that, depending on a particle size and wind speed regime, S_p values can range from 10 to 90 sr (Masonis et al., 2003; Sayer et al., 2012). Therefore, as size distribution (and chemical composition) of marine aerosol may vary over the oceans, a constant lidar ratio used in CALIOP algorithms may lead to erroneous retrievals of AOD.

In this study, we present a new method for deriving lidar ratios for individual CALIOP retrievals of single aerosol layer columns over the ocean. We have used the Synergized Optical Depth of Aerosols (SODA) product (described in Sect. 2.2) to estimate S_p for a strictly defined subset of CALIPSO data. The S_p values are calculated as a correction to achieve the best agreement between SODA and CALIPSO marine aerosol AOD values. Using CALIPSO level 2 aerosol layer data for years 2007 to 2010, we have created a 3-year averaged climatology of clean marine aerosol lidar ratio over the globe. Analyses were also carried out to assess dependence of S_p values on wind speed and estimate possible error sources in our calculations.

2 Instrumentation and methods

2.1 CALIPSO satellite

The CALIPSO mission (Winker et al., 2009), launched on 28 April 2006, has been able to provide the scientific community with vertically resolved measurements of both aerosol and cloud optical properties like depolarization ratio (a measure of particle sphericity), AOD, and ice/water phase since June 2006. The CALIPSO payload includes a high-powered digital camera, an infrared radiometer, and the two-wavelength (532 and 1064 nm) near-nadir, polarization-sensitive elastic backscatter lidar CALIOP.

The level 1 data algorithms are responsible for the geolocation and range determination of the satellite and produce profiles of attenuated backscatter coefficients. Data in this work were obtained from the 5 km level 2 operational products, version 3.01. Level 2 products have undergone various processing algorithms from the Selective Iterated BoundarY Locator (SIBYL), the Scene Classification Algorithm (SCA), and the Hybrid Extinction Retrieval Algorithm (HERA) (Vaughan et al., 2004, 2009). First, SIBYL identifies layers, then the SCA identifies the type of feature (i.e., aerosol or cloud) and the subtype (i.e., aerosol type, ice/water phase), and finally the HERA generates extinction profiles for the feature. The theoretical basis of the algorithm can be found online at www-calipso.larc.nasa.gov/resources/project_documentation.php.

The CALIPSO 5 km aerosol layer data include many operational products, only a few of which are used in this study. Among them are the integrated attenuated backscatter and its uncertainty at 532 nm, the layer features such as number found in the column and their top and bottom altitudes, and the feature classification flags.

2.2 Synergized Optical Depth of Aerosols

CloudSat was launched in 2006 with CALIPSO and was positioned in sun-synchronous orbit as part of the A-Train satellite constellation. CloudSat and CALIPSO have paved the way for new multi-sensor data products like SODA to be developed. The main instrument on CloudSat is the Cloud Profiling Radar (CPR), a nearly nadir-looking (0.16°) 94 GHz ($\approx$ 3 mm; W-band) radar. The CPR, like CALIOP, can re-

trieve information on hydrometeor microphysical properties at different heights in a vertical column. The CPR signal is mostly attenuated by water vapor; however, for cloud-free regions over the ocean, the CPR data can be used to retrieve AOD. A method developed by Josset et al. (2008) and later expanded by Josset et al. (2010a) uses a combination of CALIOP and CPR measurements of the ocean surface reflectance to derive AOD. The design of SODA utilizes the ratio of the radar-to-lidar ocean surface scattering cross section to infer column optical depth for non-cloudy atmospheric columns. Since the radar signal attenuates mostly due to water vapor and the lidar signal weakens mostly due to aerosols, after the radar signal is corrected for attenuation by water vapor and oxygen, the change in the radar-to-lidar signal ratio is directly related to aerosol abundance (Josset et al., 2008, 2010a). Therefore, by using observations from two different sensors, SODA can eliminate uncertainties induced by the CALIOP aerosol extinction algorithm over oceans. SODA AODs have been shown to be in very good agreement with MODIS AOD retrievals (Josset et al., 2008). A more detailed description of the SODA technique and its application is given in Josset et al. (2008, 2010a, b, 2011, and 2012). The SODA products that are used in this study include the quality assurance measure "qa_flag_aerosol" and the 532 nm AOD.

2.3 Lidar ratio definition

One of the biggest advantages of the SODA product is that it removes the dependence of the prescribed lidar ratio while still utilizing the active sensors to retrieve an AOD, thereby providing a means for independent evaluation of the lidar ratio. In the current study we use Eq. (4) from Josset et al. (2011) to estimate lidar ratio from Cloud-Sat/CALIOP measurements of AOD values. Following Fernald et al. (1972), the particulate two-way transmittance at height Z can be written as

$$T^2(Z) = e^{-2S_p \int_0^Z \beta_p(z)dz}, \tag{1}$$

where the lidar ratio at height Z can be defined as the ratio of the particulate extinction to backscatter ($S_p = \frac{\sigma_p(Z)}{\beta_p(Z)}$). Differentiating Eq. (1) with respect to vertical coordinate (z) gives the particulate backscatter at height Z:

$$\beta_p(Z) = -\frac{1}{2S_p T^2(Z)} \frac{dT^2(Z)}{dZ}. \tag{2}$$

Since atmospheric constituents (molecules and different particle types) can interact with the lidar beam at different heights, the lidar ratio using remotely sensed data cannot be uniquely defined for a given atmospheric column. However, the lidar ratio is a particle intensive property (i.e., dependent on particle type and not on the amount). So, if we assume that there is only a single type of aerosol that is homogeneously distributed throughout the atmospheric column and that molecular scattering is sufficiently removed by the

CALIOP level 2 algorithms, then the column lidar ratio ($\overline{S_p}$) can be expressed as the ratio of the particulate column integrated extinction ($\overline{\tau_p}$ = AOD) to the attenuated backscatter ($\overline{\Gamma_p}$). Based on these assumptions, integration of Eq. (2) with respect to vertical coordinate gives the particulate lidar ratio as

$$\overline{S_p} = -\frac{\int_{T_p^2(0)}^{T_p^2(Z)} dT^2(z)}{\int_0^Z \beta_p(z) T_p^2(z) dz}. \tag{3}$$

If we first substitute in Eq. (3) the definition for two-way transmittance as $T_p^2 = e^{-2\overline{\tau_p}}$, then substitute the total particulate attenuated backscatter signal retrieved by the lidar as $\overline{\Gamma_p} = \int_0^Z \beta_p(z) T^2(z) dz$, and finally consider that $T_p^2(0) = 1$, the equation for a columnar particulate lidar ratio is

$$\overline{S_p} = \frac{1 - e^{-2\overline{\tau_p}}}{2\overline{\Gamma_p}}. \tag{4}$$

Equation (4) allows us to calculate marine aerosol lidar ratio from two independent sources: the AOD (i.e., $\overline{\tau_p}$) from SODA and the integrated attenuated backscatter ($\overline{\Gamma_p}$) from CALIOP. It should be noted that CALIOP estimation of $\overline{\Gamma_p}$ is difficult for layers that are not bounded by clear air (Vaughan et al., 2004) and therefore require carefully designed data screening algorithms. In Sect. 4 we carry out an error analysis to verify that uncertainties in $\overline{\Gamma_p}$ have a minimal effect on the retrieved lidar ratio.

2.4 Data selection method

As different aerosol sub-types have different lidar ratios, application of Eq. (4) to episodes when aerosols other than marine aerosols are present in the atmospheric column may lead to erroneous results for the calculated $\overline{S_p}$. We developed a strict scene selection algorithm to minimize the contamination of AOD and therefore $\overline{S_p}$ by aerosol types other than marine (e.g., anthropogenic pollution, biomass burning, and dust). The algorithm first uses the feature classification flags in the CALIOP aerosol layer product. We start with clean marine aerosol that is identified based on surface type (as determined by the location of the satellite) and then retain only the data with total integrated attenuated backscatter $\gamma' > 0.01 \text{ km}^{-1} \text{ sr}^{-1}$ and volume depolarization ratio $\delta' < 0.05$. (Omar et al., 2009). As multiple types of aerosols can be found within retrieved vertical profiles (e.g., dust above marine aerosols), aerosol feature types that have been identified as marine in a given atmospheric column are not enough to carry out the analysis. Therefore, when determining the lidar ratio of marine aerosol using Eq. (4), the algorithm only retains the data in which clean marine is the only type of aerosol present in the entire cloud-free atmospheric column. To further reduce the uncertainty, we constrain the analysis to single layer profiles below 2 km and remove profiles in which

marine aerosol layers are vertically stacked within an atmospheric column. Therefore, the vertically integrated particulate attenuated backscatter $\overline{\Gamma_p}$ is replaced by Γ_p. Similarly, the column lidar ratio $\overline{S_p}$ is reduced to S_p in the remainder of the text. Note also that all quantities discussed are particulate quantities; therefore, molecular scattering is removed using gridded molecular and ozone number density profile data from the Goddard Earth Observing System Model, version 5 (GEOS-5), analysis product available from the NASA Global Modeling and Assimilation Office (GMAO) (Winker et al., 2009). Operationally, particulate scattering is determined to be where the ratio of the CALIOP 532 nm scattering profile normalized by the GEOS-5 molecular scattering profile is greater than one $\left(\frac{\beta'_{532}}{\beta_m} > 1\right)$. Errors associated with $\overline{\Gamma_p}$ are discussed in Sect. 4.

All data are for nighttime and are binned into $2° \times 5°$ (latitude and longitude, respectively) grid cells. Collocated wind speed is taken from the Advanced Microwave Scanning Radiometer – EOS (AMSR-E) observing system. To identify distinct features associated with the variability in marine aerosol lidar ratio over different parts of the oceans, the selected data are examined in relation with other variables such as season, spatial location, and wind speed.

Some additional measures were taken to target layers with a high signal-to-noise ratio and grid cells with a significant number of observations. These measures included (i) ensuring the relative error in Γ_p due to random noise in molecular backscatter was < 50 %, (ii) the collocated SODA 5 km layer was composed of at least 70 % shot-to-shot data, and (iii) the total number of retrievals per $2° \times 5°$ grid cell ranked above the first quartile of the grid cell frequency distribution. Such strict quality controls considerably increase the reliability of the analysis despite reducing the total number of data points. It should be noted that a large number (over 260 000) of data points remained for robust statistics after all the quality control and quality assurance tests. A caveat, despite such rigorous quality control criteria, remains: when interpreting data near coastlines, the CALIOP scene classification algorithm may mistakenly identify mixtures of continental pollution and marine as clean marine aerosol (Burton et al., 2013; Oo and Holz, 2011; Schuster et al., 2012), causing an overestimation in the lidar ratio inferred from Eq. (4). Further discussion of error analysis is given in Sect. 4.

3 Results

3.1 Global distribution of retrieved AOD and lidar ratio

Active detectors like CALIOP require knowledge of the lidar ratio for retrieval of aerosol optical properties. Incorrect estimates of the S_p values for a given aerosol type can lead to significant errors in the retrievals of particulate extinction and AOD. Past studies using collocated CALIOP and MODIS re-

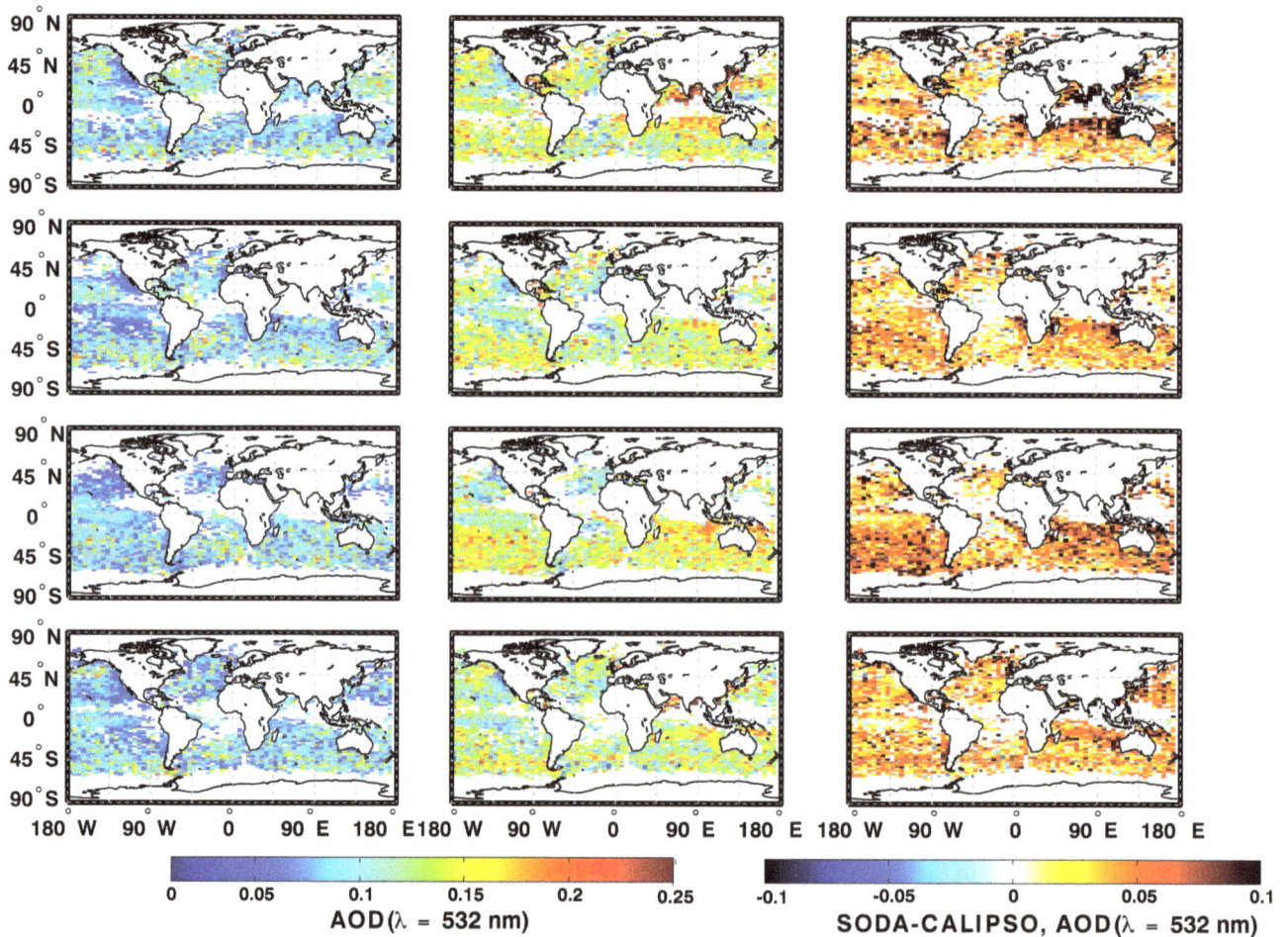

Figure 1. Seasonal median AOD values from CALIOP and SODA (columns 1 and 2) and the difference (SODA − CALIOP) plot (column 3) for December–February (row 1), March–May (row 2), June–August (row 3), and September–November (row 4) plotted on a 2° × 5° latitude longitude grid. "No data" areas are shaded white and defined as grid cells failing the quality-control algorithm (see text for details).

trievals have shown that, over the marine regions, CALIOP underestimates the AOD values relative to MODIS (Oo and Holz, 2011). As MODIS data over the ocean have been extensively evaluated with numerous field campaigns (e.g., Levy et al., 2005), it was suggested that the primary source of discrepancy between the two sensors was the low value of the marine aerosol lidar ratio used by CALIOP (Oo and Holz, 2011). Figure 1 shows seasonally averaged maps of CALIPSO and SODA marine aerosol median optical depth at 532 nm and the differences between SODA- and CALIOP-retrieved AODs. White regions in Fig. 1 represent grid cells that were rejected by the data selection algorithm and have been removed from the subsequent data analysis. Inspection of Fig. 1 reveals considerable spatial and temporal variations in marine aerosol AOD. Although the largest values of AOD seem to occur over regions with higher surface wind speed (i.e., the northern and southern oceans), elevated AOD values can also be seen over the regions downwind from dust and/or pollution sources such as the mid-latitude North Atlantic

Ocean and the Bay of Bengal and over the major oceanic gyres. The region around the Indian subcontinent and over the Bay of Bengal is believed to be just a retrieval artifact. Large disagreements between SODA- and CALIOP-reported AODs for these regions suggest that some dust/pollution aerosols might have been misclassified by CALIOP as marine aerosol. Higher S_p values for dust and pollution compared to marine aerosol would produce a higher AOD retrieval in SODA compared to CALIOP. Elevated AOD values over the oceanic regions with lower surface wind speed, however, could point to changes in marine aerosol size distribution to smaller sizes. Sub-micron sea salt aerosols (with particle diameter, $D_p < 1 \, \mu m$) are believed to have larger lidar ratios than super-micron ones (e.g., Masonis et al., 2003; Oo and Holz, 2011). In general, Fig. 1 shows positive differences between SODA- and CALIOP-retrieved seasonal median AOD values. Recalling that CALIOP-retrieved extinction is the product of the prescribed lidar ratio and the measured column integrated particulate backscatter, positive

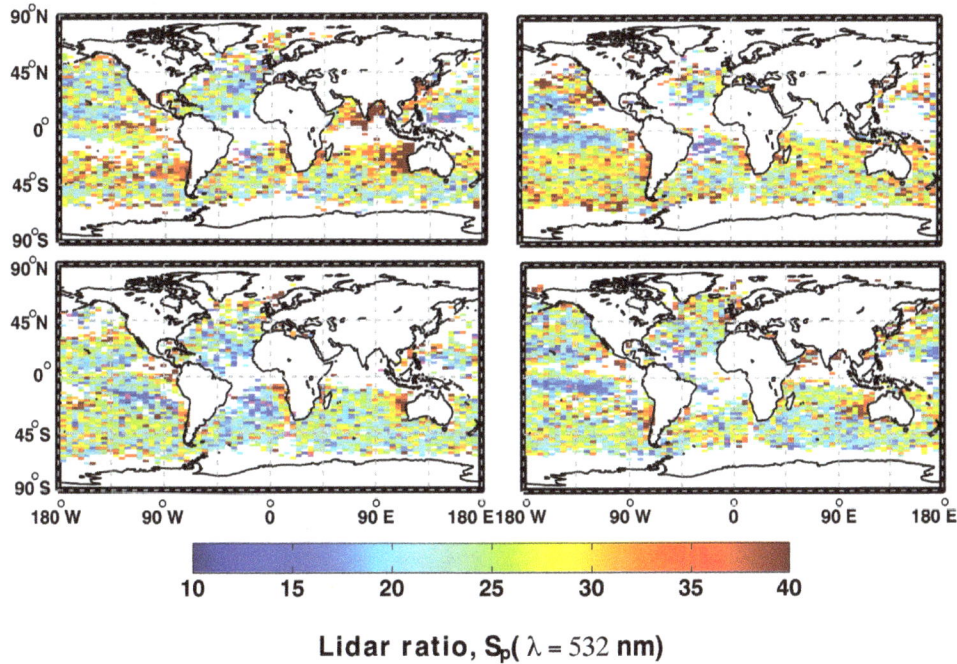

Figure 2. Seasonal lidar ratio for $2° \times 5°$ latitude longitude grid cells. Seasons are arranged as (**a**) December–February, (**b**) March–May, (**c**) June–August, (**d**) September–November.

differences between SODA and CALIOP median AODs at 532 nm over most of the oceans suggest underestimation of the marine aerosol lidar ratio prescribed in the CALIOP clean marine aerosol model. Figure 2 shows that over most of the ocean surfaces, the calculated lidar ratio is higher than the default ($S_p = 20$ sr) used in the CALIOP clean marine aerosol model. Global means and standard deviations for AOD and lidar ratio are given in Table 1. CALIOP retrievals in this study cannot be directly compared to MODIS since we only use nighttime data. Nevertheless, SODA retrievals of AOD have been shown to agree well with MODIS (Josset et al., 2008), HSRL (Fig. 7a; Josset et al., 2011) and Maritime Aerosol Network (MAN) (Smirnov et al., 2011; Fig. 8) observations, suggesting that the corrected lidar ratios will bring CALIOP retrievals close to MODIS data. Figure 2 also reveals that the value of the lidar ratio calculated using Eq. (4) changes considerably over different parts of the remote oceans, pointing to the variability in marine aerosol optical properties. It has long been known that meteorological and/or environmental factors and ocean chemical/biological composition influence marine aerosol production, entrainment, transport, and removal processes (Lewis and Schwartz, 2004) that can ultimately affect marine aerosol S_p. Moreover, due to atmospheric transport of marine aerosol, satellite-retrieved AOD values may also be related to the upwind processes. Despite the complexity of the mechanisms controlling marine aerosol mass concentration over the oceans, surface wind speed has always been considered as the major parameter governing the production, chemical composition,

and life cycle of marine aerosol (Lewis and Schwartz, 2004). Therefore, in the next section we will investigate the effect of wind speed on calculated temporal variability of marine aerosol lidar ratio.

3.1.1 Wind speed dependence

Numerous investigators have examined the effect of sea surface wind speed and sea state on marine aerosol optical properties (e.g., Smirnov et al., 2003; Sayer et al., 2012). There are two mechanisms for primary marine aerosol production: bursting of bubbles at the water surface and mechanical tearing of water drops (spume) from wave crests (for surface wind speeds $U_{10} > 9\,\mathrm{m\,s^{-1}}$, Anguelova et al., 1999). Ocean bubbles are generated by the entrainment of air due to wave action. As bubbles rise due to their buoyancy they burst at the surface, producing marine aerosol. (Blanchard and Woodcock, 1957). In this study we have selected seven different wind speed regimes (see Table 2). The lowest wind speed regime, $0 < U_{10} \leq 4\,\mathrm{m\,s^{-1}}$, was chosen to represent aerosols not generated via wind driven processes over the ocean. In general, ocean waves break at wind speed values above $\sim 4\,\mathrm{m\,s^{-1}}$ (initiating the white cap formation and bursting of the entrained bubbles) (Lewis and Schwartz, 2004). Therefore, it has been suggested that below this threshold value, there should be a weak relationship between marine aerosol optical properties and the surface wind speed (Kiliyanpilakkil and Meskhidze, 2011; Lehahn et al., 2010). Moreover, for such a low wind speed regime, most of

Table 1. Seasonal means $\pm$ 1 standard deviations for $2° \times 5°$ grid cell medians. The subscripts p, S, and C appended to τ stand for particulate, SODA, and CALIOP, respectively, where τ is the AOD.

Season	$\tau_{p,S}$	$\tau_{p,C}$	$\Gamma_p \times 10^{-3}$ (sr^{-1})	S_p (sr)
Winter	0.14 ± 0.04	0.09 ± 0.03	4.7 ± 1.2	27 ± 8
Spring	0.13 ± 0.03	0.09 ± 0.03	4.8 ± 1.2	24 ± 7
Summer	0.14 ± 0.04	0.09 ± 0.03	4.6 ± 1.2	27 ± 8
Fall	0.13 ± 0.03	0.09 ± 0.03	4.7 ± 1.1	25 ± 7

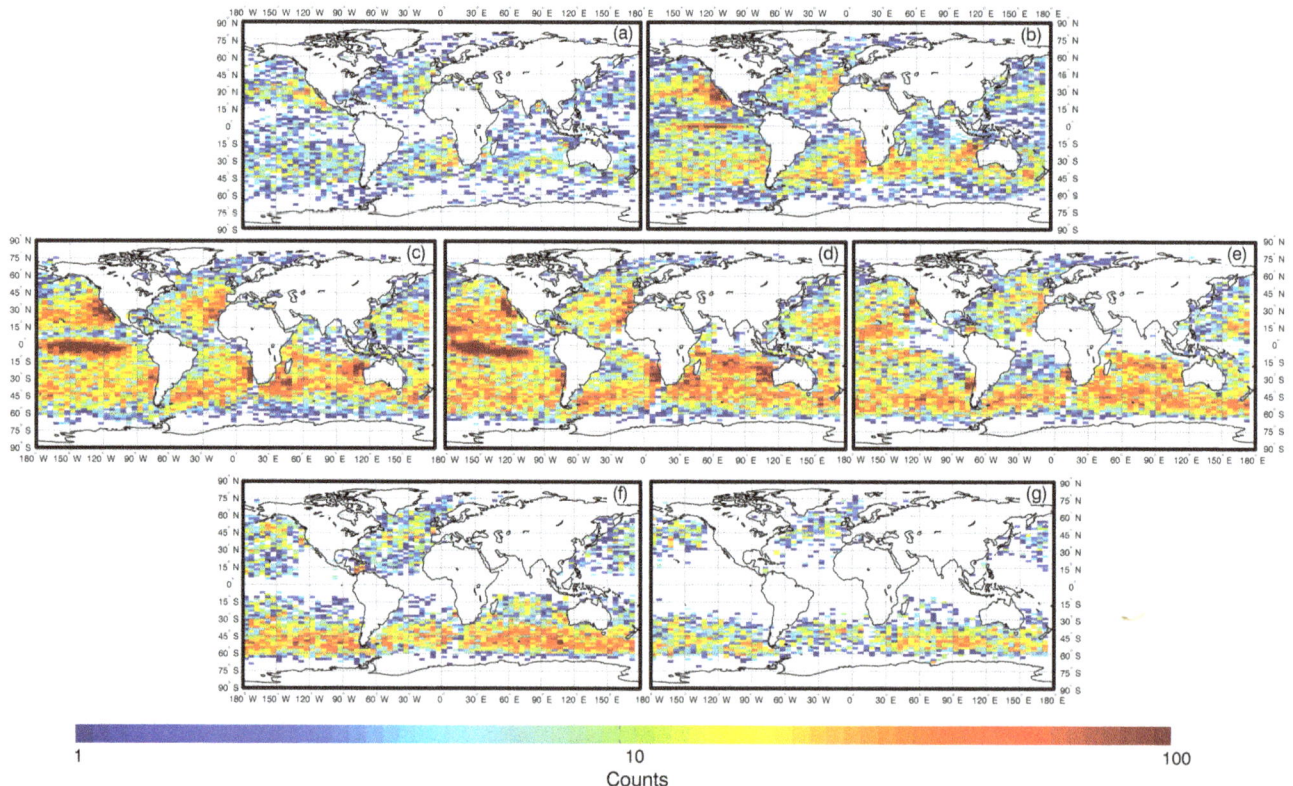

Figure 3. SODA/CALIOP retrieval counts for each $2° \times 5°$ latitude longitude grid cell and different wind speed regimes. AMSR-E wind speed regimes for figures (**a**) through (**g**), are 0–4, 4–6, 6–8, 8–10, 10–12, 12–15, and >15 m s^{-1}, respectively.

the aerosols classified as clean marine by CALIOP are either produced outside the swath and then blown into the satellite field of view or, like in cases near coastlines, mistakenly identified as marine aerosol. The highest wind speed regime, with $U_{10} > 15$ m s^{-1}, typically contributes a small fraction of CALIOP retrievals (Kiliyanpilakkil and Meskhidze, 2011) and is largely concentrated over the southern ocean and in the northern Atlantic where the highest wind speeds are observed (Bentamy et al., 2003). Figure 3 shows a spatial map of the number of retrievals for each grid cell separated by wind speed regime. According to Fig. 3, the southern ocean retrievals are dominant at the highest wind speeds and are overall consistent with the so-called "roaring forties" latitude band. Figure 3, as well as Table 2, shows that the fewest num-

ber of retrievals are found for the lowest and highest wind regimes.

The data shown in Fig. 3 are next used to generate scatter density plots for SODA- and CALIOP-retrieved AOD values for the wind speed regimes reported in Table 2 (see Fig. 4). As expected, Fig. 4 shows that increases in wind speed are typically associated with higher values of marine aerosol optical depth (note the center of the scatter distribution shifts slightly to higher AODs for larger wind speed values). However, as the majority of the SODA AODs exist above the 1 : 1 line, this figure also indicates the underestimation of CALIOP-retrieved marine aerosol optical depth values. When averaged over the entire globe, CALIOP-retrieved clean marine AOD is roughly 32 % lower compared to SODA (with an RMS error of 0.06; Supplement Fig. S2). Accord-

Table 2. Means ± 1 standard deviation for $2° \times 5°$ grid cell medians for various AMSR-E wind speed regimes. The subscripts p, S and C appended to τ stand for particulate, SODA and CALIOP, respectively, where τ is the AOD.

Wind regime (ms^{-1})	$\tau_{p,S}$	$\tau_{p,C}$	$\Gamma_p \times 10^{-3}$ (sr^{-1})	S_p (sr)	Relative number(%)
$0 < U_{10} \leq 4$	0.12 ± 0.05	0.07 ± 0.04	3.6 ± 1.4	32 ± 17	11 849 (5)
$4 < U_{10} \leq 6$	0.11 ± 0.04	0.07 ± 0.03	3.8 ± 1.1	27 ± 12	32 899 (13)
$6 < U_{10} \leq 8$	0.12 ± 0.04	0.08 ± 0.02	4.2 ± 1.0	26 ± 9	60 083 (23)
$8 < U_{10} \leq 10$	0.13 ± 0.03	0.08 ± 0.02	4.7 ± 1.0	26 ± 7	68 899 (26)
$10 < U_{10} \leq 12$	0.15 ± 0.04	0.10 ± 0.03	5.1 ± 1.0	26 ± 6	45 895 (17)
$12 < U_{10} \leq 15$	0.16 ± 0.04	0.12 ± 0.03	5.7 ± 1.2	25 ± 6	30 162 (11)
$U_{10} > 15$	0.16 ± 0.04	0.14 ± 0.04	6.4 ± 1.4	22 ± 7	12 953 (5)

ing to Fig. 4 the largest discrepancies between SODA and CALIOP retrievals are observed at lower wind speed values. One simple explanation for this is a greater chance for CALIOP misclassification over the oceanic regions where long-range continental aerosols can contribute a larger fraction of the marine boundary layer (MBL) particles (e.g., Blot et al., 2013). Terrestrial particles (e.g., mineral dust, anthropogenic pollution) are typically characterized by the larger lidar ratio values, leading to an underestimation of the CALIOP-retrieved AODs. However, measurements also show that changes in surface wind speed values can cause a considerable shift in the marine aerosol size distribution. For optically active marine aerosols, the residence time decreases considerably with increasing size. Thus the aerosol population is increasingly controlled by the smaller end of the particle size spectrum as wind speeds decrease over the ocean (Hoffman and Duce, 1974). Conversely, as wind speed increases, fine mode aerosol volume size distribution changes slightly (with mixed trends), while the coarse mode volume size distribution exhibits a large and positive response to the increase in wind speed (Lewis and Schwartz, 2004; Smirnov et al., 2003). Such variability in marine aerosol volume size distribution is expected to have an effect on the aerosol lidar ratio. As sub-micron marine aerosols are characterized with much larger lidar ratios than super-micron ones (e.g., Masonis et al., 2003; Oo and Holz, 2011), shifting marine aerosol size distribution spectra to smaller particles will cause an increase in total aerosol lidar ratio. Therefore for clean marine aerosols, AODs and lidar ratios are expected to have opposite dependences on wind speed: high wind speed regions are characteristic of high AODs and low lidar ratios while lower wind speeds favor higher lidar ratios and lower AODs (Smirnov et al., 2003; Sayer et al., 2012).

Figure 5 shows that on average, the calculated aerosol lidar ratio is weakly related to the surface wind speed. According to this figure, aerosols retrieved in the wind speed regime $0 < U_{10} \leq 4 \, \text{m s}^{-1}$ depict the largest variability in the lidar ratio as indicated by the spread of the distribution. As discussed above, aerosols in this regime likely include both marine aerosols particles produced upwind and advected into the satellite field of view (with $S_p \sim 20$ to 30 sr)

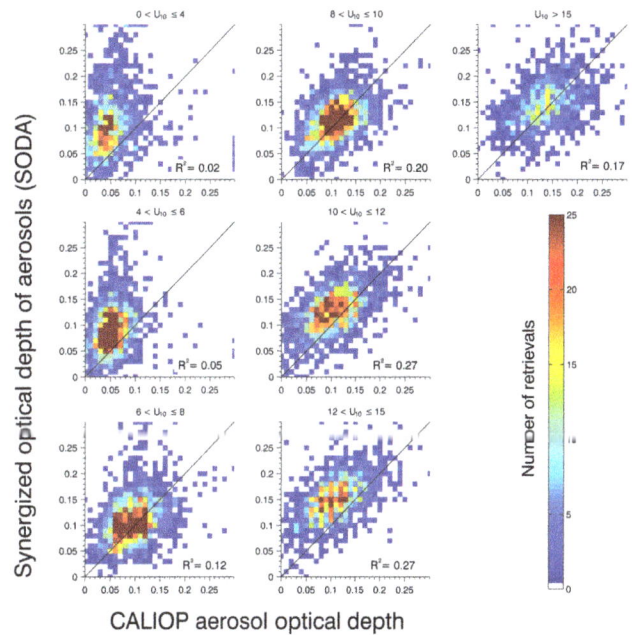

Figure 4. Scatter density plot of SODA to CALIOP AOD for each wind speed regime. Each point indicates a grid cell median, colored by frequency of occurrence. The black line is the $1:1$ relationship, with reported R^2 values.

and dust/pollution particles (with $S_p \sim 40$ to 70 sr, Omar et al., 2009) that may have been misclassified by CALIOP as marine aerosol. As shown in Table 2, the marine aerosol lidar ratio distribution in this regime is characterized by the largest standard deviation ($\sigma = 17.4$ sr), indicating that for the lowest wind speed values, a wide range of marine aerosol sizes can be present over the ocean. Since the primary marine aerosol production is minimal for the wind speed values less than $4 \, \text{m s}^{-1}$, such large spread could also indicate that under low wind conditions there is greater probability for natural continental and human-induced pollution aerosols to be miss-classified by CALIOP as clean marine.

For the higher wind speed values ($4 < U_{10} \leq 15 \, \text{m s}^{-1}$), lidar ratio generally decreases with the increase in the wind speed and approaches the lidar ratios prescribed by CALIOP

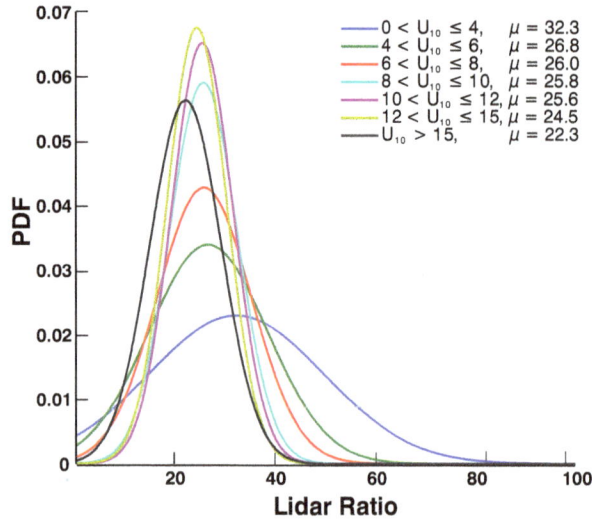

Figure 5. Probability density function of clean marine aerosol lidar ratio for selected AMSR-E wind speed regimes. The mean (μ) of each distribution is also reported.

retrieval algorithms (i.e., 20 sr) at the highest wind speed regime. According to Table 2 and Fig. 3, the most common wind values in CALIOP marine aerosol retrievals over the ocean are in the $8 < U_{10} \leq 10\,\mathrm{m\,s^{-1}}$ regime (26 % of all available data) followed by the $6 < U_{10} \leq 8\,\mathrm{m\,s^{-1}}$ regime (23 % of all available data). For the higher wind speed regimes ($U_{10} > 6\,\mathrm{m\,s^{-1}}$), surface winds play a decisive role in the determination of the lidar ratio (indicated by the narrow standard deviation, see Table 2). This is an important result, as the distributions shown in Fig. 5 may help in providing additional criteria for clean marine lidar ratio selection, yielding improved retrievals of marine aerosol AOD from CALIOP.

Analysis of data indicates that a mean lidar ratio of 26 sr is the most probable value that occurs for the majority of CALIOP retrievals over the oceans. This value compares well with those reported in the literature. Müller et al. (2007) found a marine aerosol lidar ratio of 23 ± 3 and 23 ± 5 sr using RL, and Burton et al. (2012, 2013) reported a range from 15 to 27 sr using HSRL. Bréon (2013) used a different space-based retrieval and reported S_p for marine aerosol typically on the order of 25 sr. Table S1 reports some additional values of marine aerosol S_p measured by other techniques. This new lidar ratio reduces discrepancy between CALIOP-prescribed and SODA-derived lidar ratios from about 30 to 4 %.

Previous studies reported a small decrease in marine aerosol lidar ratio with the increase in wind speed (Sayer et al., 2012). In general, wind speed alone is expected to be a poor predictor of marine aerosol lidar ratio, as aerosol volume size distribution and optical properties are likely to be influenced by a number of other parameters including relative humidity and marine boundary layer depth. Furthermore, errors increase exponentially approaching the lowest optical

depths and could be the reason for the large spread in the lidar ratio seen in Fig. 5. Untangling systematic error from real physical effects is difficult in the low (0–4 m s^{-1}) wind speed regime and highlights the need for more accurate measurements for calm wind/low AOD conditions. Despite these complications, a shift to lower lidar ratios with increasing wind speed can be seen from Fig. 5 and warrants further investigation.

4   Uncertainties, errors, and sensitivity

The method used to derive the lidar ratio in this study depends on two parameters: the CALIOP-integrated attenuated particulate backscatter (Γ_p) and the SODA aerosol optical depth (τ_p). Uncertainties in both Γ_p and τ_p retrievals are expected to propagate through the calculations of the particulate lidar ratio. Josset et al. (2008, 2010a) investigate the domain of validity for τ_p through an extensive calibration procedure. They find that for retrievals at wind speeds between 3 and 10 m s^{-1} the SODA product is in very good agreement ($R > 0.89$) with MODIS AOD, with calibration errors less than 15 %. Calibration errors in τ_p are expected to be even lower for nighttime retrievals used in this study (Josset et al., 2008). However, average uncertainty for CALIOP Γ_p retrievals has not yet been examined and is necessary for the assessment of this retrieval method. We make an estimate on this uncertainty in the following section.

Since ocean is the source of marine aerosol, clean marine aerosol layers typically extend to the ocean surface. This makes it more difficult to determine molecular and particulate backscatter components of the signal separately using satellite measurements alone. To assess the uncertainty in lidar ratio introduced for the surface connected layers (i.e., layers whose bottom bound is defined as the ocean surface), we here estimate the error in CALIOP-retrieved Γ_p values. The total attenuated backscatter signal measured by the lidar consists of molecular and particulate components:

$$\beta_{\mathrm{att}} = (\beta_p + \beta_m) e^{-2\tau_p} \cdot e^{-2\tau_m}, \tag{5}$$

with subscripts m and p representing molecular and particulate quantities, respectively. From the definition of Γ_p it follows that

$$\Gamma_p = \int_0^Z \beta_p(z) e^{-2\tau_p} \mathrm{d}z, \tag{6}$$

where the integration is from the surface to the top of the layer. β_p is the particulate backscatter and $e^{-2\tau_p}$ accounts for the attenuation of the lidar signal by the particles. Substituting Eq. (5) into Eq. (6) gives

$$\Gamma_p = \int_0^Z (\beta_{\mathrm{att}} e^{2\tau_m} - \beta_m(z) e^{-2\tau_p}) \mathrm{d}z. \tag{7}$$

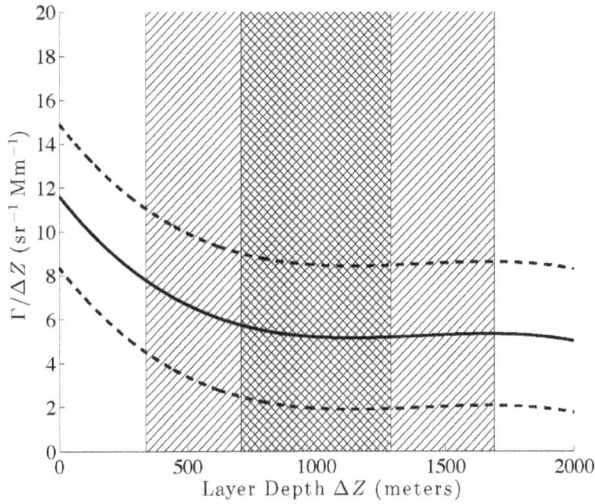

Figure 6. The normalized integrated attenuated backscatter as a function of the layer depth. The solid line shows the third-order least squares fit to the data while the dotted lines show $\pm 1\sigma$; the hatched area shows the layer depth data frequency: cross hatch between the 25th and 75th percentiles and straight hatch between 5th and 95th percentiles.

The molecular component of the signal in Eq. (7) can be derived from the GMAO modeled temperature and pressure profiles (Bloom et al., 2005). However, to solve this equation and determine the particulate attenuated backscatter value, particulate column-integrated extinction is required. To get τ_p, the CALIOP algorithm uses a prescribed value of the lidar ratio, making Eq. (4) circularly dependent. The error in CALIOP-retrieved Γ_p associated with the prescribed lidar ratio can be estimated by substituting the τ_p value from SODA. A large error would imply that the uncertainty in CALIOP-prescribed lidar ratio would introduce sizable corrections to Γ_p, making Eq. (4) unsuitable for the estimation of marine aerosol lidar ratio.

The relative error in Γ_p can be defined as

$$\text{Error} = \frac{\Gamma_{p,S} - \Gamma_{p,C}}{\Gamma_{p,C}} = \frac{(e^{-2\tau_{p,C}} - e^{-2\tau_{p,S}}) \cdot \int_0^Z \beta_m(z)\mathrm{d}z}{\Gamma_{p,C}}, \quad (8)$$

where $\Gamma_{p,S}$ and $\Gamma_{p,C}$ are columnar integrated attenuated backscatter values for SODA and CALIOP, respectively. From the theoretical basis documents for CALIOP level 1 algorithms, the molecular backscatter is estimated as $\beta_m = \frac{C_s}{S_m}\frac{T(z)}{P(z)}$, where height-dependent $T(z)$ and $P(z)$ profiles from the surface (1000 hPa) to top-of-atmosphere (0.1 hPa) pressure levels were obtained from the GMAO Modern-Era Retrospective analysis for Research and Applications data set. The molecular lidar ratio, S_m, is defined as $8\pi/3$, and C_s is a constant equal to 3.742×10^{-6} K hPa^{-1} m^{-1} (Hostetler et al., 2005). When considering all of the parameters, our analysis shows that the average error in Γ_p is approximately 1.5 %. Compared to the systematic uncertainty in the SODA prod-

uct < 15 %, the uncertainty in Γ_p is much lower, indicating that, on average, errors in Γ_p do not dominate S_p retrievals. Since an average discrepancy between CALIOP-prescribed and SODA-derived lidar ratios (~ 30 %) is more than an order of magnitude higher than uncertainty in Γ_p, we conclude that the uncertainty in the CALIOP column-integrated backscatter has a minor effect on the Eq. (4) calculated lidar ratio.

Furthermore, because in our study we use feature-integrated products for a single aerosol layer, it is also important to evaluate the relationship between Γ_p and aerosol layer thickness (ΔZ). Figure 6 shows the normalized column attenuated particulate backscatter Γ_p as a function of layer depth. For uniformly distributed aerosols throughout the column, Γ_p is likely to be proportional to ΔZ. The spread of $\Gamma_p/\Delta Z$ ratio is indicative of different amounts of marine aerosol present in the column. Two limits of very high and very low ΔZ values are of particular interest. For example, strong reduction of the $\Gamma_p/\Delta Z$ ratio at the higher ΔZ values would indicate that the lidar signal is strongly attenuated throughout the layer reaching a sensitivity limit. However, considerable increase of the ratio for the thin layers may indicate contamination of the backscattered signal by strong surface reflectance. According to Fig. 6, for the vast majority of the data, signal attenuation and surface reflectance do not seem to be major issues for the surface connected layers, suggesting that the quality control algorithm described in Sect. 2.4 was sufficient to remove the majority of erroneous measures of Γ_p.

To further assess the reliability of SODA marine aerosol product we also compared collocated HSRL (Fig. 7) and MAN (Fig. 8) AODs to SODA. Figure 7a shows results from three CALIPSO (and therefore SODA) underflights validated against HSRL. According to Fig. 7a for AODs < 0.3 (comprising the majority of marine aerosol retrievals), SODA compares reasonably well to HSRL ($R^2 = 0.82$, RMSE = 0.04; similar to the MAN comparison with RMSE = 0.03 in Fig. 8). Additionally, Fig. 7b illustrates that the relative uncertainty in the SODA-retrieved S_p is typically below 50 % for AODs > 0.05. In our study, the bulk of AODs measured by SODA (98 %) exceed this value under the quality control criteria discussed in Sect. 2.4. Errors were estimated based on Eq. (15) in Josset et al. (2012), and for AODs > 0.05 we expect lidar ratio retrieval uncertainties below 50 %. MAN and SODA collocation for Fig. 8 was determined based on a scheme in Smirnov et al. (2011) and Kleidman et al. (2011). We required that the SODA retrieval be within ± 30 min of the MAN retrieval as well as within a circle with radius of 25 km around the MAN measurement. A map of the retrieval locations and the details of the algorithm used are given in the Supplement (Fig. S1). There were 51 matching MAN data points that passed the collocation screening. The MAN data corresponding to the same SODA retrieval were averaged and used to generate the scatter plot of MAN and SODA comparison (Fig. 8). The error

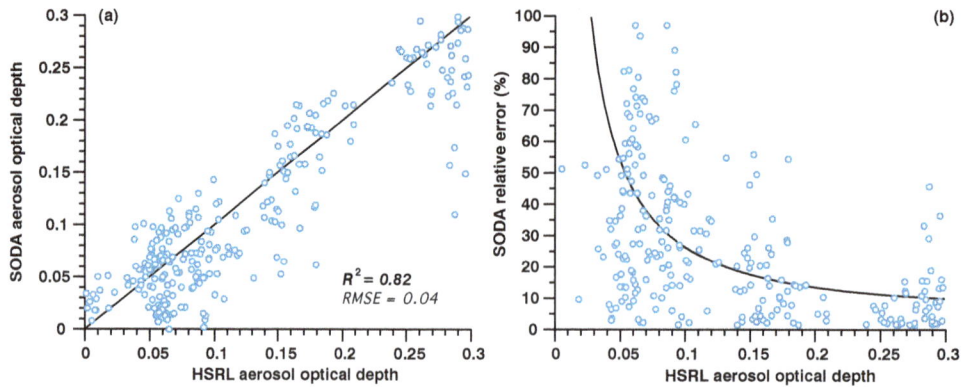

Figure 7. (a) A scatter plot of SODA AOD relative to AOD measured by HSRL at 532 nm with corresponding R^2 and RMSE. The black line illustrates the 1 : 1 line. **(b)** Relative uncertainty in the SODA column lidar ratio as a function of HSRL AOD with the black line showing the least squares exponential fit as in Josset et al. (2012), Eq. (15). All points are classified as marine plus pollution or marine plus dust and are from Table 1 in Josset et al. (2011).

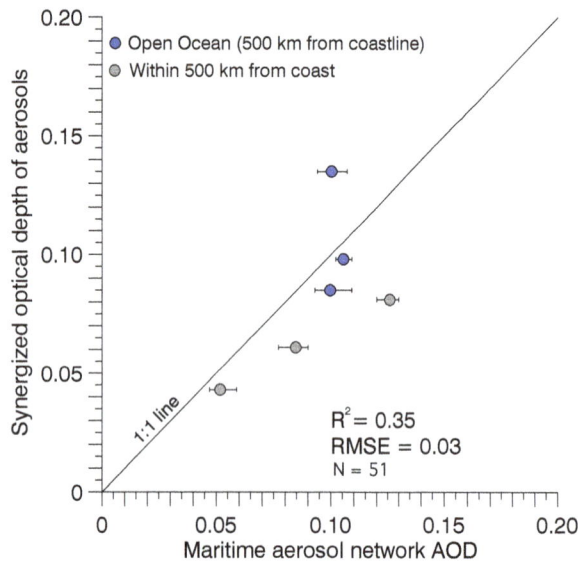

Figure 8. Scatter plot comparing the aerosol optical depth from SODA (y axis) and AERONET Maritime Aerosol Network (x axis). Blue circles represent locations that are at least 500 km from a coastline and are considered to be "open ocean".

bars in Fig. 8 indicate the maximum and minimum values of the MAN AOD reported for the closest SODA retrieval. Figure 8 shows that in general there is a good agreement between SODA and MAN retrievals with the data points located reasonably close to the 1 : 1 line. The correlation is 0.59 and the RMS error is 0.03.

5   Conclusions

A new method showing that it is possible to infer lidar ratios of marine aerosol over the ocean using two independent sources: the AOD from Synergized Optical Depth of Aerosols and the integrated attenuated backscatter from Cloud-Aerosol Lidar with Orthogonal Polarization has here been applied. The proposed equation calculates particulate lidar ratio for individual CALIOP retrievals of single aerosol layer columns as a correction to achieve the best agreement between SODA and CALIOP retrievals. The new method allows calculating marine aerosol lidar ratio and assessing its spatiotemporal variability and dependence on ocean surface wind speed. Analyses were carried out using CALIOP level 2, 5 km aerosol layer and collocated SODA nighttime data from December 2007 to November 2010. During the data analysis, over 260 000 data points passed various quality-control and quality-assurance tests to reduce errors associated with the clean marine aerosol retrievals. The calculated lidar ratios have been analyzed over the global ocean, covering a wide range of wind speed and AOD conditions. Data analysis shows that over most of the ocean surfaces, the calculated lidar ratio is higher than the default lidar ratio of 20 sr used in the CALIOP clean marine aerosol model. The calculated aerosol lidar ratios are inversely related to the surface wind speed. Increases in mean surface ocean wind speeds from 0 to >15 m s^{-1} reduces the mean lidar ratio for marine aerosol from ~ 32 sr to ~ 22 sr. Such reduction was explained by the shift in aerosol volume size distribution with the wind speed; however, it was also emphasized that future studies should explore the role of meteorological and/or environmental factors and ocean chemical/biological composition for marine aerosol intensive properties. Our data analysis showed that changes in wind speed also affect the probability density function for marine aerosol lidar ratio distribution. The largest standard deviation calculated for the lowest wind speed regime suggested that under low wind conditions, a wide range of marine aerosol sizes can be present over the ocean and there is greater probability for natural-continental and human-induced pollution aerosols to be classified by CALIOP as clean marine. We would like to mention that the role of organic aerosol at low wind speeds is still unclear. A large body of experimental data suggests that increases in the organic fraction of marine aerosol can have implications on hygroscopicity (e.g., Saxena et al., 1995; Fuentes et al., 2011; Ovadenevaite et al., 2013) and could potentially influence our results. Overall, our data analysis shows that an average value of 26 sr for clean marine aerosol lidar ratio provides the best agreement between the SODA product and CALIOP-retrieved global mean marine aerosol optical depth values. However, our study also shows large spatiotemporal variability in marine aerosol lidar ratios, suggesting that a single constant value of the lidar ratio is not suitable for a wide range of marine aerosol and can lead to large uncertainties at different locations and seasons.

We have estimated the error in CALIOP-retrieved column integrated attenuated particulate backscatter. Calculations suggest that the average uncertainty in particulate backscatter is more than an order of magnitude lower compared to the retrieved value. Data analysis also showed no clear indication for either approaching a sensitivity limit (due to strong attenuation of the lidar signal throughout the layer) or the contamination of the backscattered signal by the surface reflectance. Based on the conducted error analysis we conclude that the strict quality control criteria developed in this study is adequate to remove the majority of erroneous retrievals.

Finally, even though calculations here were carried out for marine aerosol, the technique used in this study is broad and can be used to infer lidar ratios of different species of atmospheric aerosols (i.e., mineral dust, biomass burning, etc.) advecting over the ocean. Because our data analysis shows that it is possible to derive a correction to the CALIOP prescribed marine aerosol lidar ratio, future studies should also consider conducting case studies over different oceanic regions to examine the possible effects of meteorological parameters and ocean physiochemical/biological composition on marine aerosol lidar ratio. Classification of the spatiotemporal distribution and wind speed dependence of a limited number of parameters affecting marine aerosol lidar ratios may lead to improved retrievals of AOD values over the oceans.

Acknowledgements. This research was supported by the National Aeronautics and Space Administration (NASA) through grant numbers NNX11AG72G and NNX14AL89G and by the National Science Foundation through the grant AGS-1249273. The authors gratefully acknowledge the CALIPSO, CloudSat, and NASA Langley HSRL Teams for their support and effort in making the data available. CALIPSO data were obtained from the NASA Langley Research Center Atmospheric Science Data Center. CloudSat data are produced by remote sensing systems and sponsored by the NASA Earth Science MEaSUREs DISCOVER Project and the Advanced Microwave Scanning Radiometer (AMSR-E) Science Team. The SODA product is developed at the ICARE data and services center (http://www.icare.univ-lille1.fr) in Lille (France) in the frame of the CALIPSO mission and supported by CNES.

Edited by: C. McNeil

References

Ackermann, J.: The extinction-to-backscatter ratio of tropospheric aerosol: A numerical study, J. Atmos. Ocean. Technol., 15, 1043–1050, doi:10.1175/1520-0426(1998)015<1043:TETBRO>2.0.CO;2, 1998.

Amiridis, V., Balis, D. S., Giannakaki, E., Stohl, A., Kazadzis, S., Koukouli, M. E. and Zanis, P.: Optical characteristics of biomass burning aerosols over Southeastern Europe determined from UV-Raman lidar measurements, Atmos. Chem. and Phys., 9, 2431–2440, doi:10.5194/acp-9-2431-2009, 2009.

Andreae, M. O.: Aerosols before Pollution, Science, 315, 50–51, 10.2307/20035138, 2007.

Anguelova, M., Barber Jr, R. P., and Wu, J.: Spume drops produced by the wind tearing of wave crests, J. Phys. Oceanogr., 29, 1156–1165, doi:10.1175/1520-0485(1999)029<1156:SDPBTW>2.0.CO;2, 1999.

Ansmann, A., Riebesell, M., and Weitkamp, C.: Measurement of atmospheric aerosol extinction profiles with a Raman lidar, Opt. Lett., 15, 746–748, doi:10.1364/OL.15.000746, 1990.

Ansmann, A., Wagner, F., Althausen, D., Müller, D., Herber, A., and Wandinger, U.: European pollution outbreaks during ACE 2: Lofted aerosol plumes observed with Raman lidar at the Portuguese coast, J. Geophys. Res., 106, 20725–20733, doi:10.1029/2000JD000091, 2001.

Ansmann, A. and Müller, D.: Lidar and Atmospheric Aerosol Particles, in: Lidar, edited by: Weitkamp, C., Springer New York, 105–141, 2005.

Bentamy, A., Katsaros, K. B., Mestas-Nuñez, A. M., Drennan, W. M., Forde, E. B., and Roquet, H.: Satellite estimates of wind speed and latent heat flux over the global oceans, J. Climate, 16, 637–656, doi:10.1175/1520-0442(2003)016<0637:SEOWSA>2.0.CO;2, 2003.

Blanchard, D. and Woodcock, A.: Bubble formation and modification in the sea and its meteorological significance, Tellus, 9, 145–158, doi: 10.1111/j.2153-3490.1957.tb01867.x, 1957.

Bloom, S., Da Silva, A., Dee, D., Bosilovich, M., Chern, J., Pawson, S., Schubert, S., Sienkiewicz, M., Stajner, I., and Tan, W.: Documentation and validation of the Goddard Earth Observing System (GEOS) data assimilation system – Version 4, NASA Tech. Memo., 104606, 187, 2005.

Blot, R., Clarke, A. D., Freitag, S., Kapustin, V., Howell, S. G., Jensen, J. B., Shank, L. M., McNaughton, C. S., and Brekhovskikh, V.: Ultrafine sea spray aerosol over the southeastern Pacific: open-ocean contributions to marine boundary layer CCN, Atmos. Chem. Phys., 13, 7263–7278, doi:10.5194/acp-13-7263-2013, 2013.

Bréon, F. M.: Aerosol extinction-to-backscatter ratio derived from passive satellite measurements, Atmos. Chem. Phys., 13, 8947–8954, doi:10.5194/acp-13-8947-2013, 2013.

Burton, S., Ferrare, R., Hostetler, C., Hair, J., Rogers, R., Obland, M., Butler, C., Cook, A., Harper, D., and Froyd, K.: Aerosol classification using airborne High Spectral Resolution Lidar measurements–methodology and examples, Atmos. Meas. Tech., 5, 73–98, doi:10.5194/amtd-4-5631-2011, 2012.

Burton, S. P., Ferrare, R. A., Vaughan, M. A., Omar, A. H., Rogers, R. R., Hostetler, C. A., and Hair, J. W.: Aerosol classification from airborne HSRL and comparisons with the CALIPSO vertical feature mask, Atmos. Meas. Tech., 6, 1397–1412, doi:10.5194/amt-6-1397-2013, 2013.

Carslaw, K. S., Lee, L. A., Reddington, C. L., Pringle, K. J., Rap, A., Forster, P. M., Mann, G. W., Spracklen, D. V., Woodhouse, M. T., Regayre, L. A., and Pierce, J. R.: Large contribution of natural aerosols to uncertainty in indirect forcing, Nature, 503, 67, doi:10.1038/nature12674, 2013.

Cattrall, C., Reagan, J., Thome, K., and Dubovik, O.: Variability of aerosol and spectral lidar and backscatter and extinction ratios of key aerosol types derived from selected Aerosol Robotic Network locations, J. Geophys. Res., 110, D10S11, doi:10.1029/2004JD005124, 2005.

de Leeuw, G., Andreas, E. L., Anguelova, M. D., Fairall, C., Lewis, E. R., O'Dowd, C., Schulz, M., and Schwartz, S. E.: Production flux of sea spray aerosol, Rev. Geophys., 49, doi:10.1029/2010RG000349, 2011.

Doherty, S. J., Anderson, T. L., and Charlson, R. J.: Measurement of the lidar ratio for atmospheric aerosols with a 180 backscatter nephelometer, Appl. Opt., 38, 1823–1832, 1999.

Eloranta, E.: High spectral resolution lidar, in: Lidar: Range-Resolved Optical Remote Sensing of the Atmosphere, edited by: Weitkamp, K., Springer, New York, 143–163, 2005.

Fernald, F. G., Herman, B. M., and Reagan, J. A.: Determination of aerosol height distributions by lidar, J. Appl. Meteorol., 11, 482–489, 1972.

Fuentes, E., Coe, H., Green, D., and McFiggans, G.: On the impacts of phytoplankton-derived organic matter on the properties of the primary marine aerosol – Part 2: Composition, hygroscopicity and cloud condensation activity, Atmos. Chem. Phys., 11, 2585–2602, doi:10.5194/acp-11-2585-2011, 2011.

Ghan, S., Laulainen, N., Easter, R., Wagener, R., Nemesure, S., Chapman, E., Zhang, Y., and Leung, R.: Evaluation of aerosol direct radiative forcing in MIRAGE, J. Geophys. Res., 106, 5295–5316, doi:10.1029/2000JD900502, 2001.

Groß S., Esselborn, M., Weinzierl, B., Wirth, M., Fix, A., and Petzold, A.: Aerosol classification by airborne high spectral resolution lidar observations, Atmos. Chem. Phys., 13, 2487–2505, doi:10.5194/acp-13-2487-2013, 2013.

Groß, S., Gasteiger, J., Freudenthaler, V., Wiegner, M., Geiß, A., Schladitz, A., Toledano, C., Kandler, K., Tesche, M., Ansmann, A., and Wiedensohler, A.: Characterization of the planetary boundary layer during SAMUM-2 by means of lidar measurements, Tellus B, 63, 695–705, doi:10.1111/j.1600-0889.2011.00557.x, 2011a.

Groß S., Tesche, M., Voker, F., Toledano, C., Wiegner, M., Ansmann, A., Althausen, D., and Seefeldner, M.: Characterization of Saharan dust, marine aerosols and mixtures of biomass-burning aerosols and dust by means of multi-wavelength depolarization and Raman lidar measurements during SAMUM 2, Tellus B, 63, 706–724, doi:10.1111/j.1600-0889.2011.00556.x, 2011b.

Grund, C. J. and Eloranta, E. W.: University of Wisconsin High Spectral Resolution Lidar, Optical Engineering, 30, 6–12, doi:10.1117/12.55766, 1991.

Hair, J. W., Hostetler, C. A., Cook, A. L., Harper, D. B., Ferrare, R. A., Mack, T. L., Welch, W., Izquierdo, L. R., and Hovis, F. E.: Airborne High Spectral Resolution Lidar for profiling aerosol optical properties, Appl. Opt., 47, 6734–6752, doi:10.1364/AO.47.006734, 2008.

Hoffman, E. J. and Duce, R. A.: The organic carbon content of marine aerosols collected on Bermuda, J. Geophys. Res., 79, 4474–4477, 1974.

Holben, B. N., Eck, T. F., Slutsker, I., Tanré, D., Buis, J. P., Setzer, A., Vermote, E., Reagan, J. A., Kaufman, Y. J., Nakajima, T., Lavenu, F., Jankowiak, I., and Smirnov, A.: AERONET-A Federated Instrument Network and Data Archive for Aerosol Characterization, Remote Sens. Environ., 66, 1, doi:10.1016/S0034-4257(98)00031-5, 1998.

Hoose, C., Kristjánsson, J., Iversen, T., Kirkevåg, A., Seland, Ø., and Gettelman, A.: Constraining cloud droplet number concentration in GCMs suppresses the aerosol indirect effect, Geophys. Res. Lett., 36, L12807, doi:10.1029/2009GL038568, 2009.

Hostetler, C., Liu, Z., Reagan, J., Vaughan, M., Winker, D., Osborn, M., Hunt, W., Powell, K., and Trepte, C.: CALIOP algorithm theoretical basis document – Part 1: Lidar level I ATBD-Calibration and level 1 data products, Rep.PC-SCI, 201, 66, 2005.

Josset, D., Pelon, J., Protat, A., and Flamant, C.: New approach to determine aerosol optical depth from combined CALIPSO and CloudSat ocean surface echoes, Geophys. Res. Lett., 35, L10805, doi:10.1029/2008GL033442, 2008.

Josset, D., Pelon, J., and Hu, Y.: Multi-instrument calibration method based on a multiwavelength ocean surface model, Geoscience and Remote Sensing Letters, IEEE, 7, 195–199, doi:10.1109/LGRS.2009.2030906, 2010a.

Josset, D., Zhai, P., Hu, Y., Pelon, J., and Lucker, P. L.: Lidar equation for ocean surface and subsurface, Opt. Express, 18, 20862–20875, doi:10.1364/OE.18.020862, 2010b.

Josset, D., Rogers, R., Pelon, J., Hu, Y., Liu, Z., Omar, A., and Zhai, P.: CALIPSO lidar ratio retrieval over the ocean, Opt. Express, 19, 18696–18706, doi:10.1364/OE.19.018696, 2011.

Josset, D., Pelon, J., Garnier, A., Hu, Y., Vaughan, M., Zhai, P., Kuehn, R., and Lucker, P.: Cirrus optical depth and lidar ratio retrieval from combined CALIPSO-CloudSat observations using ocean surface echo, J. Geophys. Res., 117, D05207, doi:10.1029/2011JD016959, 2012.

Kaufman, Y. J., Tanré, D., and Boucher, O.: A satellite view of aerosols in the climate system, Nature, 419, 215–223, doi:10.1038/nature01091, 2002.

Kiliyanpilakkil, V. P. and Meskhidze, N.: Deriving the effect of wind speed on clean marine aerosol optical properties using the A-Train satellites, Atmos. Chem. Phys., 11, 11401–11413, doi:10.5194/acp-11-11401-2011, 2011.

Kleidman, R. G., Smirnov, A., Levy, R. C., Mattoo, S., and Tanre, D.: Evaluation and Wind Speed Dependence of MODIS Aerosol

Retrievals Over Open Ocean, IEEE T. Geosci. Remote, 99, 1–7, doi:10.1109/TGRS.2011.2162073, 2011.

Lehahn, Y., Koren, I., Boss, E., Ben-Ami, Y., and Altaratz, O.: Estimating the maritime component of aerosol optical depth and its dependency on surface wind speed using satellite data, Atmos. Chem. Phys., 10, 6711–6720, doi:10.5194/acp-10-6711-2010, 2010.

Levy, R., Remer, L., Martins, J., Kaufman, Y., Plana-Fattori, A., Redemann, J., and Wenny, B.: Evaluation of the MODIS aerosol retrievals over ocean and land during CLAMS, J. Atmos. Sci., 62, 974–992, doi:10.1175/JAS3391.1, 2005.

Lewis, R. and Schwartz, E.: Sea salt aerosol production: mechanisms, methods, measurements and models – a critical review, American Geophysical Union, doi:10.1029/GM152, 2004.

Masonis, S. J., Anderson, T. L., Covert, D. S., Kapustin, V., Clarke, A. D., Howell, S., and Moore, K.: A Study of the Extinction-to-Backscatter Ratio of Marine Aerosol during the Shoreline Environment Aerosol Study, J. Atmos. Ocean. Technol., 20, 1388–1402, doi:10.1175/1520-0426(2003)020<1388:ASOTER>2.0.CO;2, 2003.

Meskhidze, N., Xu, J., Gantt, B., Zhang, Y., Nenes, A., Ghan, S., Liu, X., Easter, R., and Zaveri, R.: Global distribution and climate forcing of marine organic aerosol – Part 1: Model improvements and evaluation, Atmos. Chem. Phys., 11, 11689–11705, doi:10.5194/acp-11-11689-2011, 2011.

Müller, D., Ansmann, A., Mattis, I., Tesche, M., Wandinger, U., Althausen, D., and Pisani, G.: Aerosol-type-dependent lidar ratios observed with Raman lidar, J. Geophys. Res., 112, doi:10.1029/2006JD008292, 2007.

Omar, A. H., Winker, D. M., Kittaka, C., Vaughan, M. A., Liu, Z., Hu, Y., Trepte, C. R., Rogers, R. R., Ferrare, R. A., Lee, K., Kuehn, R. E., and Hostetler, C. A.: The CALIPSO Automated Aerosol Classification and Lidar Ratio Selection Algorithm, J. Atmos. Ocean. Technol., 26, 1994–2014, doi:10.1175/2009JTECHA1231.1, 2009.

Oo, M. and Holz, R.: Improving the CALIOP aerosol optical depth using combined MODIS-CALIOP observations and CALIOP integrated attenuated total color ratio, J. Geophys. Res., 116, D14201, doi:10.1029/2010JD014894, 2011.

Ovadnevaite, J., Ceburnis, D., Martucci, G., Bialek, J., Monahan, C., Rinaldi, M., Facchini, M. C., Berresheim, H., Worsnop, D. R., and O'Dowd, C.: Primary marine organic aerosol: A dichotomy of low hygroscopicity and high CCN activity, Geophys. Res. Lett., 38, doi:10.1029/2011GL048869, 2011.

Piironen, P. and Eloranta, E.: Demonstration of a high-spectral-resolution lidar based on an iodine absorption filter, Opt. Lett., 19, 234–236, 1994.

Redemann, J., Vaughan, M. A., Zhang, Q., Shinozuka, Y., Russell, P. B., Livingston, J. M., Kacenelenbogen, M., and Remer, L. A.: The comparison of MODIS-Aqua (C5) and CALIOP (V2 & V3) aerosol optical depth, Atmos. Chem. Phys., 12, 3025–3043, doi:10.5194/acp-12-3025-2012, 2012.

Saxena, P., Hildemann, L. M., McMurry, P. H., and Seinfeld, J. H.: Organics alter hygroscopic behavior of atmospheric particles, J. Geophys. Res., 100, 18755–18770, doi:10.1029/95JD01835, 1995.

Sayer, A., Smirnov, A., Hsu, N., and Holben, B.: A pure marine aerosol model, for use in remote sensing applications, J. Geophys. Res., 117, doi:10.1029/2011JD016689, 2012.

Schuster, G. L., Vaughan, M., MacDonnell, D., Su, W., Winker, D., Dubovik, O., Lapyonok, T., and Trepte, C.: Comparison of CALIPSO aerosol optical depth retrievals to AERONET measurements, and a climatology for the lidar ratio of dust, Atmos. Chem. Phys., 12, 7431–7452, doi:10.5194/acp-12-7431-2012, 2012.

Shipley, S. T., Tracy, D., Eloranta, E. W., Trauger, J. T., Sroga, J., Roesler, F., and Weinman, J. A.: High spectral resolution lidar to measure optical scattering properties of atmospheric aerosols. 1: Theory and instrumentation, Appl. Opt., 22, 3716–3724, 1983.

Smirnov, A., Holben, B., Eck, T., Dubovik, O., and Slutsker, I.: Effect of wind speed on columnar aerosol optical properties at Midway Island, J. Geophys. Res., 108, 4802, doi:10.1029/2003JD003879, 2003.

Smirnov, A., Holben, B. N., Giles, D. M., Slutsker, I., O'Neill, N. T., Eck, T. F., Macke, A., Croot, P., Courcoux, Y., Sakerin, S. M., Smyth, T. J., Zielinski, T., Zibordi, G., Goes, J. I., Harvey, M. J., Quinn, P. K., Nelson, N. B., Radionov, V. F., Duarte, C. M., Losno, R., Sciare, J., Voss, K. J., Kinne, S., Nalli, N. R., Joseph, E., Krishna Moorthy, K., Covert, D. S., Gulev, S. K., Milinevsky, G., Larouche, P., Belanger, S., Horne, E., Chin, M., Remer, L. A., Kahn, R. A., Reid, J. S., Schulz, M., Heald, C. L., Zhang, J., Lapina, K., Kleidman, R. G., Griesfeller, J., Gaitley, B. J., Tan, Q., and Diehl, T. L.: Maritime aerosol network as a component of AERONET – first results and comparison with global aerosol models and satellite retrievals, Atmos. Meas. Tech., 4, 583–597, doi:10.5194/amt-4-583-2011, 2011.

Smirnov, A., Holben, B. N., Slutsker, I., Giles, D. M., McClain, C. R., Eck, T. F., Sakerin, S. M., Macke, A., Croot, P., Zibordi, G., Quinn P. K., Sciare, J., Kinne, S., Harvey, M., Smyth, T. J., Piketh S., Zielinski, T., Proshutinsky A., Goes, J. I., Nelson, N. B., Larouche, P., Radionov, V. F., Goloub, P., Krishna Moorthy, K., Matarrese, R., Robertson, E. J., and Jourdin, F.: Maritime Aerosol Network as a component of Aerosol Robotic Network, J. Geophys. Res., 114, D06204, doi:10.1029/2008JD011257, 2009.

Tesche, M., Ansmann, A., Müller, D., Althausen, D., Engelmann, R., Freudenthaler, V., and Groß, S.: Vertically resolved separation of dust and smoke over Cape Verde using multiwavelength Raman and polarization lidars during Saharan Mineral Dust Experiment 2008, J. Geophys. Res., 114, D13202, doi:10.1029/2009JD011862, 2009a.

Tesche, M., Ansmann, A., Müller, D., Althausen, D., Mattis, I., Heese, B., Freudenthaler, V., Wiegner, M., Esselborn, M., Pisani, G., and Knippertz, P.: Vertical profiling of Saharan dust with Raman lidars and airborne HSRL in southern Morocco during SAMUM, Tellus B, 61, 144–164, doi:10.1111/j.1600-0889.2008.00390.x, 2009b.

Vaughan, M. A., Young, S. A., Winker, D. M., Powell, K. A., Omar, A. H., Liu, Z., Hu, Y., and Hostetler, C. A.: Fully automated analysis of space-based lidar data: an overview of the CALIPSO retrieval algorithms and data products, Proc. SPIE, 5575, 16–30, doi:10.1117/12.572024, 2004.

Vaughan, M. A., Powell, K. A., Kuehn, R. E., Young, S. A., Winker, D. M., Hostetler, C. A., Hunt, W. H., Liu Z., McGill, M. J., and Getzewich, B. J.: Fully automated detection of cloud and aerosol layers in the CALIPSO lidar measurements, J. Atmos. Ocean. Technol., 26, 2034–2050, doi:10.1175/2009JTECHA1228.1, 2009.

Wang, M. and Penner, J.: Aerosol indirect forcing in a global model with particle nucleation, Atmos. Chem. Phys., 9, 239–260, doi:10.5194/acp-9-239-2009, 2009.

Westervelt, D., Moore, R., Nenes, A., and Adams, P.: Effect of primary organic sea spray emissions on cloud condensation nuclei concentrations, Atmos. Chem. Phys., 12, 89–101, doi:10.5194/acp-12-89-2012, 2012.

Winker, D. M., Vaughan, M. A., Omar, A., Hu, Y., Powell, K. A., Liu, Z., Hunt, W. H., and Young, S. A.: Overview of the CALIPSO mission and CALIOP data processing algorithms, J. Atmos. Ocean. Technol., 26, 2310–2323, doi:10.1175/2009JTECHA1281.1, 2009.

Young, S. A., Cutten, D. R., Lynch, M. J., and Davies, J. E.: Lidar-derived variations in the backscatter-to-extinction ratio in southern hemisphere coastal maritime aerosols, Atmos. Environ., 27, 1541–1551, doi:10.1016/0960-1686(93)90154-Q, 1993.

Young, S. A. and Vaughan, M. A.: The retrieval of profiles of particulate extinction from cloud-aerosol lidar infrared pathfinder satellite observations (CALIPSO) data: algorithm description, J. Atmos. Oceanic Technol., 26, 1105–1119, doi:10.1175/2008JTECHA1221.1, 2009.

7

Spatiotemporal variability of water vapor investigated using lidar and FTIR vertical soundings above the Zugspitze

H. Vogelmann, R. Sussmann, T. Trickl, and A. Reichert

Karlsruhe Institute of Technology, IMK-IFU, Garmisch-Partenkirchen, Germany

Correspondence to: H. Vogelmann (hannes.vogelmann@kit.edu)

Abstract. Water vapor is the most important greenhouse gas and its spatiotemporal variability strongly exceeds that of all other greenhouse gases. However, this variability has hardly been studied quantitatively so far. We present an analysis of a 5-year period of water vapor measurements in the free troposphere above the Zugspitze (2962 m a.s.l., Germany). Our results are obtained from a combination of measurements of vertically integrated water vapor (IWV), recorded with a solar Fourier transform infrared (FTIR) spectrometer on the summit of the Zugspitze and of water vapor profiles recorded with the nearby differential absorption lidar (DIAL) at the Schneefernerhaus research station. The special geometrical arrangement of one zenith-viewing and one sun-pointing instrument and the temporal resolution of both instruments allow for an investigation of the spatiotemporal variability of IWV on a spatial scale of less than 1 km and on a timescale of less than 1 h. The standard deviation of differences between both instruments σ_{IWV} calculated for varied subsets of data serves as a measure of variability. The different subsets are based on various spatial and temporal matching criteria. Within a time interval of 20 min, the spatial variability becomes significant for horizontal distances above 2 km, but only in the warm season ($\sigma_{IWV} = 0.35$ mm). However, it is not sensitive to the horizontal distance during the winter season. The variability of IWV within a time interval of 30 min peaks in July and August ($\sigma_{IWV} > 0.55$ mm, mean horizontal distance = 2.5 km) and has its minimum around midwinter ($\sigma_{IWV} < 0.2$ mm, mean distance > 5 km). The temporal variability of IWV is derived by selecting subsets of data from both instruments with optimal volume matching. For a short time interval of 5 min, the variability is 0.05 mm and increases to more than 0.5 mm for a time interval of 15 h. The profile variability of water vapor is determined by analyzing subsets of water vapor profiles recorded by the DIAL within time intervals from 1 to 5 h. For all altitudes, the variability increases with widened time intervals. The lowest relative variability is observed in the lower free troposphere around an altitude of 4.5 km. Above 5 km, the relative variability increases continuously up to the tropopause by about a factor of 3. Analysis of the covariance of the vertical variability reveals an enhanced variability of water vapor in the upper troposphere above 6 km. It is attributed to a more coherent flow of heterogeneous air masses, while the variability at lower altitudes is also driven by local atmospheric dynamics. By studying the short-term variability of vertical water vapor profiles recorded within a day, we come to the conclusion that the contribution of long-range transport and the advection of heterogeneous layer structures may exceed the impact of local convection by 1 order of magnitude even in the altitude range between 3 and 5 km.

1 Introduction

Water vapor plays a key role in weather and climate phenomena and is the most important greenhouse gas (e.g., Harries, 1997; Kiehl and Trenberth, 1997; Trenberth et al., 2007). However, the feedback between the anthropogenic (CO_2-driven) temperature increase and the influence of water vapor is far from understood (e.g., Wagner et al., 2006). Furthermore, climate projections still suffer from inaccurate parameterizations of water vapor absorption processes within the radiation code of general circulation models (e.g., Turner and Mlawer, 2010). Understanding the role of water vapor in the climate system is particularly complex because water vapor is the only trace compound in the atmosphere appearing in all three states of matter. This involves a variety of factors, e.g., the possibility of latent heat transport (thereby damping latitudinal temperature gradients) and the fact that precipitation is the largest sink of atmospheric water vapor. The latter is the main reason for the strong decrease of water vapor concentration with altitude, and it is the reason why water vapor has an average lifetime in the atmosphere of just about 9 days, shorter than for any other greenhouse gas. The short lifetime is a basis of the very high spatiotemporal variability of water vapor (Trenberth, 1998).

However, the spatiotemporal variability of water vapor on the scales relevant to weather and climate is still far from being quantitatively characterized, and the underlying processes are not well understood. Variability, for instance, may be caused by local dynamics above complex mountain terrain (which changes with season), by regional meteorological effects, or by advection on larger scales. A highly interesting question is the variance of water vapor as a function of altitude on different timescales. Previous studies at our site based on ozone and aerosol lidar profiling demonstrated that the free troposphere may be affected by regional contributions, long-range transport, and stratosphere–troposphere exchange causing strongly and rapidly changing vertical structures in the concentration profile (Eisele et al., 1999; Stohl and Trickl, 1999; Trickl et al., 2003, 2010, 2011). In particular, we frequently observed very dry and sometimes very thin layers in the free troposphere, which were associated with stratospheric intrusion events. It remains open, however, how much such processes significantly contribute to the observed variability of water vapor in the middle and upper troposphere.

For understanding the long-term changes and the variability of water vapor, high-quality vertical sounding of water vapor with high temporal density is required. During the past years, a variety of optical remote sounders has been developed for this purpose in addition to the classical radiosondes (e.g., Kämpfer, 2013). Lidars, Fourier transform infrared (FTIR) spectrometers, and microwave radiometers fulfill the requirements of frequent measurements. In particular, we developed a differential absorption lidar (DIAL) for use at the Zugspitze, which allows for continuous day- and nighttime soundings of water vapor profiles up to the tropopause (Vogelmann and Trickl, 2008). For measuring integrated water vapor (IWV), the solar FTIR technique was found to be one of the most accurate and precise ground-based sounding techniques with a precision better than 0.05 mm (2.2 % of the mean) (Sussmann et al., 2009). According to a recent validation study, the lidar and FTIR water vapor sounders used for the work presented here are in excellent agreement (Vogelmann et al., 2011).

Comparing two high-precision state-of-the-art water vapor sounders, we also found that it is necessary to use very strict temporal coincidence criteria on the timescale of minutes and a spatial matching on the scale of 100 m. Otherwise, the combined precision of the instruments will be affected by

Table 1. Specifications of the FTIR and the DIAL on the Zugspitze.

	FTIR	DIAL
Geographical Coordinates	$10°59'8.7''$ E $47°25'15.6''$ N	$10°58'46.8''$ E $47°25'0''$ N
Altitude a.s.l.	2964 m	2675 m
Vertical range a.s.l.	above 2.96 km	2.95–12 km
Typ. integration time	13.3 min	17 min
Spectral range [cm^{-1}]	micro windows 839.5–840.5 849.0–850.2 852.0–853.1	ν_{on} 12 236.560 12 237.466 12 243.537

the natural variability of water vapor (Sussmann et al., 2009; Vogelmann et al., 2011). This was confirmed by Bleisch et al. (2011), who reported that in case of long distances between the locations of the intercompared instruments, atmospheric variability tends to blur out the significance of validation results. The question of co-location has also become an issue in the Global Climate Observing System (GCOS) Reference Upper Air Network (GRUAN) (Immler et al., 2010; Sun et al., 2010; Seidel et al., 2011; Fassò et al., 2014), and it was addressed when evaluating water vapor sounding validation campaigns like MOHAVE (2009), LUAMI (2008), WAVES (2006), AWEX-G (2003) (Leblanc et al., 2011; Stiller et al., 2012; Wirth et al., 2009; Adam et al., 2010; Whiteman et al., 2006). Co-location also is of relevance to ground-based validation of satellite missions and has been addressed many times (e.g., Tobin et al., 2006; Soden and Lanzante, 1996).

The goal of this paper is to derive quantitative information relating to the spatiotemporal variability of water vapor. The solar FTIR spectrometer on the summit of the Zugspitze (2962 m a.s.l.), and the DIAL located only 680 m to the southwest and about 288 m below provide a unique geometrical arrangement of two high-precision water vapor sounders, allowing for an advanced analysis of the spatiotemporal variability of integrated water vapor (IWV) on small scales ($\Delta t < 1$ h, $\Delta x < 1$ km).

After a brief description of the instrumental setup as well as of the FTIR and DIAL IWV data with their geometrical and temporal properties, we present the quantification of the spatial and temporal variability of IWV by statistical analysis of selected subsets of IWV data from the FTIR and the DIAL (Sects. 3.1 and 3.2). The profile-type variability of the vertical water vapor distribution is analyzed quantitatively by investigating selected subsets of DIAL soundings and by calculating a profile covariance matrix (Sect. 4). Different mechanisms driving the short-term variability of water vapor are investigated in four case studies (Sect. 5). Finally, major results are summarized (Sect. 6).

2 Instrumentation and geographical arrangement

2.1 Zugspitze solar FTIR system

Solar absorption FTIR spectrometry uses the direct radiation from the sun in the mid-infrared range as a light source. The FTIR provides total columns of numerous atmospheric trace gases. Additionally, information on the vertical distribution of trace gases can be derived (typically 2–3 degrees of freedom in a retrieval optimized for IWV) from the shape of the pressure-broadened infrared lines. Due to its principle, the solar FTIR points towards the actual position of the sun and measures slant columns/profiles that are angle corrected for consistency with vertical profiles. The FTIR instrument (Table 1) located on the summit of the Zugspitze is based on a Bruker IFS125HR interferometer and is described in detail by Sussmann and Schäfer (1997). The retrieval of IWV is based on the SFIT 2 algorithm (Pougatchev et al., 1995), which is the standard code of the Network for the Detection of Atmospheric Composition Change (NDACC). An FTIR retrieval optimized for IWV was developed recently by Sussmann et al. (2009). The precision of the IWV retrieval was estimated to be better than 0.05 mm (2.2 % of the mean).

2.2 Differential absorption lidar (DIAL)

DIAL is a laser-based remote sensing technique that provides number density profiles of trace gases. Measurements are based on specific molecular absorption and well-established spectroscopy. The Zugspitze DIAL is operated with single absorption lines in the 817 nm band of H_2O (Table 1) for ground-based water vapor profiling in the free troposphere. In order to keep a balanced signal-to-noise ratio, a vertical resolution (VDI Guideline 4210) of 50 to 300 m is adapted dynamically to the vertical range from 2.95 to roughly 12 km a.s.l., respectively. Thus, statistical measurement uncertainties are kept below about 5 % related to a mean humidity profile throughout the free troposphere. The sensitivity limit is roughly 18 ppm at 10 km a.s.l. which can occasionally be undercut in the upper troposphere. If this is the case, the upper end of the valid measurement range is reasonably reduced to lower altitudes. The DIAL instrument is located at the Schneefernerhaus research station (UFS) on the steep southern slope of the Zugspitze at an altitude of 2675 m a.s.l. The range of the Zugspitze DIAL starts 250 m above the laboratory, slightly below the altitude of the FTIR spectrometer. The DIAL system at Schneefernerhaus/Zugspitze and the retrieval of water vapor profiles are described in more detail by Vogelmann and Trickl (2008). Water vapor profiles from the Zugspitze DIAL allow for retrieving IWV with a precision better than 0.1 mm (Vogelmann et al., 2011).

2.3 Geographical setup and IWV data selection

The Zugspitze (47.42° N, 10.98° E, 2962 m a.s.l.) is by far the highest mountain on the northern rim of the Alps. The

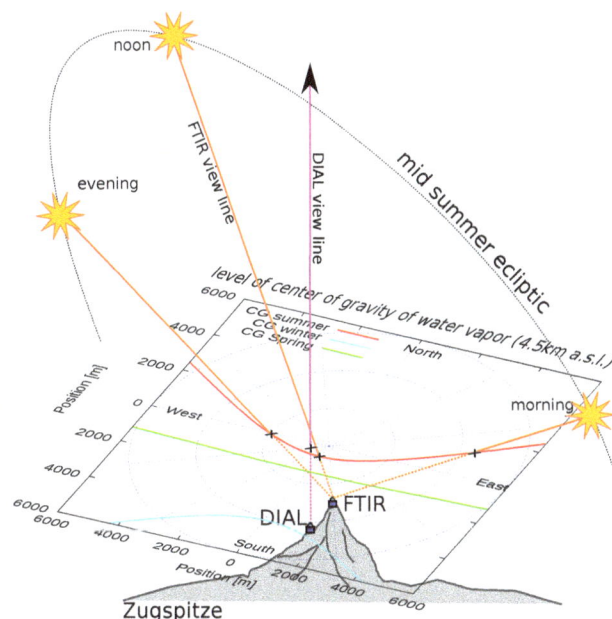

Figure 1. Geometrical setup of the IWV intercomparison between DIAL and FTIR on the Zugspitze. The DIAL is located 680 m to the southwest of the FTIR and 288 m below. The horizontal coordinate grid plane marks the mean altitude of the center of gravity of the water vapor distribution above the Zugspitze (see text) and its point of origin is vertical above the FTIR. The red, green, and blue curves in the CG plane are the trajectories of the points, where the view line (e.g., orange lines from FTIR to the sun in the case of midsummer) of the FTIR meets the CG plane in midsummer, spring, and midwinter. Consequently, the trajectories mark the horizontal position of the center of gravity of the water vapor distribution measured by the FTIR along its slanted view line. The pink line marks the fixed vertical view line of the DIAL.

free troposphere above this site is representative of central Europe. The mountain is above the moist boundary layer for most of the year. Due to reduced absorption losses this site is ideal for sensitive spectroscopic measurements of water vapor throughout the free troposphere. While the FTIR instrument is located on the summit of the Zugspitze the DIAL instrument is located at the Schneefernerhaus research station (UFS) on the steep southern slope of the Zugspitze at an altitude of 2675 m a.s.l., 680 m southwest of the FTIR instrument (Fig. 1).

The sun-pointing geometry of the FTIR instrument and the fixed zenith-pointing geometry of the DIAL allow for studying the differences of IWV values measured by both instruments with defined spatial and temporal matching (Fig. 1). According to reanalysis data from the National Center for Environmental Prediction (NCEP), the center of gravity of the water vapor vertical distribution above the Zugspitze is most frequently located at a rather constant altitude between 4300 m a.s.l. in summer and 4400 m a.s.l. in winter. For simplicity, it is assumed that the FTIR IWV is horizontally located at the point where the viewing direction of the instru-

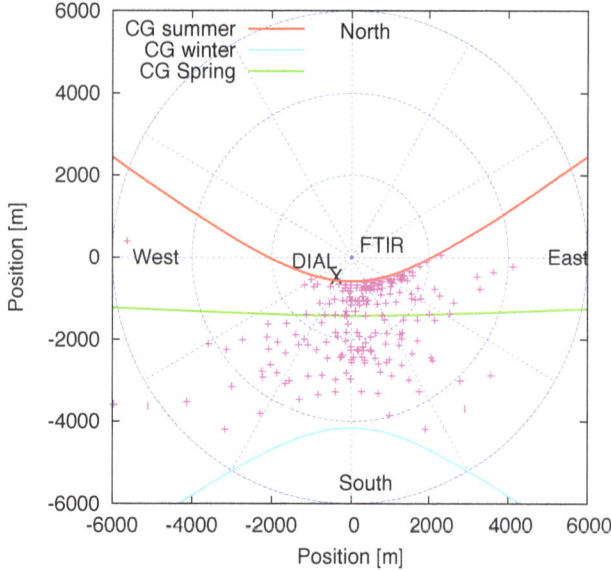

Figure 2. Trajectories of the horizontal positions of the center of gravity (CG) of the vertical water vapor distribution measured by the FTIR for IWV midsummer, spring, and midwinter. Center of gravity horizontal locations from FTIR measurements chronological coinciding with DIAL measurements ($\Delta t \leq 30$ min) are marked by crosses.

ment meets the altitude level of the center of gravity of the IWV distribution. This assumption, of course, describes the reality at high sun elevation angles better while the measured FTIR IWV is more horizontally blurred for low sun elevations close to the horizon. From this and the actual position of the sun, a rough estimate of the varying horizontal position of the IWV measured by the FTIR instrument is possible. The zenith angle of the sun defines the horizontal distance from the instrument, which may vary from less than 1 km around noon in midsummer to more than 10 km at very low sun positions. The azimuth of the FTIR IWV position is equal to the azimuth of the sun position which depends on daytime and season. In contrast to this, the horizontal position of IWV measured with the DIAL is always fixed to the location of the instrument, 680 m southwest of the FTIR site. This is illustrated in Fig. 1.

Figure 2 shows the horizontal allocation of all FTIR IWV measurements recorded concurrently ($\Delta t \leq 30$ min) with a DIAL measurement. The horizontal distance between the location of the DIAL and the horizontal position of the IWV measured by the FTIR is defined as spatial matching Δx. Figure 2 also shows the daily trajectories of the horizontal position of the center of gravity of IWV probed with the FTIR instrument for midsummer, equinox, and midwinter. In the summer season, the mean horizontal distance Δx is obviously smaller than during winter (see dashed curve in Fig. 4).

3 Variability of integrated water vapor in space and time

Of more than 350 lidar profiles recorded in the years 2007–2009, more than 250 profiles were measured during daytime (i.e., between 05:00 and 19:00 LT). In the same period, more than 3500 column measurements were made by the FTIR instrument. The systems operate with a typical integration time of 13 min (FTIR) and 17 min (DIAL). In order to obtain a quantitative measure of the water vapor variability, we analyzed certain measurement samples recorded by the two different instruments under certain spatiotemporal matching criteria for Δx and Δt. The centers of the integration time of both FTIR and DIAL were used to determine the temporal matching. We retrieved σ_{IWV} by calculating the standard deviation of the differences of IWV values from a linear model $y = a \cdot x + b$:

$$\sigma_{\mathrm{IWV}} = \sqrt{\frac{1}{n-2}\sum_{i=1}^{k}(y_i - (ax_i + b))^2}, \tag{1}$$

where y_i and x_i are the IWV values from the DIAL and the FTIR, respectively, within one sample, and n is the sample size. a and b were calculated by a regression analysis using the method of least squares. Thus,

$$\sigma_{\mathrm{IWV}} = \sqrt{\frac{1}{n-2}\sum_{i=1}^{n}\left(y_i - \bar{y} - \frac{(x_i - \bar{x})\sum_j (x_j - \bar{x})y_j}{\sum_j (x_j - \bar{x})^2}\right)^2}. \tag{2}$$

The matching criteria, amongst others, define the sample size n, which influences the uncertainty of σ_{IWV} itself. The uncertainty of σ_{IWV} is given by $\sigma_{\mathrm{IWV}}/\sqrt{(2(n-1))}$ and is illustrated by the error bars in Figs. 3–5. The inherent integration times of the instruments (roughly 15 min) cause a statistical underestimation of short-term variabilities on the scale of minutes. For the shortest time intervals investigated here (4 min), variations are statistically underestimated by factor of about 2.

3.1 Spatial variability

We decided to analyze the spatial and temporal variabilities separately for summer and winter because of two counteracting effects:

1. The special observation geometry in this study implies that the spatial overlap Δx of both soundings depends on both daytime and season. As shown in Figs. 1, 2, and 4 (dashed curve), the best spatial matching ($\Delta x < 1$ km) is achieved around midsummer in the early afternoon only (between 12:00 and 14:00 UTC), while Δx is always larger during the winter season.

2. Due to heat-driven convective dynamics in complex mountain surroundings, spatial and temporal variabilities of IWV are expected to be higher during the summer season. The convection above alpine terrain can

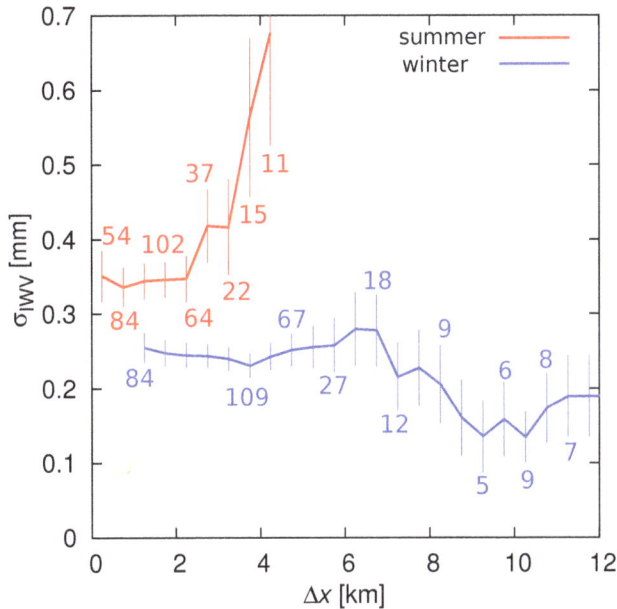

Figure 3. σ_{IWV} as a function of the horizontal distance Δx between the center of gravity of FTIR IWV and DIAL IWV in the summer season (red) and in the winter season (blue). The coincidence time interval Δt is 60 min for the blue curve and 30 min for the red curve. For geometrical reasons, the shortest distance in the winter season is 1 km. The number of measurement pairs from which σ_{IWV} was calculated is indicated by the numbers near the curves (not for all nodes). The uncertainties ($\pm\sigma$) are indicated by the error bars (for calculation see text).

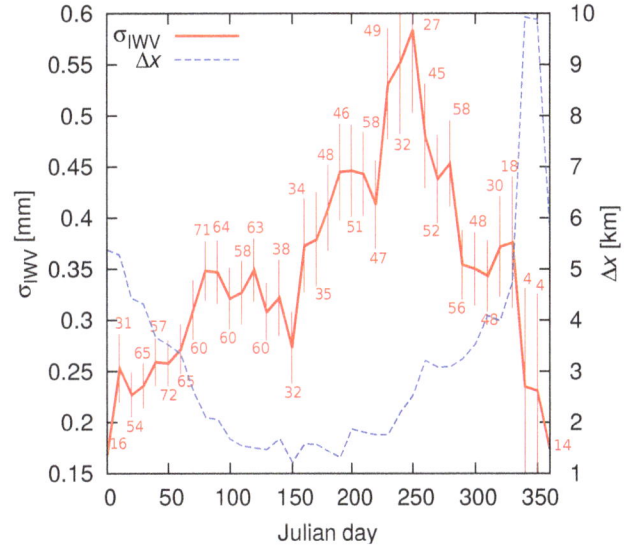

Figure 4. σ_{IWV} as a function of a Julian day. The coincidence interval is 20 min in this case; pairs within 30 days were taken into account. The quantity of measurement pairs from which σ_{IWV} was calculated is indicated by the numbers near the curve. The dashed line shows the mean horizontal distance between the pairwise soundings of IWV as a function of the season. The uncertainties ($\pm\sigma$) are indicated by the error bars (for calculation see text).

reach an altitude of about 1.5 km above the mean summit levels in summer (Carnuth and Trickl, 2000; Kreipl, 2006). During all other seasons, the convection usually does not even reach the summit of the Zugspitze and our measurement range.

For determining the spatial variability of IWV, we calculated σ_{IWV} as a function of varied spatial matching Δx by using measurement pairs within a time interval of $\Delta t = 30$ min (summer) and $\Delta t = 60$ min (winter). As mentioned above, it was shown that for a good agreement of both systems, very tight spatial and temporal matching criteria are mandatory (Vogelmann et al., 2011). Figure 3 (red curve) shows σ_{IWV} as a function of the horizontal distance of the probed volumes in the summer season. While σ_{IWV} constantly remains around 0.35 mm for $\Delta x < 2$ km, it rises to values of more than 0.65 mm at a distance of $\Delta x = 4$ km. This result shows that the variability depends on the spatiotemporal matching. Up to $\Delta x = 2$ km, the temporal variability within the selected time interval ($\Delta t = 30$ min) predominates. For larger distances, the contribution of spatial variability becomes significant.

In contrast to this, σ_{IWV} is not increasing with Δx in the winter season (Fig. 3, blue curve). This is in agreement with the assumption that local convection does not reach the ver-

tical measurement range during the winter season and that the IWV variability is probably dominated by horizontal advection of filamentary structures in the free troposphere from very different source regions. Consequently, the observed variability during winter is due to larger spatial scale processes (compared to local convection in summer), which would explain the absence of an increase with Δx in Fig. 3. Note that because IWV is much lower in winter than in summer, the relative variabilities (i.e., if σ_{IWV} were given in percent) are larger for the blue curve in Fig. 3. This means that advection of filaments (winter) leads to larger relative changes of IWV than local convection in summer. We will discuss this finding in more detail within the context of the variability of the vertical water vapor profile in Sect. 4. Figure 3 also indicates that σ_{IWV} even shows a trend towards lower values for distances above 6 km. We explain this by the fact that measurements with large horizontal mismatch ($\Delta x > 6$ km) require extraordinarily calm and clear weather conditions, because the FTIR instrument requires a cloudless field of view and a sun position close to the horizon.

Figure 4 shows σ_{IWV} as a function of the Julian day. Here, counteracting effects can be observed. While the mean horizontal distance (dashed curve) is low in the summer season ($\Delta x < 2$ km), it reaches up to almost 10 km around midwinter. The variability over the entire field of horizontal distances within a certain time interval (e.g., 20 min) reaches its maximum of almost 0.6 mm when the temperature peaks around the end of July. We assume that this is a direct effect of the

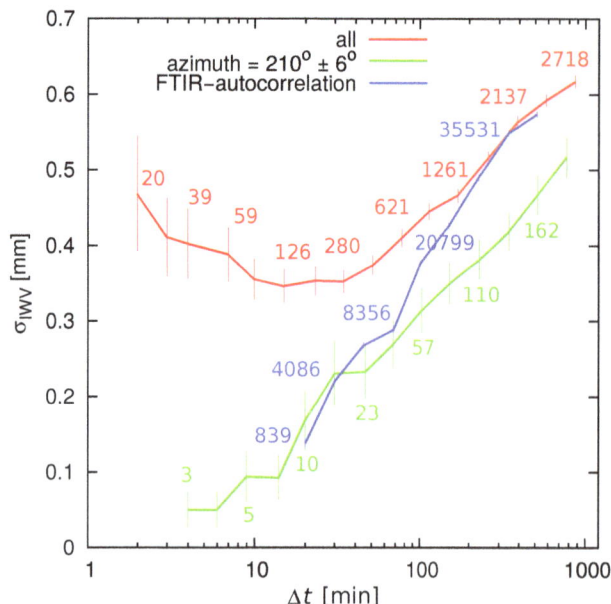

Figure 5. Variability as a function of the length of the time interval. The red curve shows σ_{IWV} from all measurements with no geometrical restrictions as a function of the length of the time interval in which data were taken into account. The green curve only includes measurements recorded in the early afternoon when the volume matching peaks with a sun azimuth of $210 \pm 6°$. The blue curve only shows σ_{IWV} of IWV values from the FTIR instrument. The quantity of measurement pairs from which σ_{IWV} was calculated is indicated by the numbers near the curves (not for all nodes). The uncertainties ($\pm \sigma$) are indicated by the error bars (for calculation see text).

heat-driven local convection, which can reach altitudes of 4.5 km at the Zugspitze site during the summer season (Reiter et al., 1983; Müller and Reiter, 1986; Carnuth and Trickl, 2000; Carnuth et al., 2002; Kreipl, 2006). The fact that the variability shows moderate values at the minimum average distance leads to the assumption that it is partially caused by local effects. As expected, the minimum variability of about 0.15 mm is observed around midwinter when temperatures are low, although the mean horizontal mismatch of both instruments is largest at this time of the year. This supports the assumption that local dynamics do not play a significant role during midwinter.

3.2 Temporal variability

For the analysis of temporal variability, we calculated the standard deviation of differences σ_{IWV} between IWV values from both instruments as a function of temporal coincidence. This was repeated for varied spatial matching criteria. When using all IWV values from both instruments without applying any geometrical matching criteria, σ_{IWV} shows a flat minimum around a coincidence interval of $\Delta t = 20$ min, see red curve in Fig. 5. About 100–300 coincident pairs contribute to

the ensembles within this minimum. At first, a minimal σ_{IWV} for the shortest interval length was expected. Two different effects are responsible for the minimum around $\Delta t = 20$ min. First of all, most FTIR and lidar measurements were carried out in the morning, because there are still few clouds. As a consequence, most of the pairs with the shortest coincidence intervals are found in the morning where the spatial matching is worst (see Figs. 1 and 2). This slightly increases σ_{IWV} on the very left hand side of the red curve in Fig. 5. Secondly, many pairs with good spatial matching can be found around noon, even for somewhat larger temporal coincidence intervals. This explains the decrease of σ_{IWV} towards the minimum (red curve in Fig. 5).

When considering measurement pairs with an FTIR sun azimuth close to the position of the DIAL instrument ($210 \pm 6°$) only, σ_{IWV} is much smaller in general and has its minimum at the shortest coincidence intervals (green curve in Fig. 5). For time intervals on the minutes scale, we find $\sigma_{IWV} = 0.05$ mm, which agrees with the validated (combined) precision of our instruments Vogelmann et al. (2011).

The temporal variability of IWV can also be estimated from the standard deviation of differences of measurements recorded by the same instrument within certain time intervals. In our case, this was possible with data from the FTIR instrument only, thanks to its more frequent and continuous operation. The result is reflected by the blue curve in Fig. 5. Due to the solar FTIR's 13.3 min integration time, the curve starts at an interval length of $\Delta t = 20$ min. The blue curve begins to deviate increasingly from the green curve beyond 30 min and converges towards the red curve for larger time intervals. This corresponds to the fact that we observe a superposition of temporal and spatial variability with the solar FTIR, i.e., for larger time intervals, the FTIR instrument produces a spatial mismatch by itself: due to its sun-pointing geometry, the FTIR instrument probes a different volume after a certain time. This spatial mismatch has a significant effect for time intervals longer than 30 min.

4 Profile variability

The variability of the vertical water vapor distribution on timescales of $\Delta t \leq 5$ h was derived from water vapor number density profiles retrieved from the DIAL measurements. We built ensembles of DIAL water vapor profiles recorded within a range of time intervals (e.g., 1–5 h). After normalizing each profile using the respective ensemble mean profile, we merged all normalized profiles into a large ensemble for statistical analysis. First, we calculated the relative variance σ^2/μ^2 (with μ = ensemble mean number density) as a function of altitude for different time intervals. This is plotted on the left hand side of Fig. 6. For the shortest time interval of this investigation (1 h), the relative variance starts with a value of about 0.02. Above 5 km, the variance continuously increases to more than 0.38 at an altitude of about 11 km a.s.l.

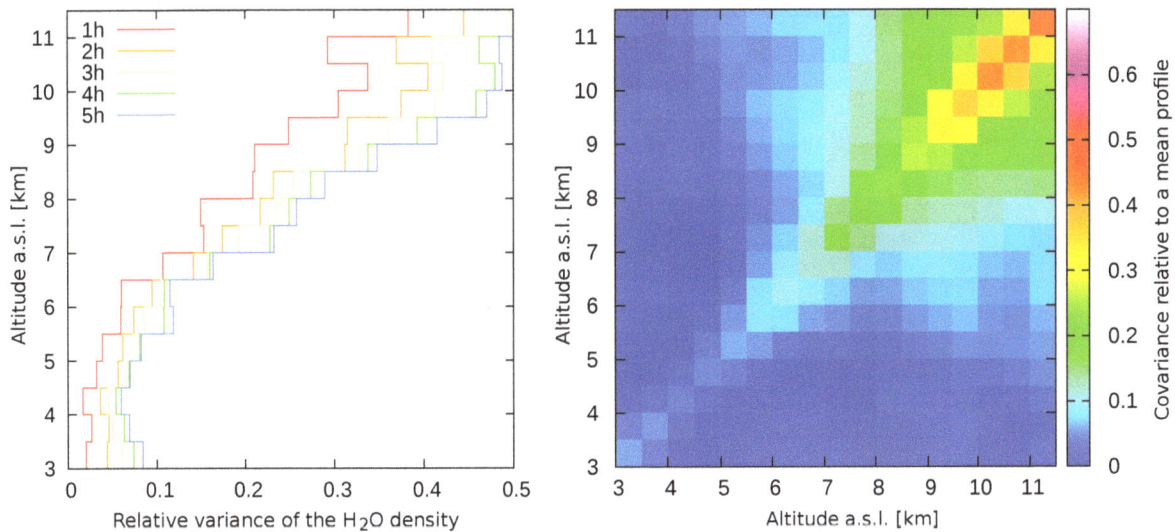

Figure 6. The short-term variability of the vertical water vapor profile is illustrated by the plot of the relative variance as a function of altitude within different time intervals (left plot). The covariance matrix (right plot) gives an idea of the interconnectivity of the variation between different altitudes.

For longer time intervals up to 5 h, the relative variance behaves quite similarly, but is shifted to higher values at all altitudes. This is in agreement with our results of IWV variability analysis, according to which longer time intervals lead to larger variabilities. In comparison to the 1 h profile, we see a more significant maximum at the lower edge at 3 km and a significant minimum at 4.5 km for longer time intervals. This enhanced increase between 3 and 4 km is, to our understanding, induced by the diurnally varying upper edge of the boundary layer during the warm season (see below).

For the lowest layer (i.e., 3–4 km), where most of the entire column above the Zugspitze site is located, we find equal relative variabilities as for IWV. This means that for a time interval $\Delta t = 1$ h, the coefficient of variation $\sigma/\mu = 0.12$. From the green curve in Fig. 5, we obtain a 1 h variability of 0.27 mm with a 60 min ensemble mean IWV of 2.33 mm, which also yields a coefficient of variation of 0.12.

In contrast to this, the relative variability increases with altitude above 5 km. This can be explained by the increasing wind speed at higher altitudes in the troposphere. The temporal variability of the water vapor density in the free troposphere at a certain altitude primarily features a horizontal variability combined with a horizontal wind velocity at this altitude. From NCEP reanalysis data, we derived an average wind speed as a function of altitude, which increases from a few meters per second near the ground to about 22 m s^{-1} in the tropopause region (Fig. 7). Similar values were reported by Birner et al. (2002) based on radiosonde data recorded above Munich (southern Germany). Depending on the pathway of the jet stream or the polar vortex, maximum wind velocities of more than 100 m s^{-1} occur occasionally (Riehl, 1962). Considering a time interval of 60 min, this means a mean horizontal spread of about 80 km around 10 km alti-

tude with a potential increase to more than 360 km in the jet stream regime.

The general increase of the relative short-term variability of water vapor above 5 km (Fig. 6, left) seems to flatten slightly at about 10 km. This can be explained by the fact that the wind speed has its maximum here and decreases at higher altitudes. Above 9 km, the contribution of measurement errors becomes significant. The DIAL is not able to measure water vapor concentrations below 18 ppm (sensitivity limit at 10 km), which may be even lower in the tropopause region. Hence, for the calculation of variances and covariances, only profiles valid in the entire range (3–12 km) are taken into account including a statistical error calculation.

The coherence of the short-term variability of water vapor at different altitudes is analyzed using the covariance matrix of the vertical profile variability (Fig. 6, right). The covariance matrix is calculated from all normalized profiles recorded from 2007 to 2011, which are contained in the sub-ensembles of profiles recorded within a 5 h time interval. Consequently, the diagonal of the covariance matrix is identical to the 5 h curve of the variability profile shown on the left hand side of Fig. 6. There are no significant off-diagonal values below 6 km. We interpret this as a sign of the lower altitudes not being dominated by a coherent air flow for most of the observations. This means that the horizontal flow at certain altitudes below 6 km is not or only weakly coupled to the flow above or below. The slight increase of off-diagonal values between 6 and 8 km indicates a partially coherent flow. The high off-diagonal values above 8 km indicate a large fraction of coherent flow of inhomogeneous air masses in this altitude region.

The weak coupling between different layers at lower altitudes is in agreement with the assumption of local convection

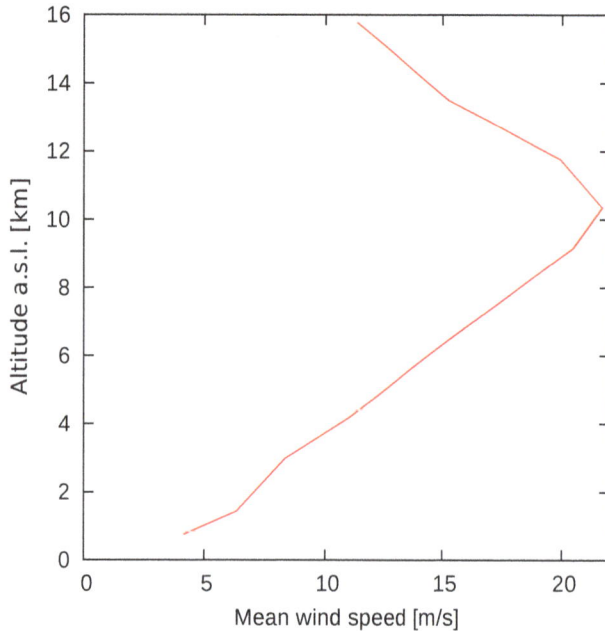

Figure 7. Mean wind speed above the Zugspitze as a function of altitude (data from the National Center for Environmental Prediction, NCEP). Under the jet stream regime, the wind velocity at 10 km can occasionally exceed $100\,\mathrm{m\,s^{-1}}$.

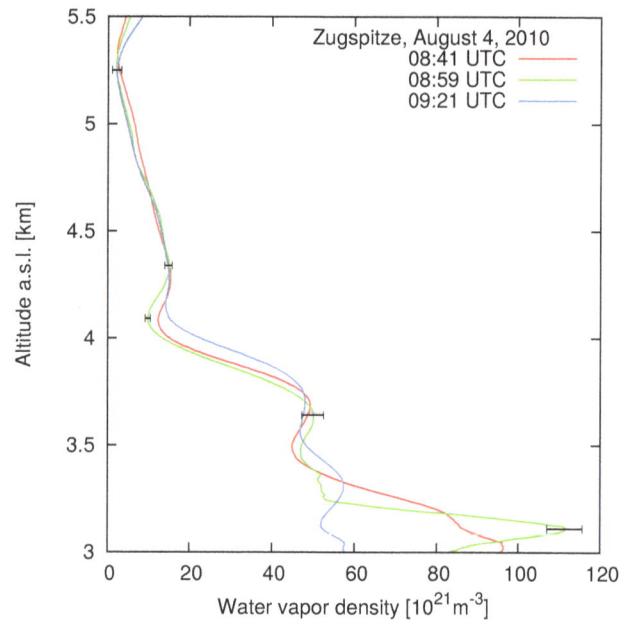

Figure 8. Short-term variability of the water vapor profile induced by local convection within a clearly confined upper edge of the boundary layer at 3.5 km under stable atmospheric conditions. The variations do not exceed a factor of 2. The example error bars ($\pm 2\sigma$) represent statistical uncertainties caused by electronic noise in the detection.

and turbulence being the dominant sources of variability in the lower part of the examined altitude range. This behavior can be described by barely interacting "bubbles" of humid air. In the upper troposphere, on the other hand, varying air masses are more coherently exchanged within the upper air flow, as a result of which layers of a wider vertical spread are affected.

5 Mechanisms driving the observed variability

In the troposphere, evaporation is the only relevant source of water vapor and precipitation the only relevant sink. Thus, water vapor is injected into the free troposphere by uplifting processes, such as local convection or large-scale warm conveyor belts. These uplifting processes cause inhomogeneity in the horizontal water vapor distribution at a certain altitude. Furthermore, air ascending to high altitudes undergoes cooling. If this air initially was humid, part of its water vapor content can be precipitated during the ascent. As a result, the absolute humidity of upper tropospheric air is low in general. Downwelling of dry air from high altitudes, in particular from the tropopause region or even the stratosphere, also produces inhomogeneity in the horizontal humidity field at the affected altitude levels. In contrast to uplifting processes, downwelling generally is not a local phenomenon. As regards the short-term variability (i.e., $\Delta t < 6\,\mathrm{h}$) of the vertical distribution of water vapor, it is reasonable to distinguish between inhomogeneity produced locally on a small scale

and inhomogeneity produced remotely and transported via long-range pathways. By analyzing the measured water vapor profiles in combination with trajectory calculations from atmospheric models, we found that the short-term variability of the profiles shows contributions from both local effects and long-range transport at the same time. The short-term variability above 5 km can be attributed to the advection of a heterogeneous layer structure in most cases. Below 5 km, on the other hand, a clear assignment is not always possible. Backward trajectories were calculated from reanalysis data with the NOAA Hybrid Single-Particle Lagrangian Integrated Trajectory (HYSPLIT) vertical velocity model (http://ready.arl.noaa.gov/HYSPLIT.php, Draxler and Hess, 1998). However, the performance of a trajectory model is also limited above complex terrain and running times of several days occasionally involve large uncertainties even in the free troposphere. Sometimes several attempts are necessary to guess the correct starting altitude due to shifts in the orographic data used by the model (Trickl et al., 2010). Thus, trajectory calculations are not considered as a proof, but as support for plausibility. Our experience in the analysis of long-range transport events suggests a high reliability of free-tropospheric trajectories. In the following subsections we highlight four different types of dynamics producing short-term variability of water vapor.

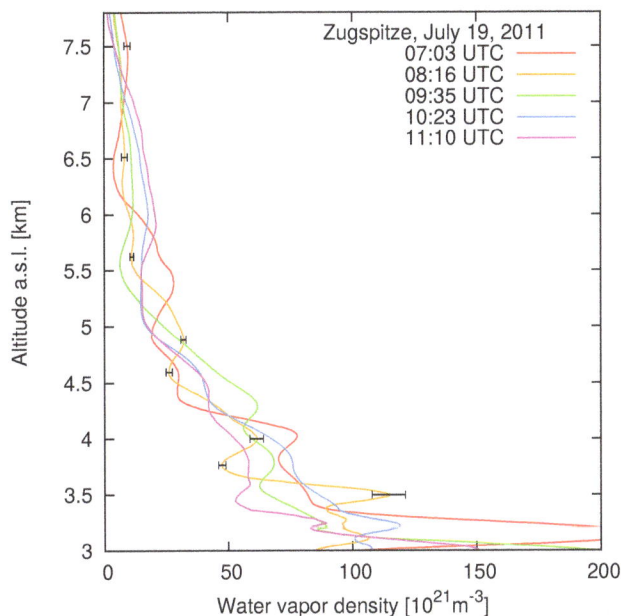

Figure 9. Short-term variability of the water vapor profile under atmospheric instability, high-reaching convection, and only a few hours before the formation of a thunderstorm. The example error bars ($\pm 2\sigma$) represent statistical uncertainties caused by electronic noise in the detection.

5.1 Local convection

5.1.1 Case studies

A case of local convection under stable atmospheric conditions (high pressure) is shown in Fig. 8. Three water vapor profiles were recorded within 40 min. The variability stops at the upper edge of the boundary layer at 3.5 km. Above this level, the water vapor distribution remains constant throughout that period. The upper edge of the boundary layer was visually verified by the upper edge of cumulus clouds located at the top of some thermals. Strongly enhanced backscatter from boundary layer aerosols was recorded up to 3.5 km. Some weaker aerosol structure that slowly moves downwards was observed above 4.5 km and even up to 7.1 km.

The situation is somewhat different under conditions of low pressure and atmospheric instability. This case is shown in Fig. 9. Five profiles were recorded within a time interval of 4 h before a heavy thunderstorm developed in the afternoon. The short-term variability of water vapor was rather high and reached far into the upper troposphere up to at least 7.5 km. Due to the travel time of upwelling air and the increasing horizontal wind speed, the variations at high altitudes (e.g., above 5 km) were less local than the variations near the ground. Cloud formation was first observed between 5.5 km and 6.5 km. However, only a few minutes later, clouds formed also above 2.5 km. Due to cloud interference, the last valid profile was recorded at 11:10 UTC

(LT − 1 h). Strongly enhanced backscatter from boundary-layer aerosols was recorded up to 4.7 km already by the morning (07:03 UTC). This altitude is rather high. The latest profile at 11:10 UTC exhibits boundary-layer aerosols up to 4.2 km only and also a lower humidity compared to the profiles recorded before. In our understanding, this indicates a downflow near but outside of the thunderstorm. This downwelling air had probably lost most of its original water content during its ascent in the thunderstorm through precipitation. At 12:37 UTC (profile not shown), the extended head of the cumulonimbus cloud of the upcoming thunderstorm led to overcast at the site above 7.7 km. In addition, strong aerosol structures appear up to 7.5 km. Backward trajectory calculations (HYSPLIT) suggest that air between 6 and 7.5 km originated from the Caribbean boundary layer.

5.1.2 General discussion

During the warm season, local convection usually reaches altitudes of up to 1.5 km above summit levels (Carnuth and Trickl, 2000; Carnuth et al., 2002; Kreipl, 2006), which is about 4.5 km a.s.l. in our case. The enhanced updraft along sunny mountain slopes is also referred to as "Alpine pumping". The slightly elevated short-term variability at lower altitudes around 3.5 km (Fig. 6, left) is attributed to local convection and the diurnal variation in the upper edge of the planetary boundary layer, which is caused by Alpine pumping. Due to the strong vertical gradient of the water vapor profile, this dominates the short-term variability of IWV in most cases when local convection significantly exceeds 3 km (which is the bottom of our measurement range). From the comparatively low mean wind speed at lower altitudes (Fig. 7), we conclude that the elevated variability here is caused by larger horizontal gradients in the water vapor concentration. This means that variations occur on smaller horizontal scales compared to higher altitudes, which underlines the fact that local processes (e.g., thermal lifts) on small scales are the dominant source. Short-term variations of the water vapor concentration at a certain altitude within the upper part of the boundary layer (i.e., 3–4.5 km a.s.l.), which are caused by local convection, are estimated to be smaller than a factor of 2. Convection penetrating into the free troposphere or even the upper troposphere can cause short-term variation factors of more than 5 at these high altitudes (e.g., Fig. 9, other observations). The presence of aerosols (enhanced backscatter) usually indicates upwelling air from the planetary boundary layer. Aerosol structures in the free troposphere are also helpful for estimating the vertical velocity of the probed air. Both cases were visually verified by the observation of cloud formation, while trajectory calculations from models are not able to resolve these small-scale local processes. However, they indicate a general downwelling for the case under stable high-pressure conditions and a general upwelling for the case under unstable low-pressure conditions.

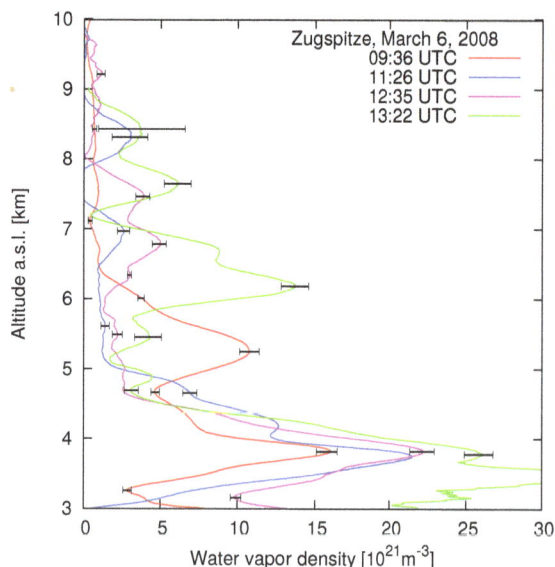

Figure 10. Example of extreme temporal variability of the vertical distribution of water vapor during a stratospheric intrusion event. Due to the advection velocity of about $11 \, \mathrm{ms^{-1}}$ between 3 and 4 km altitude (data from radiosonde at Munich at 12:00 UTC, 100 km to the north) a time shift of 1 h corresponds to a horizontal shift of about 40 km within this altitude range. The example error bars ($\pm 2\sigma$) represent statistical uncertainties caused by electronic noise in the detection.

Figure 11. Example of extreme variability of the vertical distribution of water vapor under rather humid conditions. Due to a wind speed of about $16 \, \mathrm{ms^{-1}}$ at an altitude of 4.5 km (data from radiosonde Munich at 12:00 UTC, 100 km to the north) a time shift of 2 h corresponds to a horizontal shift of about 115 km at this altitude. The two profiles were recorded within less than 2 h. The example error bars ($\pm 2\sigma$) represent statistical uncertainties caused by electronic noise in the detection.

5.2 Long-range transport

5.2.1 Case studies

Figures 10 and 11 show cases of extreme vertical variability of water vapor on a timescale of hours recorded with the DIAL. Similar scenarios have been observed many times. From these incidents we learned that the water vapor density at a certain altitude can vary by a factor of more than 30 within a few hours. Thus, the short-term variability of water vapor induced by long-range transport and the advection of very inhomogeneous layer structures can exceed the impact of local convection by 1 order of magnitude.

This is particularly pronounced for stratospheric intrusions that descend from the Arctic to central Europe. These intrusion layers occasionally become the main source of short-term variability of water vapor in the altitude range between 3 and 5 km. However, such events occur predominantly during the winter season and are accompanied by non-convective weather conditions. Under these conditions heterogeneous air masses are usually advected at a high velocity which results in a very high variability at certain altitudes even on the short timescale of 1 h. Due to the origin of these layers, stratospheric intrusion events are usually accompanied by rather dry conditions. This is illustrated by the example given in Fig. 10 where three layers of stratospheric air have been advected at the same time at different altitudes, thus cre-

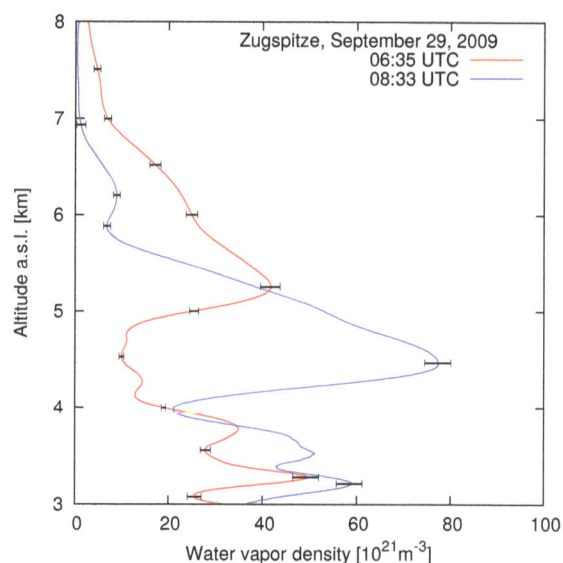

ating relative variations of the water vapor density of more than a factor of 10 at certain altitudes within 4 h. The stratospheric intrusion originated above Greenland about 2–3 days before reaching our site on 6 March 2008. It exhibited several descending filaments lying upon one other. The very complex dynamics and its accompanying heterogeneous vertical layering is discussed in great detail in a separate publication including a 4-day forward trajectory calculation for this case (Trickl et al., 2014, and references therein). Stratospheric intrusions into the lower free troposphere usually occur in the winter season with a frequency of roughly 4 to 10 times per month above the Zugspitze (Stohl et al., 2000; Trickl et al., 2010).

Also, humid air from remote boundary layers sometimes causes rather intense short-term variations of the water vapor distribution. An example is shown in Fig. 11. The humidity profile shows a significant increase between 4 and 5 km a.s.l. within 2 h. Backward trajectory calculations from reanalysis data with the HYSPLIT vertical velocity model (see above) for this case suggest a sudden change in the source region from the North American upper troposphere (dry) to the northwest Pacific and rather low altitudes of about 2 km a.s.l. within 2 h (Fig. 12). In contrast, the air at an altitude of about 3.3 km constantly originates from the subtropical North Atlantic boundary layer (moderately humid, trajectories not shown here). The trajectory starting above the northwest Pacific Ocean exhibits a fast ascent to the upper troposphere within 2 days. This behavior is attributed to a warm conveyor belt using the criteria published by Eckhardt et al. (2004). Satellite images show that the ascending part of the blue trajectory is near the warm front of a cyclone that is located about 2000 km south of the peninsula of Kamchatka (northwest Pacific Ocean). Warm conveyor belts are known to be the most important extra-tropical transport mechanism of water vapor to the free and upper troposphere, although the water vapor flux moves like a jet from a rather restricted area (Browning and Roberts, 1994; Browning et al., 1997; Eckhardt et al., 2004; Ziv et al., 2009). It is remarkable that these filamentary structures are partially preserved, while traveling around half of the hemisphere. A wind speed of $16 \, \mathrm{m \, s^{-1}}$ at an altitude of 4.5 km (Munich radiosonde, 12:00 UTC) transforms a time shift of 2 h into a horizontal shift of about 115 km. The water vapor density at this altitude changes by more than a factor of 5 within 2 h in this case.

5.2.2 General discussion

It is reasonable to assume that much of the variability in the free troposphere is caused by the rich layer structure advected along or in the vicinity of the North Atlantic storm track or from the Mediterranean basin and northern Africa. From our lidar measurements of ozone, water vapor, and aerosol, we know that the persistence of specific free-tropospheric layers above the Zugspitze can range from less than 1 h to more than 1 day (Eisele et al., 1999; Stohl and Trickl, 1999; Trickl et al., 2003, 2010, 2011). Along the jet stream, many different ascending and descending air streams merge or separate (e.g., Appenzeller et al., 1996; Stohl, 2001; Cooper et al., 2001, 2002, 2004a, b; Flentje et al., 2005). The advection of filamentary and heterogeneous layer structures affects the entire free troposphere and dominates the variability of water vapor in the upper troposphere above 5 km. The most important source regions contributing to observations above the Zugspitze are the stratosphere (very dry air), North America, the (sub)tropical Atlantic (very humid), and also Asia. Sometimes, dry and ozone-rich air flows along the northward spiraling subtropical jet streams (Trickl et al., 2011). The layers frequently possess a meridional component, leading to a transverse passage of adjacent layers across the observational site. This implies a rapid change in concentrations.

6 Summary and conclusions

The result of our studies is a quantitative description of the short-term variability of water vapor in the free troposphere above the Zugspitze, which is a location representative of central Europe. From measurement data recorded with two high-precision optical water vapor sounders arranged in a unique pointing geometry, we derived information about the

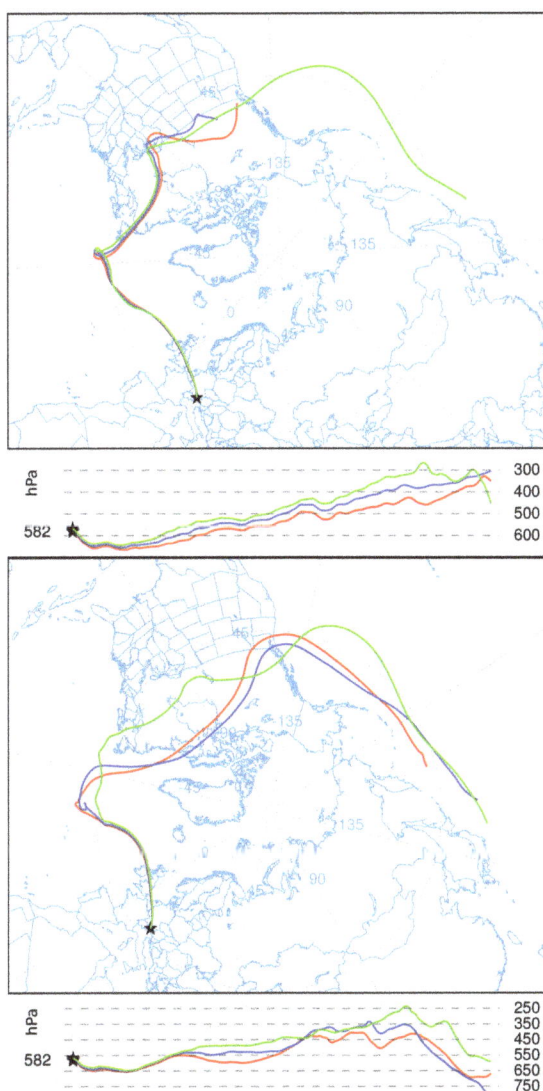

Figure 12. Backward trajectories from the NOAA HYSPLIT model (see text) ending above the Zugspitze at 4600, 4700 and 4800 m a.s.l. at 06:00 UTC (upper plot) and 08:00 UTC (lower plot), 29 July 2009 were calculated from reanalysis data with a vertical velocity model and a duration of 315 h. The vertical sections are referred to the air pressure along the pathways. The remarkable coherence of the three pathways during a longer time indicates a rather good reliability.

spatiotemporal variability of integrated water vapor (IWV) on the scale of kilometers and of minutes.

Within a time interval of 20 min, a variability of about 0.35 mm was determined in the summer season under the condition of good volume matching ($\Delta x < 2 \, \mathrm{km}$). The spatial variability became significant for horizontal distances above 2 km, but only in the warm season. The variability of IWV observed in the winter season was generally lower and did not increase with a horizontal mismatch of the probed volume ($\Delta x < 12 \, \mathrm{km}$). Its relative value, however, was larger than in the summer season. The seasonality of the IWV short-term variability and the geometrical restrictions of the measurements underline that local convection is the main source of variability during the warm season, while the variability in the winter season is driven by dynamics on a larger scale. The temporal variability of IWV was determined to be 0.05 mm on the scale of minutes (5 min) with a uniform increase to 0.5 mm on a timescale of 1 day.

The free-tropospheric profile variability of water vapor on the timescale of hours (e.g., 1–5 h) shows a broad minimum around 4.5 km a.s.l. and much larger values for higher altitudes with a constant increase up to the tropopause region. Longer time intervals generally yield larger variations at all altitudes and additionally show a more significant maximum at the lower edge of the measurement range (3 km). These findings are explained by the vertical wind profile and the heterogeneity of air masses within the upper air flow advected with a high velocity and, additionally, by the impact of local convection below 4.5 km. The covariance matrix of the profile variability yields information about the coherence of neighboring layers and shows that the air flow below 6 km is rather incoherent, while the upper air stream above 8 km is much more coherent.

We presented four case studies in which the profile variability of water vapor on the timescale of hours was attributed to specific mechanisms: local and vertically limited convection under stable conditions, high-reaching convection under unstable conditions, downwelling of a stratospheric intrusion, and long-range transport from very different source regions.

The source of the variability can be either local convection or long-range transport of inhomogeneous air masses. When reviewing all profiles of our study, we found that it is not always possible to distinguish clearly between both mechanisms of short-term variability. In particular, for altitudes below 4.5 km, which are potentially affected by local convection even under stable atmospheric conditions, we must assume a mixture of both local contributions and the advection of inhomogeneous layer structures from different remote source regions. From cases where a clear assignment was possible, we conclude that the long-range advection of very inhomogeneous layer structures can cause relative short-term variations of the water vapor concentration at a certain altitude, which are larger by 1 order of magnitude than variations in cases dominated by the impact of local convection. Due to the high altitude of the measurement site, our analysis is mostly restricted to the free troposphere. The upper edge of the Alpine boundary layer reaches our measurement range usually only during afternoons in the summer season. The consequence of measuring above a complex alpine terrain (steep mountain slopes) is that we observe the influence of local convection in our measurement range (above 3 km a.s.l.) quite frequently. The impact of local convection undercuts the possible impact of long-range transport by roughly 1 order of magnitude. This suggests, at least for the summer sea-

son, that the variability inside the boundary layer is probably reduced to values that we observe with dominating local convection reaching our measurement range. This assumption, of course, implies that the fast advection of heterogeneous air layers does not impact the boundary layer. However, the reported IWV variability during the warm season with dominating local convection, in principle, supports the findings from a recent IWV variability assessment by Steinke et al. (2015), although the underlying IWV determination started at lower altitudes and above less complex terrain. This less complex terrain assumably justifies our observations of relative short-term variations of about a factor of 2 higher of IWV in summer.

In spite of the missing convection, the relative short-term variability of water vapor (IWV and profiles) in the free troposphere is higher during the winter season. This is explained by the results of Trickl et al. (2010), according to which stratospheric air intrusions above the Zugspitze exhibit a pronounced maximum during the winter season. Roughly three-fourths of them reach the Zugspitze summit (2962 m) and were detected directly by the in situ instrumentation.

Our results for the first time provide a quantitative description of the free-tropospheric spatiotemporal variability of water vapor on the scales of minutes and kilometers (horizontal) for IWV and the scales of hours and 500 m (vertical) for profiles. This information can be useful for the parameterization of humidity in atmospheric models as well as for estimating the influence of the atmospheric variability of water vapor on the significance of water vapor measurements performed with a given integration time. In a related sense, our results also provide the information necessary for evaluating intercomparison studies of imperfectly co-located or synchronized instruments. Our findings fit perfectly with the results of our previous intercomparison study (Vogelmann et al., 2011) that indicated a high variability of water vapor, as a result of which, very tight matching criteria are required down to the scales of 10 min and several hundred meters to reduce co-location effects to a negligible level.

Acknowledgements. We thank Hans-Peter Schmid (KIT/IMK-IFU) for his continuous interest in this work and M. Rettinger (KIT/IMK-IFU) for executing the FITR measurements at the Zugspitze. We also thank Michael Sprenger (ETH Zürich) for providing the forward trajectory calculations (for the analysis of stratosphere-to-troposphere transport). We also acknowledge the team of the Schneefernerhaus research station (UFS) for maintaining our lidar measurements and the Bavarian Ministry of Environment and Consumer Protection for funding our work within the ALOMAR cooperation.

Edited by: M. Tesche

References

Adam, M., Demoz, B. B., Whiteman, D. N., Venable, D. D., Joseph, E., Gambacorta, A., Wei, J., Shephard, M. W., Miloshevich, L. M., Barnet, C. D., Herman, R. L., Fitzgibbon, J., and Connell, R.: Water Vapor Measurements by Howard University Raman Lidar during the WAVES 2006 Campaign, J. Atmos. Ocean. Tech., 27, 42–60, doi:10.1175/2009JTECHA1331.1, 2010.

Appenzeller, C., Davies, H. C., and Norton, W. A.: Fragmentation of stratospheric intrusions, J. Geophys. Res., 101, 1435–1456, doi:10.1029/95JD02674, 1996.

Birner, T., Dörnbrack, A., and Schumann, U.: How sharp is the tropopause at midlatitudes?, Geophys. Rev. Lett., 29, 1700, doi:10.1029/2002GL015142, 2002.

Bleisch, R., Kämpfer, N., and Haefele, A.: Retrieval of tropospheric water vapour by using spectra of a 22 GHz radiometer, Atmos. Meas. Tech., 4, 1891–1903, doi:10.5194/amt-4-1891-2011, 2011.

Browning, K. A. and Roberts, N. M.: Structure of a frontal cyclone, Q. J. Roy. Meteor. Soc., 120, 1535–1557, doi:10.1002/qj.49712052006, 1994.

Browning, K. A., Roberts, N. M., and Illingworth, A. J.: Mesoscale analysis of the activation of a cold front during cyclogenesis, Q. J. Roy. Meteor. Soc., 123, 2349–2375, doi:10.1002/qj.49712354410, 1997.

Carnuth, W. and Trickl, T.: Transport studies with the IFU three-wavelength aerosol lidar during the VOTALP Mesolcina experiment, Atmos. Environ., 34, 1425–1434, 2000.

Carnuth, W., Kempfer, U., and Trickl, T.: Highlights of the tropospheric lidar studies at IFU within the TOR project, Tellus, 54B, 163–185, 2002.

Cooper, O. R., Moody, J. L., Parrish, D. D., Trainer, M., Ryerson, T. B., Holloway, J. S., Hübler, G., Fehsenfeld, F. C., Oltmans, S. J., and Evans, M. J.: Trace gas signatures of the airstreams within North Atlantic cyclones: Case studies from the North Atlantic Regional Experiment (NARE 97) aircraft intensive, J. Geophys. Res., 106, 5437–5456, doi:10.1029/2000JD900574, 2001.

Cooper, O. R., Moody, J. L., Parrish, D. D., Trainer, M., Holloway, J. S., Hübler, G., Fehsenfeld, F. C., and Stohl, A.: Trace gas composition of midlatitude cyclones over the western North Atlantic Ocean: A seasonal comparison of O3 and CO, J. Geophys. Res., 107, 4057, doi:10.1029/2001JD000902, 2002.

Cooper, O., Forster, C., Parrish, D., Dunlea, E., Hübler, G., Fehsenfeld, F., Holloway, J., Oltmans, S., Johnson, B., Wimmers, A., and Horowitz, L.: On the life cycle of a stratospheric intrusion and its dispersion into polluted warm conveyor belts, J. Geophys. Res., 109, D23S09, doi:10.1029/2003JD004006, 2004a.

Cooper, O. R., Forster, C., Parrish, D., Trainer, M., Dunlea, E., Ryerson, T., Hübler, G., Fehsenfeld, F., Nicks, D., Holloway, J., de Gouw, J., Warneke, C., Roberts, J. M., Flocke, F., and Moody, J.: A case study of transpacific warm conveyor belt transport: Influence of merging airstreams on trace gas import to North America, J. Geophys. Res., 109, D23S08, doi:10.1029/2003JD003624, 2004b.

Draxler, R. R. and Hess, G. D.: An Overview of the HYSPLIT_4 Modelling System for Trajectories, Dispersion, and Deposition, Australian Meteorological Magazine, 47, 295–308, 1998.

Eckhardt, S., Stohl, A., Wernli, H., James, P., Forster, C., and Spichtinger, N.: A 15-Year Climatology of Warm Conveyor Belts., J. Climate, 17, 218–237, doi:10.1175/1520-0442(2004)017<0218:AYCOWC>2.0.CO;2, 2004.

Eisele, H., Scheel, H. E., Sládkovič, R., and Trickl, T.: High-Resolution Lidar Measurements of Stratosphere-Troposphere Exchange, J. Atmos. Sci., 56, 319–330, 1999.

Fassò, A., Ignaccolo, R., Madonna, F., Demoz, B. B., and Franco-Villoria, M.: Statistical modelling of collocation uncertainty in atmospheric thermodynamic profiles, Atmos. Meas. Tech., 7, 1803–1816, doi:10.5194/amt-7-1803-2014, 2014.

Flentje, H., Dörnbrack, A., Ehret, G., Fix, A., Kiemle, C., Poberaj, G., and Wirth, M.: Water vapor heterogeneity related to tropopause folds over the North Atlantic revealed by airborne water vapor differential absorption lidar, J. Geophys. Res., 110, D03115, doi:10.1029/2004JD004957, 2005.

Harries, J. E.: Atmospheric radiation and atmospheric humidity, Q. J. R. Meteor. Soc., 123, 2173–2186, 1997.

Immler, F. J., Dykema, J., Gardiner, T., Whiteman, D. N., Thorne, P. W., and Vömel, H.: Reference Quality Upper-Air Measurements: guidance for developing GRUAN data products, Atmos. Meas. Tech., 3, 1217–1231, doi:10.5194/amt-3-1217-2010, 2010.

Kämpfer, N., Ed.: Monitoring Atmospheric Water Vapour - Ground-Based Remote Sensing and In-situ Methods, Springer, Berlin, Heidelberg, 2013.

Kiehl, J. T. and Trenberth, K. E.: Earth's Annual Global Mean Energy Budget, B. Am. Meteorol. Soc., 78, 197–208, 1997.

Kreipl, S.: Messung des Aerosoltransports am Alpennordrand mittels Laserradar (Lidar), Dissertation (in German), Universität Erlangen, 2006.

Leblanc, T., Walsh, T. D., McDermid, I. S., Toon, G. C., Blavier, J.-F., Haines, B., Read, W. G., Herman, B., Fetzer, E., Sander, S., Pongetti, T., Whiteman, D. N., McGee, T. G., Twigg, L., Sumnicht, G., Venable, D., Calhoun, M., Dirisu, A., Hurst, D., Jordan, A., Hall, E., Miloshevich, L., Vömel, H., Straub, C., Kampfer, N., Nedoluha, G. E., Gomez, R. M., Holub, K., Gutman, S., Braun, J., Vanhove, T., Stiller, G., and Hauchecorne, A.: Measurements of Humidity in the Atmosphere and Validation Experiments (MOHAVE)-2009: overview of campaign operations and results, Atmos. Meas. Tech., 4, 2579–2605, doi:10.5194/amt-4-2579-2011, 2011.

Müller, H. and Reiter, R.: Untersuchung der Gebirgsgrenzschicht über einem großen Alpental bei Berg-Talwindzirkulation, Meteorol. Rdsch., 39, 247–256 (in German), 1986.

Pougatchev, N. S., Connor, B. J., and Rinsland, C. P.: Infrared measurements of the ozone vertical distribution above Kitt Peak, J. Geophys. Res., 100, 16689–16697, doi:10.1029/95JD01296, 1995.

Reiter, R., Müller, H., Sladkovic, R., and Munzert, K.: Aerologische Untersuchungen der tagesperiodischen Gebirgswinde unter besonderer Berücksichtigung des Windfeldes im Talquerschnitt, Meteorol. Rdsch., 36, 225–242 (in German), 1983.

Riehl, H.: Jet Streams of the Atmosphere, Tech. Rep. 32, Department of Atmospheric Science Colorado State University Fort Collins, Colorado, 1962.

Seidel, D. J., Sun, B., Pettey, M., and Reale, A.: Global radiosonde balloon drift statistics, J. Geophys. Res., 116, D07102, doi:10.1029/2010JD014891, 2011.

Soden, B. J. and Lanzante, J. R.: An Assessment of Satellite and Radiosonde Climatologies of Upper-Tropospheric

Water Vapor., J. Clim., 9, 1235–1250, doi:10.1175/1520-0442(1996)009<1235:AAOSAR>2.0.CO;2, 1996.

Steinke, S., Eikenberg, S., Löhnert, U., Dick, G., Klocke, D., Di Girolamo, P., and Crewell, S.: Assessment of small-scale integrated water vapour variability during HOPE, Atmos. Chem. Phys., 15, 2675–2692, doi:10.5194/acp-15-2675-2015, 2015.

Stiller, G. P., Kiefer, M., Eckert, E., von Clarmann, T., Kellmann, S., García-Comas, M., Funke, B., Leblanc, T., Fetzer, E., Froidevaux, L., Gomez, M., Hall, E., Hurst, D., Jordan, A., Kämpfer, N., Lambert, A., McDermid, I. S., McGee, T., Miloshevich, L., Nedoluha, G., Read, W., Schneider, M., Schwartz, M., Straub, C., Toon, G., Twigg, L. W., Walker, K., and Whiteman, D. N.: Validation of MIPAS IMK/IAA temperature, water vapor, and ozone profiles with MOHAVE-2009 campaign measurements, Atmos. Meas. Tech., 5, 289–320, doi:10.5194/amt-5-289-2012, 2012.

Stohl, A.: A 1-year Lagrangian "climatology" of airstreams in the Northern Hemisphere troposphere and lowermost stratosphere, J. Geophys. Res., 106, 7263–7280, doi:10.1029/2000JD900570, 2001.

Stohl, A. and Trickl, T.: A textbook example of long-range transport: Simultaneous observation of ozone maxima of stratospheric and North American origin in the free troposphere over Europe, J. Geophys. Res., 104, 30445–30462, doi:10.1029/1999JD900803, 1999.

Stohl, A., Spichtinger-Rakowsky, N., Bonasoni, P., Feldmann, H., Memmesheimer, M., Scheel, H. E., Trickl, T., Hübener, S., Ringer, W., and Mandl, M.: The influence of stratospheric intrusions on alpine ozone concentrations, Atmos. Env., 34, 1323–1354, doi:10.1016/S1352-2310(99)00320-9, 2000.

Sun, B., Reale, A., Seidel, D. J., and Hunt, D. C.: Comparing radiosonde and COSMIC atmospheric profile data to quantify differences among radiosonde types and the effects of imperfect collocation on comparison statistics, J. Geophys. Res., 115, D23104, doi:10.1029/2010JD014457, 2010.

Sussmann, R. and Schäfer, K.: Infrared spectroscopy of tropospheric trace gases: combined analysis of horizontal and vertical column abundances, Appl. Opt., 36, 735–741, doi:10.1364/AO.36.000735, 1997.

Sussmann, R., Borsdorff, T., Rettinger, M., Camy-Peyret, C., Demoulin, P., Duchatelet, P., Mahieu, E., and Servais, C.: Technical Note: Harmonized retrieval of column-integrated atmospheric water vapor from the FTIR network – first examples for long-term records and station trends, Atmos. Chem. Phys., 9, 8987–8999, doi:10.5194/acp-9-8987-2009, 2009.

Tobin, D. C., Revercomb, H. E., Knuteson, R. O., Lesht, B. M., Strow, L. L., Hannon, S. E., Feltz, W. F., Moy, L. A., Fetzer, E. J., and Cress, T. S.: Atmospheric Radiation Measurement site atmospheric state best estimates for Atmospheric Infrared Sounder temperature and water vapor retrieval validation, J. Geophys. Res., 111, D09S14, doi:10.1029/2005JD006103, 2006.

Trenberth, K., Jones, P., Ambenje, P., Bojariu, R., Easterling, D., Tank, A., Parker, D., Rahimzadeh, F., Renwick, J., Rusticucci, M., Soden, B., and Zhai, P.: Observations: Surface and Atmospheric Climate Change. In Climate Change 2007: The Physical Science Basis. Contribution of Working Group I to the Fourth Assessment Report of the Intergovernmental Panel on Climate Change, chap. 3, 235–336, Cambridge, United Kingdom and New York, N.Y., USA, Cambridge University Press, 2007.

Trenberth, K. E.: Atmospheric Moisture Residence Times and Cycling: Implications for Rainfall Rates and Climate Change, Clim. Change, 39, 667–694, doi:10.1023/A:1005319109110, 1998.

Trickl, T., Cooper, O. R., Eisele, H., James, P., Mücke, R., and Stohl, A.: Intercontinental transport and its influence on the ozone concentrations over central Europe: Three case studies, J. Geophys. Res., 108, 8530, doi:10.1029/2002JD002735, 2003.

Trickl, T., Feldmann, H., Kanter, H.-J., Scheel, H.-E., Sprenger, M., Stohl, A., and Wernli, H.: Forecasted deep stratospheric intrusions over Central Europe: case studies and climatologies, Atmos. Chem. Phys., 10, 499–524, doi:10.5194/acp-10-499-2010, 2010.

Trickl, T., Eisele, H., Bärtsch-Ritter, N., Furger, M., Mücke, R., Sprenger, M., and Stohl, A.: High-ozone layers in the middle and upper troposphere above Central Europe: potential import from the stratosphere along the subtropical jet stream, Atmos. Chem. Phys, 11, 9343–9366, doi:10.5194/acp-11-9343-2011, 2011.

Trickl, T., Vogelmann, H., Giehl, H., Scheel, H.-E., Sprenger, M., and Stohl, A.: How stratospheric are deep stratospheric intrusions?, Atmos. Chem. Phys., 14, 9941–9961, doi:10.5194/acp-14-9941-2014, 2014.

Turner, D. D. and Mlawer, E. J.: The Radiative Heating in Underexplored Bands Campaigns, Bulletin of the American Meteorological Society, 91, 911–923, doi:10.1175/2010BAMS2904.1, 2010.

Vogelmann, H. and Trickl, T.: Wide Range Sounding of Free Tropospheric Water Vapor with a Differential Absorption Lidar (DIAL) at a High Altitude Station, Appl. Opt., 47, 2116–2132, doi:10.1364/AO.47.002116, 2008.

Vogelmann, H., Sussmann, R., Trickl, T., and Borsdorff, T.: Intercomparison of atmospheric water vapor soundings from the differential absorption lidar (DIAL) and the solar FTIR system on Mt. Zugspitze, Atmos. Meas. Tech., 4, 835–841, doi:10.5194/amt-4-835-2011, 2011.

Wagner, T., Beirle, S., Grzegorski, M., and Platt, U.: Global trends (1996-2003) of total column precipitable water observed by Global Ozone Monitoring Experiment (GOME) on ERS-2 and their relation to near-surface temperature, J. Geophys. Res., 111, D12102, doi:10.1029/2005JD006523, 2006.

Whiteman, D. N., Russo, F., Demoz, B., Miloshevich, L. M., Veselovskii, I., Hannon, S., Wang, Z., Vömel, H., Schmidlin, F., Lesht, B., Moore, P. J., Beebe, A. S., Gambacorta, A., and Barnet, C.: Analysis of Raman lidar and radiosonde measurements from the AWEX-G field campaign and its relation to Aqua validation, J. Geophys. Res., 111, D09S09, doi:10.1029/2005JD006429, 2006.

Wirth, M., Fix, A., Ehret, G., Reichardt, J., Begie, R., Engelbart, D., Vömel, H., Calpini, B., Romanens, G., Apituley, A., Wilson, K. M., Vogelmann, H., and Trickl, T.: Intercomparison of Airborne Water Vapour DIAL Measurements with Ground Based Remote Sensing and Radiosondes within the Framework of LU-AMI 2008, in: Proceedings of the 8th International Symposium on Tropospheric Profiling, edited by: Apituley, A., Russchenberg, H., and Monna, W., Delft, the Netherlands, poster presentation, 2009.

Ziv, B., Saaroni, H., Romem, M., Heifetz, E., Harnik, N., and Baharad, A.: Analysis of conveyor belts in winter Mediterranean cyclones, Theor. Appl. Climatol., 99, 441–455, doi:10.1007/s00704-009-0150-9, 2009.

8

Comparison of mercury concentrations measured at several sites in the Southern Hemisphere

F. Slemr[1], H. Angot[2], A. Dommergue[2,3], O. Magand[3], M. Barret[2,3], A. Weigelt[4], R. Ebinghaus[4], E.-G. Brunke[5], K. A. Pfaffhuber[6], G. Edwards[7], D. Howard[7], J. Powell[8], M. Keywood[8], and F. Wang[9]

[1]Max-Planck-Institute for Chemistry, Hahn-Meitner-Weg 1, 55128 Mainz, Germany
[2]Université Grenoble Alpes, LGGE, 38041 Grenoble, France
[3]CNRS, LGGE, 38041 Grenoble, France
[4]Helmholtz-Zentrum Geesthacht (HZG), Institute of Coastal Research, Max-Planck-Strasse 1, 21502 Geesthacht, Germany
[5]South African Weather Service c/o CSIR, P.O. Box 320, Stellenbosch 7599, South Africa
[6]Norwegian Institute for Air Research (NILU), P.O. Box 100, 2027 Kjeller, Norway
[7]Macquarie University, Environmental Science, Sydney, NSW, Australia
[8]CSIRO Ocean and Atmosphere Flagship Research, Aspendale, VIC, Australia
[9]Centre for Earth Observation Science, Department of Environment and Geography, University of Manitoba, Winnipeg, MB, R3T 2N2, Canada

Correspondence to: F. Slemr (franz.slemr@mpic.de)

Abstract. Our knowledge of the distribution of mercury concentrations in air of the Southern Hemisphere was until recently based mostly on intermittent measurements made during ship cruises. In the last few years continuous mercury monitoring has commenced at several sites in the Southern Hemisphere, providing new and more refined information. In this paper we compare mercury measurements at several remote sites in the Southern Hemisphere made over a period of at least 1 year at each location. Averages of monthly medians show similar although small seasonal variations at both Cape Point and Amsterdam Island. A pronounced seasonal variation at Troll research station in Antarctica is due to frequent mercury depletion events in the austral spring. Due to large scatter and large standard deviations of monthly average median mercury concentrations at Cape Grim, no systematic seasonal variation could be found there. Nevertheless, the annual average mercury concentrations at all sites during the 2007–2013 period varied only between 0.85 and 1.05 ng m^{-3}. Part of this variability is likely due to systematic measurement uncertainties which we propose can be further reduced by improved calibration procedures. We conclude that mercury is much more uniformly distributed throughout the Southern Hemisphere than the distributions suggested by measurements made onboard ships. This finding implies that smaller trends can be detected in shorter time periods. We also report a change in the trend sign at Cape Point from decreasing mercury concentrations in 1996–2004 to increasing concentrations since 2007.

1 Introduction

Our knowledge of the distribution of mercury in air over the Southern Hemisphere is mostly based on measurements made during ship cruises. According to the most comprehensive review of shipboard measurements made between 1990 and 2009 by Soerensen et al. (2012) mercury concentrations varied between 0.72 ng m^{-3} reported by Kuss et al. (2011) for the southern Atlantic Ocean and 2.20 ng m^{-3} observed by Xia et al. (2010) over the southeastern Indian Ocean. These data were collected in different areas during different seasons, typically over a period of 1 or 2 months. Only a few of these measurements were accompanied by measurements of tracers specific to anthropogenic pollution and the influence from the ship such as CO, nitrogen oxides, and particles. Consequently, the influence of sources such as from biomass burning, regional pollution, and pollution from the ship itself could not be properly filtered out from the data. Part of the reported variability may also be due to the use of frequently undeclared and non-uniform standard conditions under which these concentrations are reported. Mercury concentrations in $ng\,m^{-3}$ are usually reported at a standard pressure of 1013 hPa and a standard temperature of 273.14 K. However, some researchers and organisations use 293.14 K or 298.14 K. Since the same concentrations reported at 273.14 and 298.14 K differ by almost 10 %, the non-uniform standard conditions alone would prevent the detection of the statistically significant decrease in annual median mercury concentrations at Cape Point from $\sim 1.3\,ng\,m^{-3}$ in 1996 to below $1.2\,ng\,m^{-3}$ in 2004 (Slemr et al., 2008). Lastly, averages and standard deviations are quite frequently quoted without the number of measurements on which they are based. This means that the averages or medians cannot be weighed by the number of the measurements. It also makes statistical tests for the differences of averages impossible. It is not surprising that, using such data, Soerensen et al. (2012) concluded that no significant trend in the Southern Hemisphere could be detected so far. While we agree with this conclusion, a qualification is required: the quality of the data used by Soerensen et al. (2012) does not allow detection of trends smaller than their variability, i.e. some 50 % or even more. Consequently, with trends of up to ~ 2 % per year (Slemr et al., 2008; Ebinghaus et al., 2011), it would take several decades to detect trends from measurements onboard ships.

Recently, mercury has been measured continuously at several remote sites in the Southern Hemisphere over periods of a year or more. In this paper we will compare these measurements in terms of their monthly and annual statistics. We selected stations which are either baseline stations (Amsterdam Island, Troll research station in Antarctica) or where additional measurements (e.g. CO, ^{222}Rn, wind direction, aerosol) allow us to filter out baseline conditions (Cape Point and Cape Grim). The results show that atmospheric mercury is more uniformly distributed over the Southern Hemisphere than the measurements onboard ships suggest. Stationary sites with continuous and reproducible measurements of higher quality over longer periods allow for the detection of smaller trends in shorter time periods.

2 Experimental

Figure 1 shows the location of the sites whose data are used in this paper: Amsterdam Island, Cape Grim, Cape Point, Troll research station, and Galápagos Archipelago.

The Cape Point site (CPT, $34°21'$ S, $18°29'$ E) is operated as one of the Global Atmospheric Watch (GAW) baseline monitoring observatories of the World Meteorological Orga-

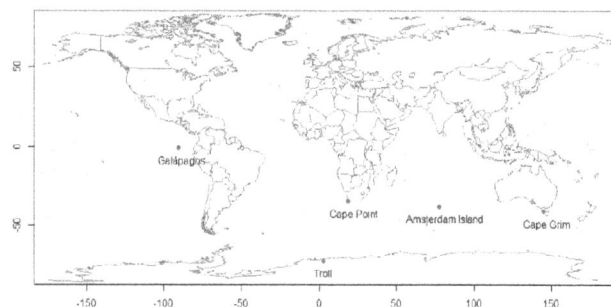

Figure 1. The location of the sites whose data are reported in this paper.

nization (WMO). The station is located on the southern tip of the Cape Peninsula within the Cape Point National Park on top of a peak 230 m a.s.l. and about 60 km south of Cape Town. The station has been in operation since the end of the 1970s and its current continuous measurement portfolio includes Hg, CO, O_3, CH_4, N_2O, ^{222}Rn, CO_2, several halocarbons, particles, and meteorological parameters. The station receives clean marine air masses for most of the time. Occasional events with continental and polluted air can easily be filtered out using a combination of the CO and ^{222}Rn measurements (Brunke et al., 2004). Gaseous elemental mercury (GEM) was measured by a manual amalgamation technique (Slemr et al., 2008) between September 1995 and December 2004 and has been measured by the automated Tekran 2537B instrument (Tekran Inc., Toronto, Canada) since March 2007. Only the Tekran data are reported here. These data were obtained in compliance with the standard operating procedures of the GMOS (Global Mercury Observation System, www.gmos.eu) project. The instrument has been run with a 15 min sampling frequency. For data analysis, 30 min averages were used. On average, 30 % of the data were classified as baseline using the ^{222}Rn $\leq 250\,mBq\,m^{-3}$ criterion.

Amsterdam Island (AMS, $37°48'$ S, $77°33'$ E) is a small isolated island ($55\,km^2$) located in the Indian Ocean 3400 km east of Madagascar. AMS is a GAW global station established in 1967. The climate of Amsterdam Island is mild oceanic, with frequent presence of clouds. Measurements are performed at Pointe Bénédicte station, which is located 2 km west of the Saint Martin de Viviès base on the edge of a cliff 55 m a.s.l. (GPS coordinates: $37°48'$ S, $77°33'$ E). GEM has been measured using a Tekran 2537B connected to a Tekran 1130/1135 speciation unit since January 2012 with a 5 min sampling frequency. For data analysis, 1 h averages were used. Details on operation and calibration procedures are given in Angot et al. (2014) and follow GMOS standard operating procedures. The station receives clean marine air masses almost all the time.

The Norwegian Antarctic Troll research station (TRS) is located in Queen Maud Land at $72°01'$ S and $2°32'$ E at an elevation of 1275 m and about 220 km from the Antarctic coast.

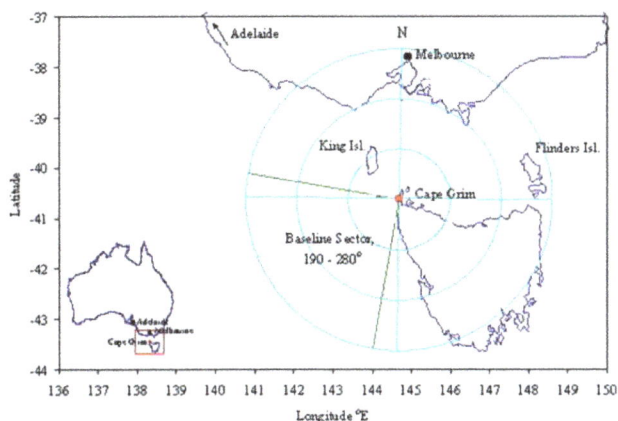

Figure 2. Location of the Cape Grim station and definition of the baseline sector.

The station has been in operation since January/February 2007 and its current continuous measurements include mercury, CO, O_3, particles, greenhouse gases, hydrocarbons, persistent organic compounds (POPs) and meteorological parameters (Hansen et al., 2009; Pfaffhuber et al., 2012). Mercury has been measured using the Tekran 2537B instrument since February 2007 with a 5 min sampling frequency. For data analysis, 1 h averages were used. The original mercury concentrations were reported at a standard temperature of 293.14 K and were converted to the standard temperature of 273.14 K to be comparable with all other data reported here.

The Cape Grim Baseline Air Pollution Station is located on the north-western coast of Tasmania, Australia (40°41′ S, 144°41′ E, Fig. 2). The Cape Grim Baseline Air Pollution Station was established in 1976 to monitor and study global atmospheric composition and is part of the WMO GAW programme. Measurements at Cape Grim include greenhouse gases such as CO_2, CH_4, N_2O, O_3, reactive nitrogen oxides, stratospheric ozone depleting chemicals such as chlorofluorocarbons (CFCs), radon, and GEM. The Tekran 2537A instrument was run with 5 min sampling time. For data analysis, 15 min averages were used. Additionally, meteorological parameters are measured, such as wind speed and direction, rainfall, temperature, humidity, air pressure, solar radiation, along with condensation nuclei (CN) concentration (particles greater than 10 nm), ultrafine condensation nuclei concentration (greater than 3 nm), aerosol absorption, aerosol scattering, cloud condensation nuclei concentration and rainfall chemical composition. Baseline conditions are defined as those with wind directions at 50 m altitude lying between 190 and 280°. In addition, CN should be less than a threshold concentration determined from 5 years of CN data for the current month based on the 90th percentile of CN hourly medians for this period, interpolated using cubic splines to give daily values (Fig. 2). During 2011–2013, the station received baseline marine air for 33 % of the time.

All mercury measurements reported here were made by an automated dual channel, single amalgamation, cold vapour atomic fluorescence analyser (Tekran-Analyzer model 2537 A or B, Tekran Inc., Toronto, Canada). The instrument features two gold cartridges. While one is adsorbing mercury during a sampling period, the other is being thermally desorbed using argon as a carrier gas. Mercury is detected using cold vapour atomic fluorescence spectroscopy (CVAFS). The functions of the cartridges are then interchanged, allowing continuous sampling of the incoming air stream. The instrument can be combined with a speciation unit (Tekran 1130/1135) consisting of a denuder, aerosol filter and pyrolyser that enables a determination of GEM, gaseous oxidised mercury (GOM), and particle-bound mercury (PM, < 2.5 μm) typically every 2–3 h (Landis et al., 2002). Operation and calibration of the instruments follow established and standardised procedures (e.g. Steffen and Schroeder, 1999). All mercury concentrations reported here are given in ng m^{-3} at 273.14 K and 1013 hPa.

In this paper we compare measurements at different sites in terms of monthly and annual average and median concentrations. Random uncertainties of individual measurements will average out and all we have to discuss are thus the systematic uncertainties, i.e. biases. The Tekran analyser is a complex instrument and the systematic uncertainties of its measurements depend on the operation procedure, the performance of the instrument, and the experience of its operators. All instruments used in this study are equipped with an internal mercury permeation source that is used to check and adjust periodically the instrument span and zero, typically every 25–72 h depending on the standard operating procedures that are used. This periodical internal calibration removes drifts both in span and zero that are caused mostly by the temperature and ageing of the fluorimeter lamp. The permeation rate of $\sim$ 1 pg Hg s^{-1} is, however, too low to allow a gravimetric determination of the permeation rate within a reasonable time period, as is usually done when certifying permeation devices for other gases (Barratt, 1981). Consequently, the permeation rate is calibrated every 6–12 months by repeated injection (at least 10 injections) of known volumes of gas saturated with mercury vapour at a known temperature. A skilled operator can achieve an individual injection precision of $\sim$ 3 %, resulting in an uncertainty of $\sim$ 1 % for 10 injections. The flow rate uncertainty of $\sim$ 1 % represents the second major contribution to the overall systematic uncertainty (Widmer et al., 1982). Adding smaller contributions from uncertainties associated with the injected volume and the temperature of the mercury vapour saturating device yields an overall systematic uncertainty of $\sim$ 3 %. We consider this to be the lower limit of the overall systematic uncertainty because this estimate assumes ideal performances of the instrument, its internal permeation device, the calibration mercury vapour saturating device, the injection syringes, as well as of the instrument operators.

A comprehensive analysis of all random and systematic uncertainties involved in a single manual determination of mercury concentration in air is given by Brown et al. (2008), who estimated the combined relative uncertainty to be 16.7 % at the concentration of 1.2 ng m^{-3}. This uncertainty includes the uncertainty from different published mercury vapour pressure curves and can be reduced to 12.6 % when one vapour pressure curve is accepted to be correct, as is the case here. This uncertainty analysis, however, is not directly applicable to measurements with the Tekran instrument because most items in the uncertainty budget are random rather than systematic. The combined systematic uncertainty (square root of the sum of uncertainties in quadrature) from uncertainties in flow calibration (2 %) and detector calibration (7 %) would be $\sim$7 %. Since one vapour pressure curve was used, the 5.5 % uncertainty in the saturated mercury concentration can be neglected. The overall systematic uncertainty would then be $\sim$3 % and is comparable to our estimate.

Contributions of deviations from an ideal performance, such as slow deactivation of the traps, difference between the concentrations from the two traps, contamination of the switching valves and traps, and leaks (Steffen et al., 2012), are difficult to quantify. Thus we take published results of Tekran instrument intercomparisons as a measure of practically achievable systematic uncertainty. In an intercomparison described by Ebinghaus et al. (1999) three Tekran instruments that were operated side by side at Mace Head were biased by 0.02–0.11 ng m^{-3} (median 0.01–0.13 ng m^{-3}) against each other. With an average concentration of 1.75 ng m^{-3}, this represents the highest systematic uncertainty of $\sim$6 %. Two Tekran instruments were run side by side for 4 days at a site in Tuscany in June 1998 (Munthe et al., 2001) with an average bias of 9 %. Mercury was measured by five Tekran instruments for 28 days within a 6-week period in May and June 2006 at German EMEP station Waldhof (Aas et al., 2006). The median concentrations were 2.02, 1.88, 1.77, 1.70, and 1.69 ng m^{-3}, and their average was 1.81 $\pm$ 0.14 ng m^{-3}. The average bias was thus $\sim$8 % and the bias between the instruments with the lowest and highest readings was $\sim$18 % (related to the average concentrations). In summary, based on experimental evidence, we can expect an average systematic uncertainty of $\sim$10 %, in extreme cases up to 20 %.

Despite using the same instrumentation, the measurements may target different mercury species at different sites, depending on their configuration and/or local conditions. At Amsterdam Island the instrument was operated with the Tekran 1130/1135 speciation unit. It showed GOM concentrations of less than 5 pg m^{-3} representing less than 1 % of the total gaseous mercury (TGM) concentrations of $\sim$1 ng m^{-3} (Angot et al., 2014). The data for Amsterdam Island presented here are stated explicitly as GEM. The instruments at Cape Point, Cape Grim, and Troll research station are operated without speciation units but with PTFE

(Teflon) filters to protect the instrument from sea salt and other particles. Although not proven, we assume that the surface active GOM in the humid air of the marine boundary layer at Cape Point and Cape Grim will be filtered out together with PM, partly by the salt particle loaded PTFE filter (denuders coated with KCl are used to adsorb GOM (Landis et al., 2002)) and partly on the walls of the inlet tubing. Consequently, we assume that measurements at Cape Point and Cape Grim represent GEM only and are thus directly comparable to those at Amsterdam Island. Although at Troll research station the same configuration with a PTFE filter is used, measurements by Temme et al. (2003) showed that at the low temperature and humidity prevailing at this site, GOM passed the inlet tubing and the PTFE filter. The measurements at Troll research station are thus assumed to represent TGM. As the GOM concentrations at Amsterdam Island in particular and in the marine boundary layer in general are below 10 pg m^{-3} (Soerensen et al., 2010; Angot et al., 2014), the difference between TGM and GEM at Amsterdam Island, Cape Grim and Cape Point is usually less than 1 %, which is insignificant when compared with the uncertainties discussed above. Consequently, GEM measurements at Cape Point, Cape Grim and Amsterdam Island are comparable to TGM measured at Troll research station. We caution, however, that recent studies have shown that the KCl-coated denuder in the Tekran speciation technique does not efficiently collect all GOM (Gustin et al., 2013; Huang et al., 2013; Ambrose et al., 2013). The bias between the TGM measurements at Troll research station and GEM measurements at all other stations can thus be larger.

The pair data difference tests were done using a t test (Kaiser and Gottschalk, 1972). A Mann–Kendal test for trend detection and the estimate of Sen's slope were made using the program by Salmi et al. (2002).

3 Results and discussion

3.1 Comparison of seasonal variations

Figure 3 shows seasonal variation of median mercury concentrations at Amsterdam Island, Cape Point, Cape Grim, and Troll research station in Antarctica during 2011–2013. Plotted are the averages of monthly median mercury concentrations and their standard deviations. We prefer here the use of monthly medians because they are less influenced by extreme values. The medians for Cape Point and Cape Grim were calculated both from unfiltered data and data filtered using the ^{222}Rn $\leq$ 250 mBq m^{-3} criterion for Cape Point and the baseline criteria mentioned above for Cape Grim. Pair tests for systematic differences between the monthly medians of filtered and unfiltered data (Kaiser and Gottschalk, 1972) did not show any significant difference (significance level <95 %) at both sites. Thus pollution events occasionally observed at Cape Point (Brunke et al. 2012; Slemr et

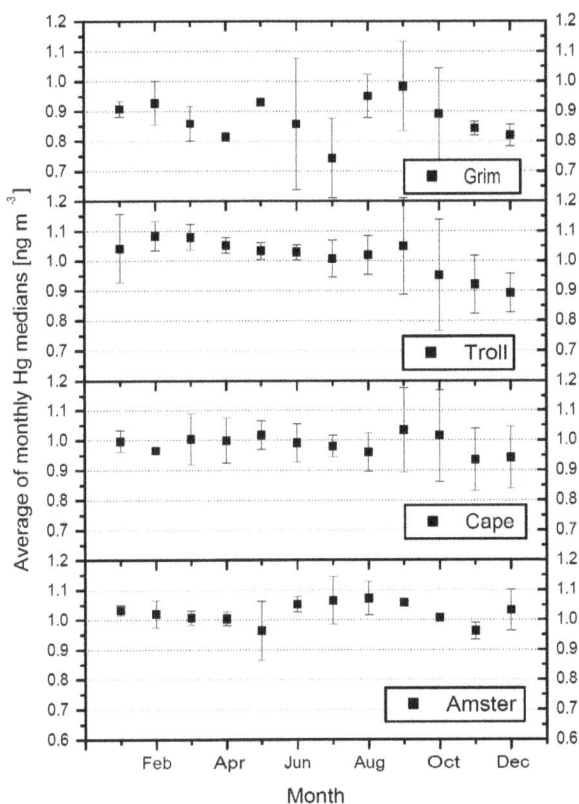

Figure 3. Seasonal variation of average monthly medians of mercury concentrations in 2011–2013 at Cape Point (no data in February 2011) and Troll research station (no data in September and October 2011). At Amsterdam Island the data cover only the 28 January 2012 to 31 December 2013 period and, at Cape Grim, data from January to August and November 2011, and April, May and October 2013 are missing. Bars denote the standard deviation of the monthly averages.

al., 2013) and at Cape Grim have no substantial influence on the monthly medians of mercury concentrations. This finding also has implications for the data from Amsterdam Island: if the influence of continental air masses is unimportant at Cape Point located on the coast of South Africa and at Cape Grim near the Australian continent, even less influence can be expected at Amsterdam Island, an isolated island in the middle of the Indian Ocean. Consequently, medians of unfiltered data from all sites were used when constructing this figure.

The smallest seasonal variation, within ~ 0.1 ng m^{-3}, is observed at Cape Point and Amsterdam Island, and the data which vary around 1 ng m^{-3} are very similar. In fact, a pair test for the differences in monthly medians (23 months) revealed no significant difference (significance level < 95 %) between the measurements at Amsterdam Island and Cape Point. Standard deviations of monthly medians averaged over 3 years (2011–2013) at Cape Point tend to be somewhat larger than those averaged over 2 years at Amsterdam Island, possibly due to inter-annual variations. Taking the standard

deviations into account, there is no seasonal variation discernible at both sites.

The seasonal variation at Troll research station is, at ~ 0.2 ng m^{-3}, substantially larger, whereas the monthly standard deviations are comparable to those at Cape Point. Minimum values are observed in October, November, and December, which are the months with frequent mercury depletion events in Antarctica (Temme et al., 2003; Pfaffhuber et al., 2012), and maximum values tend to occur in February and March and are, at ~ 1.1 ng m^{-3}, somewhat higher than at Cape Point and Amsterdam Island. In November and December the monthly average concentrations are, at ~ 0.9 ng m^{-3}, somewhat lower than at Cape Point and Amsterdam Island but comparable when averaged over the whole year (see Table 1). A pair test for differences in monthly medians at Cape Point, Amsterdam Island, and Troll research station revealed no statistically significant difference between them in the 2011–2013 period (33 months for Cape Point vs. Troll, 24 months for Amsterdam Island vs. Troll). There is a significant difference (> 99 %, 79 months) between medians at Cape Point and Troll research station over the period 2007–2013, which might be due to different trends at both sites.

Cape Grim data show the largest seasonal variation of ~ 0.25 ng m^{-3}, the largest monthly standard deviations, and the lowest annual average concentration of ~ 0.85 ng m^{-3} of all four sites, some 15 % below the annual mean concentrations at all other sites. Large standard deviations in September and October coincide with similar variability at Troll research station and Cape Point. Large and random scatter of the monthly values in other months suggests that the data from Cape Grim are not as homogeneous as those from other sites. Pair tests for differences in monthly medians detected a highly significant systematic difference between data from Cape Point and Amsterdam Island on the one hand and those from Cape Grim on the other (Cape Point vs. Cape Grim: > 99.9 %, 23 months; Amsterdam Island vs. Cape Grim: > 99.9 %, 21 months). Without additional QA/QC effort we cannot find out how many of these differences between the data from Cape Grim and from the other three sites are due to regional differences and/or due to the systematic uncertainties discussed in the experimental section.

3.2 Comparison of annual averages

The annual averages and medians for the Amsterdam Island, Cape Point, Cape Grim, and Troll research stations are given in Table 1. The table also contains an average of monthly medians for March, April, May, June, and October 2011 for Galápagos Archipelago (Wang et al., 2014). Located just south of the Equator, Galápagos Archipelago may be influenced by northern hemispheric air, especially in January, when the intertropical convergence zone (ITCZ) is at its southernmost position (Wang et al., 2014). The band of mixed northern and southern hemispheric air at ITCZ in the

Table 1. Comparison of annual average and median mercury concentrations at Amsterdam Island, Cape Point, Cape Grim, Troll research station, and Galápagos Archipelago. Hourly data were available for Amsterdam Island and Troll research station, half-hourly data for Cape Point, 5–15 min data for Cape Grim, and monthly averages for Galápagos Archipelago. All concentrations are given in ng m^{-3} at 273.14 K and 1013 hPa.

Site	2011		2012		2013	
	Average and standard deviation	Median, number of measurements	Average and standard deviation	Median, number of measurements	Average and standard deviation	Median, number of measurements
Cape Point	0.923 ± 0.106	0.934, 13 918	1.017 ± 0.095	1.018, 15 040	1.052 ± 0.160	1.040, 7809
Amsterdam Island	No data	No data	1.025 ± 0.065[a]	1.028, 6164[a]	1.028 ± 0.096	1.027, 7410
Cape Grim	0.959 ± 0.146[b]	0.976, 3692[b]	0.872 ± 0.130	0.854, 35 097	0.848 ± 0.112[c]	0.858, 36 310[c]
Troll	1.032 ± 0.192	1.061, 5876	1.052 ± 0.160	1.040, 7809	0.970 ± 0.162	1.000, 8196
Galápagos Archipelago	1.054 ± 0.087[d]	1.041, 5 months[d,e]	No data	No data	No data	No data

[a] Temporal coverage 28 January 2012–31 December 2012.
[b] Only September, October and December covered by measurements.
[c] No data in April, May and October.
[d] Only March, April, May, June, and October data were considered; February eliminated because of ITCZ proximity.
[e] Average of monthly medians.

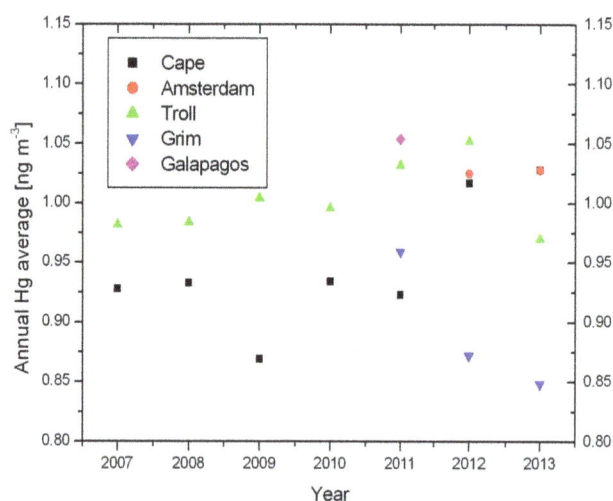

Figure 4. Annual average mercury concentrations at Cape Point, Amsterdam Island, Cape Grim, Troll research station and Galapagos Archipelago (Wang et al., 2014). Note that the 2013 annual averages at Cape Point and Amsterdam Island fall together.

marine boundary layer over the Atlantic Ocean tends to be quite narrow, usually less than 500 km broad (Slemr et al., 1985). If the same applies for the region around Galápagos Archipelago, then data from December, January and February could have been influenced by northern hemispheric air. Thus data for February 2011, although available, were not included.

Figure 4 shows an overview of the average mercury concentrations measured at different southern hemispheric sites during 2007–2013. It does not show the average mercury concentration of 1.32 ± 0.23 ng m^{-3} measured at a coastal site in Suriname for the season when the ITCZ is located north of the site and air originates from the South Atlantic (Müller et al., 2012). As the ITCZ moves seasonally over the

site in Suriname, the influence of northern hemispheric air is greater than at Galápagos Archipelago. Moreover, this site is also influenced by emissions from large-scale biomass burning in the Amazonas region (Ebinghaus et al., 2007; Müller et al, 2012). And last but not least, no annual statistics for southern hemispheric air can be made for Suriname because only seasonal concentrations are available. For these reasons, the measurements at Suriname are not included in further discussion.

Most of the annual medians and averages for individual sites in Table 1 differ by less than 0.02 ng m^{-3}, implying that the data are nearly normally distributed. Only at the Troll research station do the differences between annual medians and averages tend to be larger, while the medians tend to exceed the averages (in 6 of the 7 years). This is probably due to the extremely low values during the depletion events which occur during the Antarctic spring.

The annual averages and medians at Amsterdam Island and Cape Point differ by 0.01 and 0.01 ng m^{-3}, respectively, in 2012, and by 0.02 and 0.01 ng m^{-3}, respectively, in 2013. When compared over the overlapping period in 2012 (28 January–31 December), the averages and medians at both sites differed merely by 0.00 and 0.01 ng m^{-3}, respectively. The differences between Troll research station and the two other stations (Amsterdam Island and Cape Point) are substantially larger, by as much as 0.11 and 0.13 ng m^{-3} for 2011 averages and medians, respectively. In 2012 and 2013 the differences are below 0.1 ng m^{-3}. Annual averages over the period of 2007–2013 show that the difference between Cape Point and Troll research station never exceeded 0.14 ng m^{-3}, reached in 2009, and the average difference was 0.06 ng m^{-3}. The highest difference in medians was 0.20 ng m^{-3}, also in 2009, and the average difference was 0.08 ng m^{-3}.

Larger concentration differences are observed between Cape Grim and all other sites in 2011–2013. The annual averages and medians at Cape Grim were lower than at Amsterdam Island by 0.15 and $0.17 \, \text{ng m}^{-3}$, respectively, in 2012, and by 0.18 and $0.17 \, \text{ng m}^{-3}$, respectively, in 2013. The differences in annual averages and medians at Cape Grim and Cape Point were somewhat lower in 2012 and somewhat higher in 2013 than the corresponding differences between Cape Grim and Amsterdam Island. In 2011, data for Cape Grim and Cape Point overlap only for the period from September 6 to October 19. In this period, the average and median concentrations at Cape Grim were, at 1.03 ± 0.11 ($n = 2328$) and $1.04 \, \text{ng m}^{-3}$, respectively, substantially higher than 0.86 ± 0.07 ($n = 1474$) and $0.86 \, \text{ng m}^{-3}$, respectively, at Cape Point.

Figure 4 shows that the annual average mercury concentrations at all sites vary within $\sim 0.2 \, \text{ng m}^{-3}$ from $0.85 \, \text{ng m}^{-3}$ (Cape Grim in 2013) to $\sim 1.05 \, \text{ng m}^{-3}$ (Galápagos Archipelago in 2011 and Troll research station in 2012). It is not clear how much of this variability is real or due to systematic uncertainty issues discussed in the experimental chapter. We believe that both components contribute and that the real variability of the annual average or median mercury concentrations at southern hemispheric sites not influenced by local and regional pollution is lower. Assuming a systematic uncertainty of $\sim 10\%$ (see Experimental), the real variability at $1 \, \text{ng m}^{-3}$ in the Southern Hemisphere would be $\sim 0.1 \, \text{ng m}^{-3}$. This number can be viewed as a preliminary threshold for judging how representative the trends observed at any background site in the Southern Hemisphere are. With this threshold, much smaller trends at shorter time periods can be detected by long-term measurements at several sites when compared to shipboard measurements as reviewed by Soerensen et al. (2012).

3.3 Trend at Cape Point

Figure 4 shows an overall tendency of annual average mercury concentrations for Cape Point to increase with time. The Mann–Kendall test applied to annual averages and medians for 2007–2013 does not reveal a significant trend. However, when applied to monthly medians and averages, the trend is highly significant (at 99.99 % significance level for averages and at 99.96 % for medians). Senn's slope calculated from monthly averages is $0.018 \, \text{ng m}^{-3} \, \text{yr}^{-1}$ (0.008–$0.026 \, \text{ng m}^{-3} \, \text{yr}^{-1}$ at a significance level of 95 %) and, from monthly medians, $0.016 \, \text{ng m}^{-3} \, \text{yr}^{-1}$ (0.007–$0.025 \, \text{ng m}^{-3} \, \text{yr}^{-1}$). This is the first analysis suggesting that mercury concentrations are increasing, as would be expected based on increasing worldwide anthropogenic emissions (Streets et al., 2009; Muntean et al., 2014). A decreasing trend of $-0.015 \, \text{ng m}^{-3} \, \text{yr}^{-1}$ was derived from annual medians at Cape Point in the years 1996–2004 (Slemr et al., 2008), implying that the turning point was located between 2004 and 2007.

No trend could be detected in annual and monthly data from Troll research station over the same period: seven an-

nual averages and medians are not sufficient for trend detection as they were for Cape Point, and the trend in monthly averages and medians is probably masked by the strong seasonal variation. All other southern hemispheric data sets are too short for any trend detection.

Over 7 years of measurements at Cape Point the concentrations had increased by $0.12 \, \text{ng m}^{-3}$ when calculated from the trend of the monthly averages and $0.11 \, \text{ng m}^{-3}$ from the trend of the monthly medians. The changing trend from a decrease during the 1996–2004 period to an increase during 2007–2013 at Cape Point is not the only sign that the hemispheric trends in mercury concentrations are changing. An analysis of 1996–2013 data from Mace Head, classified according to the geographical origin of the air masses, showed a) that the downward trend of mercury concentration in air masses originating from over the Atlantic Ocean south of $28°$ N is substantially lower than for all other classes originating north of $28°$ N and b) that all downward trends for air masses originating from north of $28°$ N are decelerating (Weigelt et al., 2015). The apparent inconsistency that no decelerating trend for air masses from south of $28°$ N was found can be explained by the fact that the changes of a smaller trend are likely to be more difficult to detect.

4 Conclusions

We compared mercury concentrations measured at Cape Point, Amsterdam Island, Cape Grim, and Troll research station in Antarctica. Amsterdam Island and Troll research station are background stations per se, and at Cape Point and Cape Grim the influence of local and regional pollution can be eliminated by using filters such as CO and ^{222}Rn or wind direction and aerosol concentrations. No systematic difference was found between the unfiltered and filtered monthly median mercury concentrations at Cape Point and Cape Grim. We find that in terms of annual averages and medians the gradients of background mercury concentrations within the Southern Hemisphere are small and do not exceed $0.2 \, \text{ng m}^{-3}$. Taking into account a systematic measurement uncertainty of $\sim 0.1 \, \text{ng m}^{-3}$, the real variability could be as low as $0.1 \, \text{ng m}^{-3}$. This is much lower than the variability of shipboard mercury measurements on which the discussions of secular trends of mercury concentrations have relied so far. Consequently, smaller trends at shorter time periods can be detected by increasingly available long-term measurements at background sites in the Southern Hemisphere. The preliminary threshold of $\sim 0.1 \, \text{ng m}^{-3}$ for trend detection will further decrease when the comparability of the data sets improves.

The discussion of the measurement uncertainties shows a large difference between a small theoretical uncertainty and the much larger uncertainty achieved experimentally during several intercomparisons. Sampling flow rate can be precisely calibrated, and thus we believe that most of the "surplus" uncertainty comes from the behaviour and calibration of the Tekran internal permeation source. The issues related to the injection of known amounts of mercury are relatively well known (for example, not all syringes and replacement needles are suitable) and the uncertainty caused by them can be reduced by meticulous work. To the best of our knowledge we could not find any information about the dynamical behaviour of the internal permeation source that would enable one to calculate how much time is needed to stabilise the permeation rate (Barratt, 1981). Working practice, however, suggests that the time needed to stabilise the permeation rate increases with the decreasing permeation rate. We surmise that the very small permeation rate of the device in the Tekran instrument needs days rather than hours to stabilise within a 1 % margin required for precision measurements (Barratt, 1981). We thus conclude that the limited time of the cruises and the field conditions onboard ships are at least partly responsible for the large spread of the data from shipborne measurements.

We also report here an increasing trend for mercury concentrations at Cape Point for the period 2007–2013. No significant trend could be detected in mercury concentrations measured at Troll research station in Antarctica over the same period, but this is at least partly due to pronounced seasonal variations at Troll. As mercury concentrations at Cape Point decreased over the period 1996–2004, we conclude that the trend must thus have changed in direction between 2004 and 2007. Such change is qualitatively consistent with the trend changes observed at Mace Head in the Northern Hemisphere (Weigelt et al., 2014).

Acknowledgements. This work contributes to European Community FP7 project Global Mercury Observation System (GMOS). For Amsterdam Island, logistical support and financial support were provided by French Polar Institute IPEV (program 1028, GMOStral). Financial support was also provided by a grant from Labex OSUG@2020 (ANR10 LABX56) and LEFE CNRS/INSU (program SAMOA). We deeply thank the overwintering staff: B. Bouillard, J. Chastain, E. Coz, A. Croguennoc, M. Le Dréau, and V. Lucaire. Aurélien Dommergue acknowledges the Institut Universitaire de France. The Australian Bureau of Meteorology in Australia and CSIRO are also thanked for their continuous support of Cape Grim station. We also sincerely thank the staff at Cape Grim, S. Cleland, J. Ward, N. Sommerville and S. Baley. We acknowledge the support of the Cape Grim Science program student scholarship program. For Troll, financial support to sustain measurements is given through the Norwegian Antarctic Research Expeditions (NARE) programme administered by the Norwegian Polar Institute (NPI). We are very grateful for the technical support offered by the overwintering NPI staff at Troll. The NILU field team, but especially Jan H. Wasseng, is thanked for annual maintenance of the equipment at the Troll station.

Edited by: A. Dastoor

References

Aas, W. (Ed.): Data Quality 2004, Quality Assurance, and Field Comparisons, EMEP/CCC-Report 4/2006, NILU, Kjeller, Norway 2006.

Ambrose, J. L., Lyman, S. N., Huang, J., Gustin, M. S., and Jaffe, D. A.: Fast time resolution oxidized mercury measurements during the Reno Atmospheric Mercury Intercomparison Experiment (RAMIX), Environ. Sci. Technol. 47, 7285–7294, 2013.

Angot, H., Barret, M., Magand, O., Ramonet, M., and Dommergue, A.: A 2-year record of atmospheric mercury species at a background Southern Hemisphere station on Amsterdam Island, Atmos. Chem. Phys., 14, 11461–11473, doi:10.5194/acp-14-11461-2014, 2014.

Barratt, R. S.: The preparation of standard gas mixtures, Analyst, 106, 817–849, 1981.

Brown, R. J. C., Brown, A. S., Yardley, R. E., Corns, W. T., and Stockwell, P. B.: A practical uncertainty budget for ambient mercury vapour measurement, Atmos. Environ. 42, 2504–2517, 2008.

Brunke, E.-G., Labuschagne, C., Parker, B., Scheel, H. E., and Whittlestone, S.: Baseline air mass selection at Cape Point, South Africa: Application of ^{222}Rn and other filter criteria to CO_2, Atmos. Environ., 38, 5693–5702, 2004.

Brunke, E.-G., Ebinghaus, R., Kock, H. H., Labuschagne, C., and Slemr, F.: Emissions of mercury in southern Africa derived from long-term observations at Cape Point, South Africa, Atmos. Chem. Phys., 12, 7465–7474, doi:10.5194/acp-12-7465-2012, 2012.

Ebinghaus, R., Jennings, S. G., Schroeder, W. H., Berg, T., Donaghy, T., Guentzel, J., Kenny, C., Kock, H. H., Kvietkus, K., Landing, W., Mühleck, T., Munthe, J., Prestbo, E. M., Schneeberger, D., Slemr, F., Sommar, J., Urba, A., Wallschläger, D., and Xiao, Z.: International field intercomparison measurements of atmospheric mercury species, Atmos. Environ. 33, 3063–3073, 1999.

Ebinghaus, R., Jennings, S. G., Kock, H. H., Derwent, R. G., Manning, A. J., and Spain, T. G.: Decreasing trend in total gaseous mercury observations in baseline air at Mace Head, Ireland, from 1996 to 2009, Atmos. Environ., 45, 3475–3480, 2011.

Ebinghaus, R., Slemr, F., Brenninkmeijer, C. A. M., van Velthoven, P., Zahn, A., Hermann, M., O' Sullivan, D. A., and Oram, D. E.: Emission of gaseous mercury from biomass burning in South America in 2005 observed during CARIBIC flights, Geophys. Res. Lett. 34, L08813, doi:10.1029/2006GL028866, 2007.

Gustin, M. S., Huang, J., Miller, M. B., Peterson, C., Jaffe, D. A., Ambrose, J., Finley, B. D., Lyman, S. N., Call, K., Talbot, R., Feddersen, D., Mao, H., and Lindberg, S. E.: Do we understand what the mercury speciation instruments are actually measuring? Results of RAMIX, Environ. Sci. Technol. 47, 7295–7306, 2013.

Hansen, G., Aspmo, K., Berg, T., Edvardsen, K., Fiebig, M., Kallenborn, R., Krognes, T., Lunder, C., Stebel, K., Schmidbauer, N.,

Solberg, S., Espen Yttri, K.: Atmospheric monitoring at the Norwegian Antarctic station Troll: measurement programme and first results, Polar. Res., 28, 353–363, 2009.

Huang, J., Miller, M. B., Weiss-Penzias, P., and Gustin, M. S.: Comparison of gaseous oxidized Hg measured by KCl-coated denuders, and nylon and cation exchange membranes, Environ. Sci. Technol., 47, 7307–7316, 2013.

Kaiser, R. and Gottschalk, G.: Elementare Tests zur Beurteilung von Meßdaten, Bibliographisches Institut, Mannheim, 1972.

Kuss, J., Zülicke, C., Pohl, C., and Schneider, B.: Atlantic mercury emission determined from continuous analysis of the elemental mercury sea-air concentration difference within transects between 50° N and 50° S, Global Biogeochem. Cy., 25, GB3021, doi:10.1029/2010GB003998, 2011.

Landis, M. S., Stevens, R. K., Schaedlich, F., and Prestbo, E. M.: Development and characterization of an annular denuder methodology for the measurement of divalent inorganic reactive mercury in ambient air, Environ. Sci. Technol. 36, 3000–3009, 2002.

Müller, D., Wip, D., Warneke, T., Holmes, C. D., Dastoor, A., and Notholt, J.: Sources of atmospheric mercury in the tropics: continuous observations at a coastal site in Suriname, Atmos. Chem. Phys., 12, 7391–7397, doi:10.5194/acp-12-7391-2012, 2012.

Muntean, M., Janssens-Maenhout, G., Song, S., Selin, N. E., Olivier, J. G. J., Guizzardi, D., Maas, R., and Dentener, F.: Trend analysis from 1970 to 2008 and model evaluation of EDGARv4 global gridded anthropogenic mercury emissions, Sci. Tot. Environ., 494–495, 337–350, 2014.

Munthe, J., Wängberg, I., Pirrone, N., Iverfeldt, A., Ferrara, R., Ebinghaus, R., Feng, X., Gardfeldt, K., Keeler, G., Lanzillotta, E., Lindberg, S. E., Lu, J., Mamane, Y., Prestbo, E., Schmolke, S., Schroeder, W. H., Sommar, J., Sprovieri, F., Stevens, R. K., Stratton, W., Tuncel, G., and Urba, A.: Intercomparison of methods for sampling and analysis of atmospheric mercury species, Atmos. Environ. 35, 3007–3017, 2001.

Pfaffhuber, K. A., Berg, T., Hirdman, D., and Stohl, A.: Atmospheric mercury observations from Antarctica: seasonal variation and source and sink region calculations, Atmos. Chem. Phys., 12, 3241–3251, doi:10.5194/acp-12-3241-2012, 2012.

Salmi, T., Määttä, A., Anttila, P., Ruoho-Airola, T., and Amnell, T.: Detecting trends of annual values of atmospheric pollutants by the Mann-Kendall test and Sen's slope estimates – the Excel template application Makesens, Finnish Meteorological Institute, Helsinki, Finland, 2002.

Slemr, F., Schuster, G., and Seiler, W.: Distribution, speciation, and budget of atmospheric mercury, J. Atmos. Chem. 3, 407–434, 1985.

Slemr, F., Brunke, E.-G., Labuschagne, C., and Ebinghaus, R.: Total gaseous mercury concentrations at the Cape Point GAW station and their seasonality, Geophys. Res. Lett. 35, L11807, doi:10.1029/2008GL033741, 2008.

Slemr, F., Brunke, E.-G., Ebinghaus, R., and Kuss, J.: Worldwide trend of atmospheric mercury since 1995, Atmos. Chem. Phys., 11, 4779–4787, doi:10.5194/acp-11-4779-2011, 2011.

Slemr, F., Brunke, E.-G., Whittlestone, S., Zahorowski, W., Ebinghaus, R., Kock, H. H., and Labuschagne, C.: ^{222}Rn-calibrated mercury fluxes from terrestrial surface of southern Africa, Atmos. Chem. Phys., 13, 6421–6428, doi:10.5194/acp-13-6421-2013, 2013.

Soerensen, A. L., Skov, H., Jacob, D. J., Soerensen, B. T., and Johnson, M. S.: Global concentrations of gaseous elemental mercury and reactive gaseous mercury in the marine boundary layer, Environ. Sci. Technol. 44, 7425–7430, 2010.

Soerensen, A. L., Jacob, D. J., Streets, D. G., Witt, M. L. I., Ebinghaus, R., Mason, R. P., Andersson, M., and Sunderland, E. M.: Multi-decadal decline of mercury in the North-Atlantic atmosphere explained by changing subsurface seawater concentrations, Geophys. Res. Lett. 39, L21810, doi:10.1029/2012GL053736, 2012.

Steffen, A. and Schroeder, W.: Standard operation procedures manual for total gaseous mercury measurements, Canadian Mercury Measurement Network (CAMNet), Version 4.0, March 1999.

Steffen, A., Scherz, T., Olson, M., Gay, D., and Blanchard, P.: A comparison of data quality control protocols for atmospheric mercury speciation measurements, J. Environ. Monitor., 14, 752–765, 2012.

Streets, D. G., Zhang, Q., and Wu, Y.: Projections of global mercury emissions in 2050, Environ. Sci. Technol. 43, 2983–2988, 2009.

Temme, C., Einax, J. W., Ebinghaus, R., and Schroeder, W. H.: Measurements of atmospheric mercury species at a coastal site in the Antarctic and over the South Atlantic Ocean during polar summer, Environ. Sci. Technol. 37, 22–31, 2003.

Wang, F., Saiz-Lopez, A., Mahajan, A. S., Gómez Martín, J. C., Armstrong, D., Lemes, M., Hay, T., and Prados-Roman, C.: Enhanced production of oxidised mercury over the tropical Pacific Ocean: a key missing oxidation pathway, Atmos. Chem. Phys., 14, 1323–1335, doi:10.5194/acp-14-1323-2014, 2014.

Weigelt, A., Ebinghaus, R., Manning, A. J., Derwent, R. G., Simmonds, P. G., Spain, T. G., Jennings, S. G., and Slemr, F.: Analysis and interpretation of 18 years of mercury observations since 1996 at Mace Head at the Atlantic Ocean coast of Ireland, Atmos. Environ. 100, 85–93, 2015.

Widmer, A. E., Fehlmann, R., and Rehwald, W.: A calibration system for calorimetric mass flow devices, J. Phys. E: Sci. Instrum., 15, 213–220, 1982.

Xia, C., Xie, Z., and Sun, L.: Atmospheric mercury in the marine boundary layer along a cruise path from Shanghai, China, to Prydz Bay, Antarctica, Atmos. Environ., 44, 1815–1821, 2010.

9

Energetic particle induced intra-seasonal variability of ozone inside the Antarctic polar vortex observed in satellite data

T. Fytterer[1], M. G. Mlynczak[2], H. Nieder[1], K. Pérot[3], M. Sinnhuber[1], G. Stiller[1], and J. Urban[3,†]

[1]Institute for Meteorology and Climate Research, Karlsruhe Institute of Technology, Eggenstein-Leopoldshafen, Germany
[2]Atmospheric Sciences Division, NASA Langley Research Center, Hampton, VA, USA
[3]Department of Earth and Space Sciences, Chalmers University of Technology, Göteborg, Sweden
[†]deceased, 14 August 2014

Correspondence to: T. Fytterer (tilo.fytterer@kit.edu)

Abstract. Measurements from 2002 to 2011 by three independent satellite instruments, namely MIPAS, SABER, and SMR on board the ENVISAT, TIMED, and Odin satellites are used to investigate the intra-seasonal variability of stratospheric and mesospheric O_3 volume mixing ratio (vmr) inside the Antarctic polar vortex due to solar and geomagnetic activity. In this study, we individually analysed the relative O_3 vmr variations between maximum and minimum conditions of a number of solar and geomagnetic indices (F10.7 cm solar radio flux, Ap index, ≥ 2 MeV electron flux). The indices are 26-day averages centred at 1 April, 1 May, and 1 June while O_3 is based on 26-day running means from 1 April to 1 November at altitudes from 20 to 70 km. During solar quiet time from 2005 to 2010, the composite of all three instruments reveals an apparent negative O_3 signal associated to the geomagnetic activity (Ap index) around 1 April, on average reaching amplitudes between -5 and -10% of the respective O_3 background. The O_3 response exceeds the significance level of 95 % and propagates downwards throughout the polar winter from the stratopause down to ~ 25 km. These observed results are in good qualitative agreement with the O_3 vmr pattern simulated with a three-dimensional chemistry-transport model, which includes particle impact ionisation.

1 Introduction

Energetic particles (keV–MeV), mainly originating from the sun but also from the Earth's magnetospheric radiation belts and the aurora region, penetrate the atmosphere down to mesospheric and stratospheric regions, depending on their energy. The particles are guided by the Earth's magnetic field lines and therefore mostly precipitate at auroral and radiation belt areas (~ 55–$70°$ geomagnetic latitudes), depositing energy and directly influencing the chemical composition of the stratosphere and mesosphere. Due to the air compounds, precipitating particles mainly produce large abundances of O_2^+ as well as $N(^2D)$ and N_2^+. $N(^2D)$ and N_2^+ lead to increased concentrations of odd nitrogen ($NO_x = N + NO + NO_2$) through a number of reactions, including dissociative recombination of N_2^+ and ion-neutral chemistry with species of the oxygen family (e.g. Rusch et al., 1981). Additionally, O_2^+ and water vapour initialise chain reactions associated with water cluster ion formation and accompanied recombination reactions, which eventually lead to the production of odd hydrogen ($HO_x = H + OH + HO_2$; e.g. Solomon et al., 1981).

Both HO_x and NO_x play an important role in destroying O_3 in the mesosphere and stratosphere (e.g. Lary, 1997). However, HO_x is short-lived (seconds–hours) and therefore more important near its source region in the mesosphere, while NO_x has a relatively long lifetime (days–months), at least during night-time conditions. Consequently, NO_x can be transported downwards inside the polar vortex (e.g. Solomon et al., 1982) from the upper mesosphere/lower thermosphere down to the stratosphere, resulting in stratospheric

O_3 depletion through catalytic chemical reactions in combination with solar radiation. Thus, energetic particle precipitation (EPP) indirectly affects O_3 during polar winter. Since O_3 is the major radiative heating source in the stratosphere, variations of this gas will also influence the stratospheric temperature field and eventually lead to altered atmospheric dynamics. However, the atmospheric response to EPP is not fully understood so far. The current knowledge is discussed in more detail by Sinnhuber et al. (2012).

Observations of the EPP indirect effect on stratospheric polar O_3 are relatively rare, at least compared to other latitudes, due to a lack of long-term O_3 measurements in these regions. However, a hint for this mechanism was presented by Randall et al. (1998) which analysed the Polar Ozone and Aerosol Measurement instrument data, revealing a close anticorrelation between NO_2 and O_3 mixing ratios in winter/spring from 1994 to 1996 in the Antarctic stratosphere (~ 25–35 km). They suggested that the relationship cannot originate from downwards transported O_3-deficient air but is due to photochemical destruction of O_3 by NO_2. Further observations from several satellite instruments from 1992 to 2005 show that the stratospheric NO_x enhancement in the Southern Hemisphere is caused by EPP (Randall et al., 2007). More recent satellite observations from 2002 to 2012 reported by Funke et al. (2014) reveal that particle induced NO_x is indeed transported downwards to the middle stratosphere at polar latitudes, while further model studies suggest that the subsiding of NO_x leads to strongly reduced stratospheric O_3 concentrations (~ 30 %) down to altitudes ~ 30 km (e.g. Reddmann et al., 2010). Thus, it appears promising to search for a link between EPP and O_3 in actual data sets, because the downwards propagating signal of the EPP indirect effect on stratospheric and mesospheric O_3 throughout the polar winter has not been explicitly observed so far. Note that NO_x can be only transported downwards inside a stable large-scale dynamical structure, which provides sufficient subsidence and prevents NO_x removal/dilution by horizontal transport. These conditions are found primarily inside the Antarctic polar vortex, because the Arctic vortex is strongly disrupted by planetary waves, leading to its weakening or temporary breakdown. This large dynamical variability eventually causes high variations in O_3 volume mixing ratios (vmr), superposing the EPP indirect effect.

Therefore our study is focused on O_3 vmr observations inside the Antarctic polar vortex from ~ 20–70 km, derived from environmental satellite/Michelson Interferometer for passive atmospheric sounding (ENVISAT/MIPAS), thermosphere ionosphere mesosphere energetics and dynamics/sounding of the atmosphere using broadband emission radiometry (TIMED/SABER), and Odin/sub-millimetre radiometer (SMR) measurements. The intra-seasonal variability of the O_3 vmr values has been investigated and the relation to a number of solar and geomagnetic indices, namely the F10.7 cm solar radio flux, the Ap index, and the ≥ 2 MeV electron flux is analysed.

2 Data analysis and numerical modelling

2.1 Approximation of the Antarctic polar vortex

The position and the extension of the Antarctic polar vortex were estimated by using the gradient of the potential vorticity (PV) on isentropic surfaces (Nash et al., 1996). Assuming a dry atmosphere at altitudes ≥ 20 km, the PV was calculated from temperature, pressure, relative vorticity, and the corresponding latitude taken from ERA-Interim (https://ecaccess.ecmwf.int/ecmwf), the latest version of global atmospheric reanalysis data produced by the European Centre for Medium-Range Weather Forecasts (ECMWF). The reference pressure was set to 1000 hPa and the gravitational constant was considered to be dependent on latitude and height. The PV was calculated for all height intervals between 20 and 70 km which were adapted from the MIPAS retrieval grid (see Sect. 2.2.1). Note that ERA-Interim data are primarily model-driven at mesospheric altitudes but the individual PV results look reasonable at each height interval. As an example, Fig. 1 shows the PV, depending on time and equivalent latitude (EQL), during the Antarctic winter 2011 at ~ 40 km. The EQLs assigned to an individual PV isoline enclose the same area as the geographical latitudes of equivalent values. However, this area is located around an estimate of the vortex centre position, rather than around the geographical pole. In general, the EQL of the strongest PV gradient indicates the estimated location of the vortex edge; however, in most cases, there are at least two locations revealing gradients of similar magnitude. Therefore, Nash et al. (1996) also considered the zonal wind to locate the real vortex edge, but here we added a visual analysis instead of the zonal wind to divide the Southern Hemisphere into three non-overlapping zones: deep inside the Antarctic polar vortex (CORE) and the corresponding outermost edge (EDGE), covering all EQLs poleward of the respective borders, as well as an area not influenced by the vortex (OUTSIDE), which extends from the equator to the respective OUTSIDE border. In this study we will consider the EDGE region as the Antarctic vortex area, but the CORE and OUTSIDE region are still necessary to determine whether the observed features inside the EDGE zone are actually originating from the vortex itself. The limits of the three regions of each height interval revealed no strong variation from 2002 to 2011, therefore holding for every winter (Table 1). Note that the ECMWF ERA-Interim data only cover heights up to ~ 63 km. However, considering the behaviour of the Antarctic vortex at altitudes between 60 and 70 km (Preusse et al., 2009, their Fig. 2a), it seems reasonable to assume that the estimated limits of the three regions at ~ 63 km are also valid up to 70 km.

Table 1. Limits, derived from the potential vorticity ($10^{-6}\,\mathrm{K\,m^2\,s^{-1}\,kg^{-1}}$), of the southern hemispheric regions CORE, EDGE, and OUTSIDE at the individual heights. The altitudes are adapted from the MIPAS retrieval grid and the shown potential vorticity values hold for 2002–2011.

Nominal height (km)	CORE	EDGE	OUT-SIDE	Nominal height (km)	CORE	EDGE	OUT-SIDE
20	−60	−50	−30	37	−3600	−1900	−1000
21	−70	−50	−30	38	−6000	−2500	−1500
22	−90	−70	−40	39	−6000	−3000	−2000
23	−150	−90	−40	40	−9000	−3500	−2000
24	−200	−100	−60	41	−9000	−4000	−2000
25	−180	−110	−60	42	−9000	−4000	−2000
26	−280	−160	−100	43	−15000	−5000	−3000
27	−360	−220	−120	44	−15000	−5500	−3000
28	−600	−250	−120	46	−22000	−8000	−4000
29	−900	−300	−150	48	−18000	−10000	−2000
30	−800	−400	−200	50	−36000	−12000	−4000
31	−1000	−400	−200	52	−32000	−16000	−4000
32	−1600	−600	−300	54	−36000	−16000	−4000
33	−2000	−800	−400	56	−60000	−30000	−5000
34	−1800	−1100	−400	58	−60000	−30000	−10000
35	−2800	−1400	−800	60	−60000	−30000	−10000
36	−4000	−1600	−1000	62–70	−18000	−90000	−30000

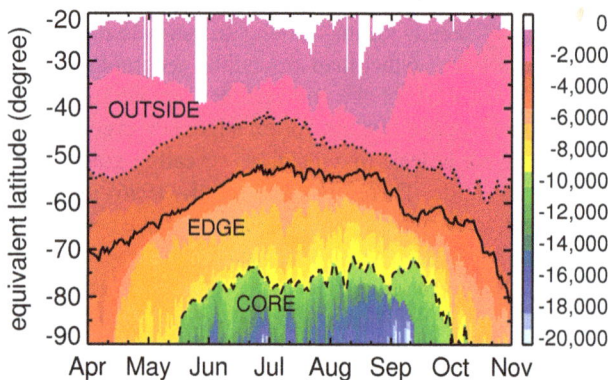

Figure 1. Potential vorticity ($10^{-6}\,\mathrm{K\,m^2\,s^{-1}\,kg^{-1}}$, colour scale) at $\sim 40\,\mathrm{km}$ during the Antarctic winter 2011 as a function of time and equivalent latitude. The thresholds of the regions OUTSIDE (dotted line), EDGE (solid line) and CORE (dashed line) are included. Potential vorticity was calculated from ECMWF Era-Interim data.

2.2 Ozone measurements

2.2.1 MIPAS

MIPAS (Fischer et al., 2008) was a limb sounder on board ENVISAT, which had a sun-synchronous orbit. The main advantages of MIPAS measurements are the global coverage from 87° S to 89° N and the availability of observations during both day and night, crossing the equator at $\sim 10{:}00$ LT and $\sim 22{:}00$ LT, respectively. MIPAS was a Fourier transform infrared (4.15–$14.6\,\mu m$) emission spectrometer, allowing simultaneous observations of several atmospheric trace gases, including O_3. MIPAS was operational from July 2002 to April 2012, but due to an instrument failure in March 2004, the entire observation period is divided into two subintervals from July 2002 to March 2004 and January 2005–April 2012 (referred to as P1 and P2 here, respectively). During P1 an almost continuous time series is available, while larger data gaps are present during P2 before October 2006. Here, we use the complete data set of the most frequent observation mode (nominal mode), covering the altitudes from the upper troposphere up to $\sim 70\,\mathrm{km}$ at the poles which was derived from the MIPAS level-2 research processor developed by IMK/IAA. Details of the retrievals are described in von Clarmann et al. (2003), Glatthor et al. (2006), and von Clarmann et al. (2009). Note that the number of tangent heights is constant during P1 (17) and P2 (23), and that the actually available altitudes (cloud contaminated observations are disregarded) only slightly differ from day to day. The corresponding vertical resolution becomes coarser at higher altitudes (independent of the geographical location), increasing from 3.5 to 8 km (Steck et al., 2007) and from 2.5 to 5 km (Eckert et al., 2014) in P1 and P2, respectively. However, the retrieval grid in all MIPAS O_3 data versions used here (V3O_O3_9, V5R_O3_220, V5R_O3_221) is independent of the tangent heights, with a grid width of 1 km below 44 and 2 km above. During P1/P2 O_3 was measured at two different wavelength intervals, ranging from 9.0–9.4/9.6–9.7 μm and 12.5–13.5/12.7–13.2 μm in particular. However, the full spectral ranges were not used, but sub-intervals (microwindows). These were selected to minimise the computing time and to optimise the relation between the measurement-noise induced random error and other errors. These other errors originate, among further error sources, from spectral contributions of further atmospheric constituents of unknown abundances. It should also be noted that there is a bias in MIPAS O_3 data between the two periods, which was estimated using a multi-linear parametric trend model (Eckert et al., 2014). To accept an O_3 data point, the recommended filter criteria for MIPAS O_3 data were applied by using an averaging kernel diagonal value > 0.03 as well as the visibility flag equal to 1 which indicates spectral available data.

At least 10 accepted data points inside the Antarctic polar vortex at a certain grid level were required to calculate the arithmetic average of one day, while at least 13 days were arithmetically averaged to a 26-day running mean from 1 April to 1 November, repeating this algorithm for each height interval and all years from 2002 to 2011. The time interval of 26 days was chosen to minimise a possible influence of the 27-day cycle of the sun, also ensuring that each time interval includes only one 27-day solar rotation maximum at most. The analysis was repeated for NO_2 (V5R_NO2_220, V5R_NO2_221) and the corresponding retrieval is described in Funke et al. (2005 and 2011).

2.2.2 SABER

The SABER instrument on board the TIMED Satellite has been nearly continuously operating since January 2002, measuring vertical profiles of several atmospheric parameters and minor constituents (e.g. O_3) from the surface up to altitudes > 100 km. The SABER measurements are governed by a periodic quasi 60-day cycle, each time changing from the Southern Hemisphere mode (83° S–52° N) to the Northern Hemisphere mode (52° S–83° N) and vice versa. Note that the "switching day" is only varying a few days from year to year. To consider both day and night O_3 observations, SABER Level 2A Ozone96 data v2.0 and v1.07 (http://saber.gats-inc.com/custom.php, Rong et al., 2009) measured at $\sim 9.6\,\mu m$ are used. However, v1.07 was only used to fill v2.0 data gaps, which seemed reasonable because the data fit quite well the results of the performed analysis during the respective periods (15–31 May, 7–31 August, not shown here). Consequently, the combined data set of both versions shows no larger data gap and the measurements of both versions were restricted to values < 20 ppm to exclude outliers. Comparisons with the results of an increased threshold to < 100 ppm revealed only minor differences (not shown here). The investigated height interval, ranging from 20 to 70 km, is divided in 38 non-overlapping subintervals and binned at the same altitudes as MIPAS data. The algorithm used to calculate the running means is also identical to the one applied for the "accepted" MIPAS data points. However, SABER needs approximately 60 days to cover all local times, leading to a quasi 60-day wave like oscillation in O_3 if 26-day running means are used. This behaviour becomes evident at altitudes > 50 km, where the averaging interval was consequently extended from 26 to 60 days. Note that the calculation of the 60-day running means required at least 30 days.

2.2.3 SMR

The Odin satellite mission started in February 2001 and is a joint project between Sweden, Canada, France and Finland (Murtagh et al., 2002). Odin was launched into a sun-synchronous polar orbit, carrying the SMR instrument and nominally covering the latitude range from 82.5° S to 82.5° N. The SMR makes vertical profile measurements during both day and night, while passing the equator at $\sim 06{:}00/18{:}00$ LT in the descending/ascending node. The O_3 data were extracted from the Odin/SMR Level 2 data product, version 2.0 (http://odin.rss.chalmers.se/, Urban et al., 2005), only using measurements of the frequency band centred around ~ 544.6 GHz, providing vertical O_3 profiles in the ~ 15–70 km altitude range. The filtering criterion used for SMR is the measurement response, which corresponds to the sum of the rows of the averaging kernel matrix. The profiles characterised by a measurement response lower than 0.9 are not reliable enough, and are therefore excluded. The algorithm to calculate the 26-day running means is identical to the one applied to MIPAS data. Note that Odin/SMR was a two-discipline satellite until April 2007, switching between atmospheric (aeronomy mode) and astronomy observations, and is entirely dedicated to aeronomy since this date. Consequently, measurements in the relevant mode are roughly performed one day out of three before April 2007 and every other day afterwards. However, the calculation of the 26-day running means is still possible because the data gaps occur in a regular way, so they do not essentially worsen the 26-day averages. The vertical resolution of the data version used here is better than 3 km below 45 km, but increases to 5–6 (50–60 km) and 7–10 km (60–70 km), leading to noisy results at altitudes > 50 km compared to the other two instruments.

2.3 Solar data and geomagnetic indices

The data of the indices were obtained from two different websites provided by the National Geophysical Data Center. In detail the flux of the 10.7 cm radio emission from the sun (F10.7) and the geomagnetic Ap index (Ap), commonly used proxies for solar variation and geomagnetic activity, respectively, were downloaded from http://spidr.ngdc.noaa.gov/spidr/. The ≥ 2 MeV electron flux (2 MeV), including the flux of all electrons with energy levels above 2 MeV, was measured by the Geostationary Operational Environmental Satellites (GOES) and the corresponding time series were downloaded from ftp://ftp.ngdc.noaa.gov/STP/SOLAR_DATA/SATELLITE_ENVIRONMENT/Daily_Fluences/. Note that the 2 MeV data set also considers contamination effects on the electron detectors on the spacecrafts due to protons > 32 MeV. Furthermore the 2 MeV data are obtained from geostationary satellites which perform in-situ measurements in the radiation belts and consequently do not directly provide observations of precipitating particles. However, it is very likely that there is at least a positive relation between 2 MeV and precipitating relativistic radiation belt particles. Thus, the 2 MeV is not used as a proxy of precipitating particles but as an indicator of the influence from the magnetosphere. Precipitating particle integral fluxes in polar regions are observed by sun-synchronous Polar orbiting Operational Environmental Satellite (POES) detectors and the corresponding data correlate better with geomagnetic indices than the GOES electron fluxes (Sinnhuber et al., 2011). However, the respective measurements of the POES instruments tend to underestimate the fluxes from ground-based observations during weak geomagnetic activity (Rodger et al., 2013). Since this study focus on 2002–2011 and an essential part of this time interval overlaps with low geomagnetic activity, GOES data and Ap are used instead of POES measurements. The time series of all data sets are based on daily values, which were arithmetically averaged to 26-day means centred at 1 April, 1 May, and 1 June. The means were separately calculated for each index for the individual years from 2002 to 2011; however, 2 MeV data are only available until 2010.

2.4 Numerical modelling

The three-dimensional chemistry and transport model (3dCTM; Sinnhuber et al., 2012, Appendix 1) used here is based on the Bremen 3dCTM (e.g. Wissing et al., 2010), extending on 47 pressure levels from the tropopause up to the lower thermosphere (~ 10–$140\,\mathrm{km}$) with a latitude/longitude resolution of $2.5° \times 3.75°$. The model was recently updated with a variable H_2 and O_2 distribution, leading to proper HO_x and consequently night time O_3 values at altitudes $> 60\,\mathrm{km}$ (see Sect. 1). The 3dCTM is driven by meteorological data obtained from simulations of the three-dimensional dynamical model LIMA (Berger, 2008) and the advection is calculated by applying the second-order moments scheme reported by Prather (1986). In the stratosphere, a family approach for the chemical families: O_x ($O + O(^1D) + O_3$), NO_x ($N + NO + NO_2$), HO_x ($H + OH + HO_2$), BrO_x ($Br + BrO$), ClO_x ($Cl + ClO + 2Cl_2O_2$), and CHO_x ($CH_3 + CH_3O_2 + CH_3OOH + CH_3O + HCO$) is used, but was not used for O_x, HO_x, and NO_x in the mesosphere/lower thermosphere region.

In this study the 3dCTM was used to investigate the impact of precipitating particles on O_3 inside the Antarctic polar vortex at altitudes from 20 to 70 km. After a multi-year two-dimensional model spin-up, two simulations from 2003 to 2009 were performed. The first run (base run) does not consider any energetic particles, while the second run (EP run) includes ionisation effects by both protons and electrons, using the ionisation rates provided by the Atmospheric Ionisation Module Osnabrück (AIMOS; Wissing and Kallenrode, 2009). The resulting NO_x production per created ion

pair includes various ionic and neutral reactions depending on the atmospheric background state (Nieder et al., 2014). Simple parameterisations are used for the production of HO_x (Solomon et al., 1981) and O (Porter et al., 1976). Note that heterogeneous chemistry was not included, which only becomes important during spring in the lower stratosphere. Both model runs considered constant solar minimum conditions (F10.7 = $70 \times 10^{-22}\,\mathrm{W\,m^{-2}\,Hz^{-1}}$) to exclude O_3 variations due to solar activity. The obtained O_3 model results of both runs were separately selected according to the vertical MIPAS retrieval grid for direct comparisons to the observations, repeating the described algorithm to calculate the 26-day running means. Finally, in order to derive the O_3 vmr variations solely originating from precipitating particles, the obtained averages of the base run were subtracted from the corresponding O_3 values of the EP run. The results were divided by the arithmetic mean of both runs and eventually multiplied by 100 %.

3 Results and discussion

3.1 Satellite observations

3.1.1 O_3 response from 2002 to 2011

The 26-day O_3 vmr averages from 2002 to 2011 of each altitude-time interval (1 April–1 November, 20–70 km) were individually grouped into years of high and low index activity. For this purpose the index median of the corresponding time series of the 26-day average of an index (F10.7, Ap, 2 MeV) centred around 1 April was calculated, only including years of actually available O_3 observations. Therefore the median of an index time series works as a threshold, dividing the entire time interval from 2002 to 2011 in years of high (above the median) and low (below the median) index activity. Note that the classification of the years does not only depend on the chosen index, but due to data gaps also on the considered height-time interval as well as the instrument used. Afterwards the arithmetic O_3 mean of the years of low index activity was subtracted from the O_3 mean of the years of high index activity, eventually dividing this absolute O_3 difference by the arithmetic O_3 average of the entire observation period and multiplying the results by 100 % for more handy values. Thus the calculated relative O_3 difference (referred to as O_3 amplitude here) represents the impact of the respective index on the O_3 background. To reduce the measurement noise of the individual instruments, the results of all three instruments were merged by simply calculating the arithmetic average but only if the corresponding O_3 amplitude of all three instruments was available. Note that due to the major sudden stratospheric warming centred around 27 September (Azeem et al., 2010) the O_3 observations from 1 September to 1 November 2002 were excluded. In contrast, the solar proton event in the end of October 2003 (Jackman

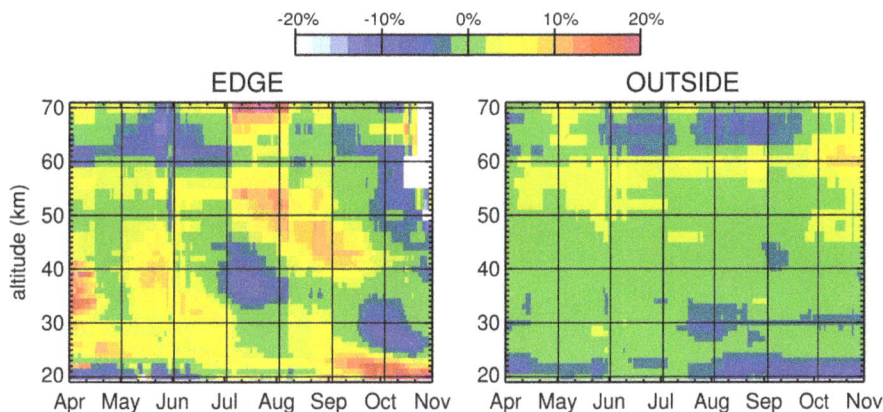

Figure 2. Example of the O_3 amplitude (see Sect. 3.1.1 for definition) observed by MIPAS from 2002 to 2011 between years of high Ap index and years of low Ap index centred around 1 April, for the regions EDGE (left) and OUTSIDE (right).

et al., 2005) was neglected due to its late occurrence. The performed analyses with O_3 observations, considering the indices from 1 May and 1 June (not shown here), revealed no essential differences compared to 1 April or the structures became less obvious. Comparisons with earlier periods are not reasonable because the vortex first builds up in April. Therefore the focus is set on the O_3 response to indices centred around 1 April. The O_3 amplitude was calculated for all three regions (CORE, EDGE, and OUTSIDE) which were introduced in Sect. 2.1. The corresponding results reveal that the pattern found inside the EDGE region are fairly similar and less noisy compared to the features observed in the CORE area (not shown here). In contrast the O_3 amplitudes outside the Antarctic polar vortex are fundamentally different. An example for the O_3 response associated to 1 April Ap in the EDGE and the OUTSIDE region derived from MIPAS measurements is presented in Fig. 2, showing considerably disagreeing structures and essentially weaker amplitudes, especially below 50 km. Thus comparison between the individual regions of the Southern Hemisphere ensures, that, the pattern found in the EDGE region are actually originating from the Antarctic polar vortex.

Figure 3 displays the corresponding results of the O_3 amplitude from 2002 to 2011, but only for values above the significance level of 95 % while shaded areas show regions between the significance level of 95 and 99 %. The significance was calculated according to a Student's t test, based on the error of the mean of the 26-day running O_3 means and assuming the worst case scenario of absolute error propagation. The MIPAS O_3 measurements (left column) reveal a high negative response to Ap (upper row) in early Antarctic winter > 60 km, on average ranging around -10 %. Further striking negative O_3 amplitudes occur in July between 30 and 40 km as well as around 1 October at ~ 30 km, at least weakly indicating the downward transport of the Ap signal in stratospheric O_3 due to NO_x predicted by model studies (e.g. Reddmann et al., 2010). In contrast, a positive O_3 am-

plitude is found at the beginning of the winter between ~ 25 and ~ 55 km (~ 10–20 %), as well as at altitudes < 30 km throughout the winter (up to ~ 20 % in October at ~ 20 km) and above the indicated subsiding layer of negative amplitudes. But considering that most of these features drop below the significance level of 95 % by combining the data of all three instruments (right column), a more detailed investigation of these patterns is not reasonable. However, the results of the merged data set show a well pronounced subsiding negative Ap signal from ~ 50 km in June down to ~ 25 km in October, which is disrupted in August, while the generally positive structures below 30 km are also still present.

The O_3 response to 2 MeV (middle row) derived from MIPAS observations also indicates a downwelling of negative O_3 amplitudes, descending from ~ 60 km in June down to ~ 30 km in late August. Additionally, the MIPAS O_3 response to 2 MeV in early winter is reversed compared to the corresponding influence from Ap on O_3, which does not originate from missing 2 MeV data from 2011. Strong positive O_3 amplitudes are generally observed throughout the winter below 30 km, exceeding values of ~ 20 % in April and October, as well as during October between 30 and 50 km where the maximum amplitude is lower (~ 10 %). The positive features can be validated with the composite results even if they are damped in the region below 30 km. However, this is not the case for the negative response, except for a small area in June in the lower mesosphere. Considering that the Ap responds to lower particle energy levels compared to 2 MeV and that the behaviour of both indices is essentially different from 2002 to 2010 (see Fig. 4), the different O_3 amplitudes associated to Ap and 2 MeV are still reasonable.

The O_3 response to F10.7 (lowermost row) is fairly similar between MIPAS and the merged measurements, and both also agree with the respective pattern observed for Ap, including the indicated downwelling of negative O_3 amplitudes during midwinter from 50 to 25 km. The composite O_3 shows strong positive amplitudes in May > 55 km which originate

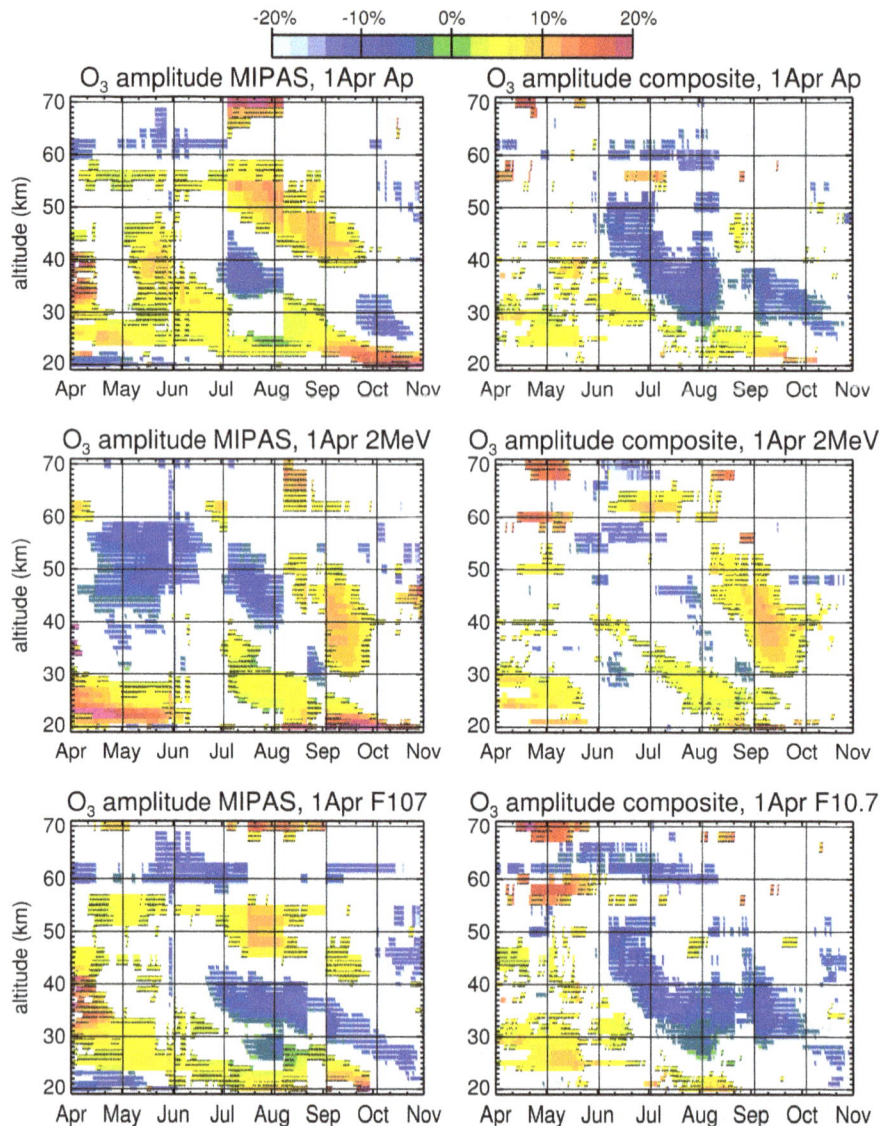

Figure 3. O_3 amplitude (see Sect. 3.1.1 for definition) inside the Antarctic polar vortex between years of high index values and years of low index values, namely Ap index (upper row), ≥ 2 MeV electron flux (middle) as well as F10.7 cm solar radio flux (lowermost row) centred around 1 April, derived from MIPAS (left column) and composite (MIPAS+SMR+SABER, right column) observations from 2002 to 2011. Shown are only values above the significance level of 95 %. Additionally, regions between the significance level of 95 and 99 % are shaded in black or white, according to a Student's t test.

from SMR measurements and are most likely due to the low vertical resolution of the SMR instrument at these altitudes (see Sect. 2.2.3). The high agreement between the results of Ap/O_3 and F10.7/O_3 might originate from the coupling of both indices during solar maximum years (Gray et al., 2010, their Fig. 1). In order to investigate a possible cross-correlation between solar radiation and geomagnetic disturbances, the analysis was repeated for years of moderate solar activity, only including 2005–2010 (Fig. 4). Similar analyses to extract a more distinct solar signal during times of approximately constant geomagnetic activity were not reasonable, because the respective years of nearly constant Ap values

(2002, 2005, 2006, 2008, 2010) do not provide a sufficient amount of data in MIPAS and SMR measurements.

3.1.2 O_3 behaviour during solar minimum activity (2005–2010)

Figure 5 displays the obtained O_3 amplitudes for solar quiet times (2005–2010) associated to 1 April Ap, again only showing values above 95 % significance level and shading the area of regions between 95 and 99 %. The MIPAS O_3 response to Ap indicates a subsiding negative signal (~ -10 to -15 %), starting in late June slightly below 50 km and prop-

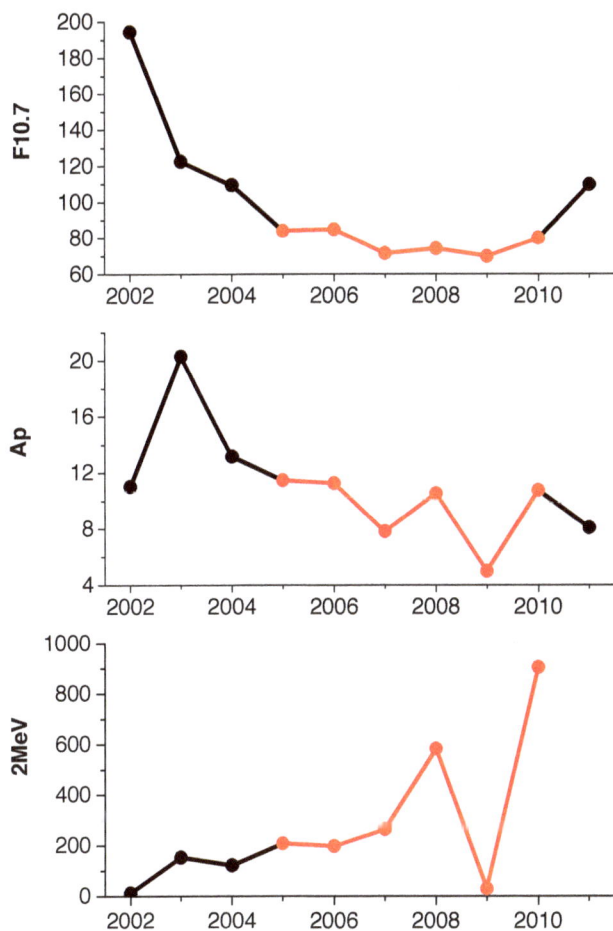

Figure 4. Time series from 2002 to 2011 of the 26-day averages centred around 1 April of the F10.7 cm solar radio flux (10^{-22} W m^{-2} Hz^{-1}, top), Ap index (middle), and ≥ 2 MeV electron flux (electrons cm^{-2} day^{-1} sr^{-1}, bottom). The period of low solar activity from 2005 to 2010 is marked in red. Note the different scaling.

agating downwards to ~ 25 km throughout the winter. However, the middle part of the downwelling between late July and late September is below the significance level of 95 % and therefore not shown here. Furthermore, the hinted subsidence is closely surrounded by well pronounced positive O_3 amplitudes, especially below ~ 30 km which maximise in September (> 20 %). There is also a negative structure centred at 1 June at ~ 60 km, which cannot be caused by NO_x but most likely results from HO_x formation (see Sect. 1). Considering the composite results, the downwelling Ap signal in O_3 becomes apparent and robust but slightly weaker (~ -10 %) while the positive features are also damped but still present. The mesospheric response is generally weak and the high positive O_3 amplitudes in May are again caused by the SMR measurements.

The 2 MeV impact on MIPAS O_3 shows generally agreeing features with the influence of the geomagnetic activity and is also of similar magnitude, however, the downwelling negative signal is hinted to already start in late May at ~ 55 km. In contrast to the O_3 response to Ap, the downwards propagating 2 MeV signal is less robust and can be only guessed in the composite O_3 amplitude, while the positive structures (~ 10–15 %) in August below 30 km and in September between 30 and 50 km are still present. In general, the 2 MeV features are less obvious in the O_3 composite, except for the positive O_3 amplitudes above the hinted downward transport. Nevertheless, the agreement between Ap and 2 MeV pattern is quite strong, in MIPAS observations in particular, although both parameters are only indirectly related to O_3. However, the O_3 structure associated to both indices is far too similar and additionally found in all three instruments to be a coincidence, even if the descending O_3 response to 2 MeV is weaker. Since Ap represents lower particle energy levels compared to the 2 MeV and both indices are only moderately correlated (see Fig. 4), the similar results strongly indicate a related source mechanism, suggesting solar wind variability.

Considering the entire process, that energetic particles produce NO_x which eventually destroys stratospheric O_3, the Ap impact observed in O_3 (see Fig. 5) is expected to be reversed in NO_x, at least in the stratosphere. In order to investigate this in more detail, the analysis was repeated for 1 April Ap and NO_x. Here NO_x, is represented only by NO_2 from MIPAS observations, because the respective NO measurements are quite noisy compared to NO_2, especially below 30 km. This is still reasonable because NO is converted to NO_2 during night and therefore NO_2 is the major fraction of NO_x inside the Antarctic polar vortex. The corresponding results include the years 2005–2010 and are displayed in Fig. 6, supporting that the stratospheric O_3 depletion can be indeed associated to the catalytic NO_x/O_3 cycle. The Ap signal in NO_2 is stronger by the factor of 2–5, compared to the respective O_3 amplitudes. The sharp gradient in mid July originates from 2005 NO_2 data, which are not available afterwards. However, the general structure of the subsiding Ap signal in NO_2 is still similar with and without 2005 observations. Note that the essentially smaller NO_2 amplitudes in October below the significance level of 95 % are not in conflict with the respective well pronounced negative O_3 response, because the latter one results from an accumulation effect from the NO_2 above. Furthermore, large negative NO_2 amplitudes throughout the entire winter below ~ 30 km are observed, matching the high positive O_3 response to Ap. A possible reason for this behaviour might be that NO_2 is stored in reservoir species, like $ClONO_2$, HNO_3, and N_2O_5, due to reactions with ClO, OH, and NO_3, respectively. However, N_2O_5 is converted to HNO_3 via water ion cluster chemistry (López-Puertas et al., 2005, their reactions 1 and 8–12), which was also investigated with respect to EPP for conditions without solar proton events by Stiller et al. (2005). These reactions eventually lead to lower NO_x concentrations, consequently slowing down the catalytic O_3 depletion. Based on the corresponding MIPAS

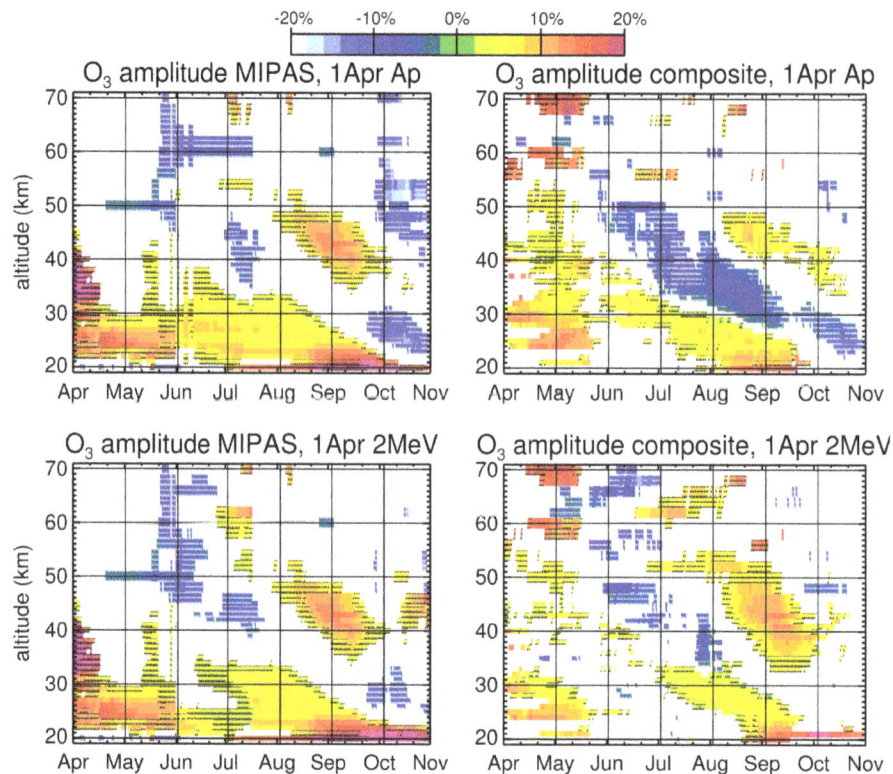

Figure 5. Same as Fig. 3 but only for Ap index (upper row) and $\geq 2\,\text{MeV}$ electron flux (lower row) from 2005 to 2010.

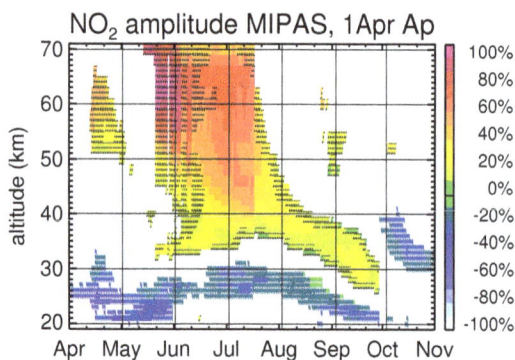

Figure 6. Same as Fig. 3, but only for the NO_2 amplitude associated to 1 April Ap index. The NO_2 was derived from MIPAS observations from 2005 to 2010.

climatologies (not shown here), HNO_3 is more important until mid July, while $ClONO_2$ is dominating afterwards and its influence becomes essentially crucial in spring due to heterogeneous chemistry which has taken place before. This suggested NO_2-$ClONO_2$ mechanism is supported by Whole Atmosphere Community Climate Model results reported by Jackman et al. (2009, their Figs. 6 and 7), who simulated the impact of the SPE in July 2000 on stratospheric O_3 and NO_y ($= NO_x + NO_3 + N_2O5 + HNO_3 + HO_2NO_2 + ClONO_2 + BrONO_2$).

Furthermore, the positive O_3 amplitudes below $\sim 30\,\text{km}$ could be also partly explained by the self healing effect of O_3 (Jackman and McPeters, 1985). Altitude regions of reduced O_3 will lead to increased solar UV radiation in the layers directly below. This is accompanied by a higher production of atomic oxygen and would consequently increase the formation of O_3. However, this proposed mechanism would only have an additional effect, contributing to the formation of O_3 in the atmospheric layer right below the subsidence, but cannot account for the entire region. Note that this layer is also present throughout the entire winter, and thus an influence from the vortex above is unlikely but any further investigations are beyond the scope of this study.

Additionally, the area of high positive Ap / O_3 structure between 35 and 50 km from August to September cannot be completely explained by the NO_x / O_3 cycle. In detail, the respective Ap influence of NO_2 is close to 0 and consequently well below the 95 %, while the respective MIPAS $ClONO_2$ amplitude (not shown here) reveals positive values, which are also mostly below the 95 % significance level. These results are at least not in conflict with a higher O_3 amplitude. Furthermore, this positive Ap impact on O_3 is essentially less visible in the composite results than in MIPAS data, and a corresponding composite analysis for Ap / NO_2 is necessary for a more detailed investigation. But this is not possible due to non-existing NO_2 measurements from SABER and SMR. Thus no definite explanation can be given at this stage and

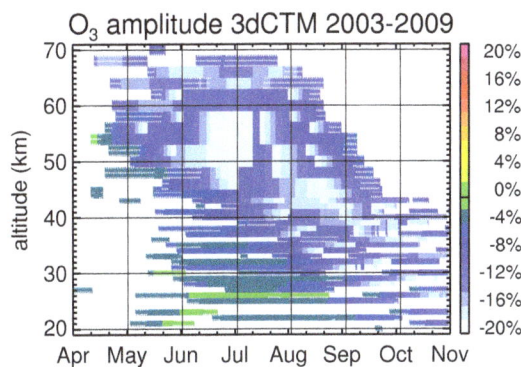

Figure 7. O_3 amplitude (see Sect. 2.4 for definition), simulated by the 3dCTM from 2003 to 2009. Shown are values above 95 % significance level, according to a Student's t test, and areas between 95 and 99 % significance level are shaded.

3.2 Comparison with 3dCTM

The simulated O_3 amplitude between the EP run and base run, representing high and low geomagnetic activity, respectively, is displayed in Fig. 7. Note that the modelled O_3 amplitude is also referred to as O_3 amplitude here, which is justified because "observed" and "modelled" O_3 amplitude still hold the same physical meaning, even if the calculation algorithm is slightly different. It is reasonable to investigate the complete simulated time interval from 2003 to 2009, because the model runs represent solar minimum conditions similar to the years 2005–2010. The results reveal apparent negative O_3 amplitudes propagating downwards throughout the winter with maximum negative values during midwinter between 45 and 60 km. The subsidence shows larger negative O_3 amplitudes compared to the measurements and is also much broader, which might be due to the constant F10.7 and the prescribed dynamics, both reducing the inter-annual variability of O_3. Furthermore, we performed an on/off experiment, while in reality the EPP indirect effect is a persistent feature. Below 30 km the observed high positive O_3 amplitudes associated with Ap are only indicated in the model results by essentially weaker and additionally negative amplitudes. However, the model amplitudes are at least less negative compared to the values above. The second positive region above the downwelling is completely missing. Further note that the strong positive response during late winter/early spring below 30 km might not be reproduced by the model due to missing heterogeneous chemistry. The proposed self healing effect of O_3 (see Sect. 3.1.2) was also tested, using O^1D as a proxy for the O_3 photolysis rate in the Lyman-alpha band and calculating the O^1D amplitude (not shown here).

this feature is a subject of a future work. However, it should be pointed out that this structure does not harm the underlying mechanism proposed to explain the identified negative O_3 amplitude and subsequent downward transport.

However, the expected positive response directly below the downwelling is only partly visible and even below the 67 % significance level.

The qualitative agreement between model results and observations in the stratosphere suggests that the subsiding Ap signal found in O_3 is actually originating from particle precipitation. However, the simulated downwelling starts at altitudes > 60 km while observations reveal no obvious structures in the mesosphere, possibly caused by satellite sampling. As already stated in Sect. 3.1.2, the mesospheric behaviour cannot be caused by NO_x, because the NO_x / O_3 cycle is not efficiently working at these altitudes. Thus the O_3 depletion > 50 km could be accounted to OH production, which is most likely overestimated in the model and consequently leads to an increased O_3 depletion not observed by the satellite instruments.

4   Conclusions

We have investigated the O_3 behaviour inside the Antarctic polar vortex from 2002 to 2011, observed by three independent satellite based instruments ENVISAT/MIPAS, Odin/SMR, and TIMED/SABER. These O_3 vmr measurements, based on 26-day running means from 1 April to 1 November covering altitudes from 20 to 70 km, were individually grouped into high and low index activity according to the 26-day averages centred around 1 April, 1 May, and 1 June of different solar and geomagnetic indices (F10.7, Ap, 2 MeV). After minimising the direct influence of the solar radiation by only considering the period of solar minimum activity from 2005 to 2010 we found a negative O_3 response caused by geomagnetic activity (Ap) from 1 April in all three instruments, ranging from −5 to −10 % and propagating downwards throughout the Antarctic winter from ~50 km down to ~25 km. This subsiding negative signal in O_3 is above the significance level of 95 % and overlaps with the corresponding positive NO_2 response to 1 April Ap, supporting that NO_x is indeed the cause of the O_3 depletion. We could also show that the high positive O_3 response below 30 km, which is present during the entire winter, is in agreement with respective negative NO_2 structures. The cause of the NO_2 behaviour is possibly related to the formation of the reservoir species $ClONO_2$ and HNO_3, slowing down the catalytic destruction of O_3 by Cl. The O_3 pattern induced by the magnetosphere (2 MeV) from 1 April are similar but weaker, compared to the respective geomagnetic activity, still suggesting a related source mechanism between 2 MeV and Ap like solar wind variability. The composite observations of all three instruments are in good qualitative agreement with 3dCTM simulation, revealing similar O_3 pattern induced by the geomagnetic activity from 1 April while the simulated O_3 response is larger but still in the same order of magnitude.

However, we have to point out that the validity of the subsiding O_3 depletion associated to geomagnetic activity and

NO_x is not ensured due to the short time series of only 6 years at most. Thus, we conclude that precipitating particles are strongly indicated as a factor contributing to stratospheric O_3 during Antarctic winter, but we cannot prove the link unambiguously.

Author contributions. T. Fytterer, analysed the satellite and indices data and wrote the final script. G. Stiller, J. Urban and K. Pérot, and M. Mlynczak provided the O_3 data from ENVISAT/MIPAS, Odin/SMR, and TIMED/SABER, respectively, and all of them contributed to interpretation. H. Nieder performed the 3dCTM simulations. M. Sinnhuber initiated the study and contributed to interpretation.

Acknowledgements. T. Fytterer, H. Nieder, and M. Sinnhuber gratefully acknowledge funding by the Helmholtz Association of German Research Centres (HGF), grant VH-NG-624. The authors also acknowledge support by Deutsche Forschungsgemeinschaft and Open Access Publishing Fund of Karlsruhe Institute of Technology. Odin is a Swedish-led satellite project funded jointly by the Swedish National Space Board (SNSB), the Canadian Space Agency (CSA), the National Technology Agency of Finland (Tekes), the Centre National d'Etudes Spatiales (CNES) in France and the third party mission program of the European Space Agency (ESA). We further like to thank the ERA-Interim for free provision of data and related support.

Edited by: Q. Errera

References

Azeem, S. M. I., Talaat, E. R., Sivjee, G. G., and Yee, J.-H.: Mesosphere and lower thermosphere temperature anomalies during the 2002 Antarctic stratospheric warming event, Ann. Geophys., 28, 267–276, doi:10.5194/angeo-28-267-2010, 2010.

Berger, U.: Modeling of middle atmosphere dynamics with LIMA, J. Atmos. Sol.-Terr. Phys., 70, 1170–1200, doi:10.1016/j.jastp.2008.02.004, 2008.

Eckert, E., von Clarmann, T., Kiefer, M., Stiller, G. P., Lossow, S., Glatthor, N., Degenstein, D. A., Froidevaux, L., Godin-Beekmann, S., Leblanc, T., McDermid, S., Pastel, M., Steinbrecht, W., Swart, D. P. J., Walker, K. A., and Bernath, P. F.: Drift-corrected trends and periodic variations in MIPAS IMK/IAA ozone measurements, Atmos. Chem. Phys., 14, 2571–2589, doi:10.5194/acp-14-2571-2014, 2014.

Fischer, H., Birk, M., Blom, C., Carli, B., Carlotti, M., von Clarmann, T., Delbouille, L., Dudhia, A., Ehhalt, D., Endemann, M., Flaud, J. M., Gessner, R., Kleinert, A., Koopman, R., Langen, J., López-Puertas, M., Mosner, P., Nett, H., Oelhaf, H., Perron, G., Remedios, J., Ridolfi, M., Stiller, G., and Zander, R.: MIPAS: an instrument for atmospheric and climate research, Atmos. Chem. Phys., 8, 2151–2188, doi:10.5194/acp-8-2151-2008, 2008.

Funke, B., López-Puertas, M., von Clarmann, T., Stiller, G. P., Fischer, H., Glatthor, N., Grabowski, U., Höpfner, M., Kellmann, S., Kiefer, M., Linden, A., Mengistu Tsidu, G., Milz, M.,

Steck, T., and Wang, D. Y.: Retrieval of stratospheric NO_x from 5.3 and 6.2 μm nonlocal thermodynamic equilibrium emissions measured by Michelson Interferometer for Passive Atmospheric Sounding (MIPAS) on Envisat, J. Geophys. Res., 110, D09302, doi:10.1029/2004JD005225, 2005.

Funke, B., Baumgaertner, A., Calisto, M., Egorova, T., Jackman, C. H., Kieser, J., Krivolutsky, A., López-Puertas, M., Marsh, D. R., Reddmann, T., Rozanov, E., Salmi, S.-M., Sinnhuber, M., Stiller, G. P., Verronen, P. T., Versick, S., von Clarmann, T., Vyushkova, T. Y., Wieters, N., and Wissing, J. M.: Composition changes after the "Halloween" solar proton event: the High Energy Particle Precipitation in the Atmosphere (HEPPA) model versus MIPAS data intercomparison study, Atmos. Chem. Phys., 11, 9089–9139, doi:10.5194/acp-11-9089-2011, 2011.

Funke, B., López-Puertas, M., Stiller, G. P., and von Clarmann, T.: Mesospheric and stratospheric NO_y produced by energetic particle precipitation during 2002–2012, J. Geophys. Res-Atmos., 119, 4429–4446, doi:10.1002/2013JD021404, 2014.

Glatthor, N., von Clarmann, T., Fischer, H., Funke, B., Gil-López, S., Grabowski, U., Höpfner, M., Kellmann, S., Linden, A., López-Puertas, M., Mengistu Tsidu, G., Milz, M., Steck, T., Stiller, G. P., and Wang, D.-Y.: Retrieval of stratospheric ozone profiles from MIPAS/ENVISAT limb emission spectra: a sensitivity study, Atmos. Chem. Phys., 6, 2767–2781, doi:10.5194/acp-6-2767-2006, 2006.

Gray, L. J., Beer, J., Geller, M., Haigh, J. D., Lockwodd, M., Matthes, K., Cubasch, U., Fleitmann, D., Harrison, G., Hood, L., Luterbacher, J., Meehl, G. A., Shindell, D., van Geel, B., and White, W.: Solar influences on climate, Rev. Geophys., 48, RG4001, doi:10.1029/2009RG000282, 2010.

Jackman, C. H. and McPeters, R. D.: The Response of Ozone to Solar Proton Events During Solar Cycle 21: A Theoretical Interpretation, J. Geophys. Res., 90, 7955–7966, 1985.

Jackman, C. H., DeLand, M. T., Labow, G. J., Fleming, E. L., Weisenstein, D. K., Ko, M. K. W., Sinnhuber, M., and Russell III, J. M.: Neutral atmospheric influences of the solar proton events in October–November 2003, J. Geophys. Res., 110, A09S27, doi:10.1029/2004JA010888, 2005.

Jackman, C. H., Marsh, D. R., Vitt, F. M., Garcia, R. R., Randall, C. E., Fleming, E. L., and Frith, S. M.: Long-term middle atmospheric influence of very large solar proton events, J. Geophys. Res., 114, D11304, doi:10.1029/2008JD011415, 2009.

Lary, D. J.: Catalytic destruction of stratospheric ozone, J. Geophys. Res., 102, 21515–21526, doi:10.1029/97JD00912, 1997.

López-Puertas, M., Funke, B., Gil-López, S., von Clarmann, T., Stiller, G. P., Höpfner, M., Kellmann, S., Mengistu Tsidu, G., Fischer, H., and Jackman, C. H.: HNO_3, N_2O_5, and $ClONO_2$ enhancements after the October–November 2003 solar proton events, J. Geophys. Res., 110, A09S44, doi:10.1029/2005JA011051, 2005.

Murtagh, D., Frisk, U., Merino, F., Ridal, M., Jonsson, A., Stegman, J., Witt, G., Eriksson, P., Jiménez, C., Megie, G., de la Noë, J., Ricaud, P., Baron, P., Pardo, J. R., Hauchcorne, A., Llewellyn, E. J., Degenstein, D. A., Gattinger, R. L., Lloyd, N. D., Evans, W. F. J., McDade, I. C., Haley, C. S., Sioris, C., von Savigny, C., Solheim, B. H., McConnell, J. C., Strong, K., Richardson, E. H., Leppelmeier, G. W., Kyrölä, E., Auvinen, H., and Oikarinen, L.: An overview of the Odin atmospheric mission, Can. J. Phys., 80, 309–319, doi:10.1139/p01-157, 2002.

Nash, E. R., Newman, P. A., Rosenfield, J. E., and Schoeberl, M. R.: An objective determination of the polar vortex using Ertel's potential vorticity, J. Geophys. Res., 101, 9471–9478, doi:10.1029/96JD00066, 1996.

Nieder, H., Winkler, H., Marsh, D. R., and Sinnhuber, M.: NO_x production due to energetic particle precipitation in the MLT region: Results from ion chemistry model studies, J. Geophys. Res.-Space., 119, 2137–2148, doi:10.1002/2013JA019044, 2014.

Porter, H. S., Jackman, C. H, and Green, A. E. S.: Efficiencies for production of atomic nitrogen and oxygen by relativistic proton impact in air, J. Chem. Phys., 65, 154–167, doi:10.1063/1.432812, 1976.

Prather, M.: Numerical advection by conservation of second-order moments, J. Geophys. Res., 91, 6671–6681, doi:10.1029/JD091iD06p06671, 1986.

Preusse, P., Eckermann, S. D., Ern, M., Oberheide, J., Picard, R. H., Roble, R. M., Riese, M., Russell III, J. M., and Mlynczak, M. G.: Global ray tracing simulations of the SABER gravity wave climatology, J. Geophys. Res., 114, D08126, doi:10.1029/2008JD011214, 2009.

Randall, C. E., Rusch, D. W., Bevilacqua, R. M., and Hoppel, K. W.: Polar Ozone And Aerosol Measurement (POAM) II stratospheric NO_2, 1993–1996, J. Geophys. Res., 103, 28361–28371, doi:10.1029/98JD02092, 1998.

Randall, C. E., Harvey, V. L., Singleton, C. S., Bailey, S. M., Bernath, P. F., Codrescu, M., Nakajima, H., and Russell III, J. M.: Energetic particle precipitation effects on the Southern Hemisphere stratosphere in 1992–2005, J. Geophys. Res., 112, D08308, doi:10.1029/2006JD007696, 2007.

Reddmann, T., Ruhnke, R., Versick, S., and Kouker, W.: Modeling disturbed stratospheric chemistry during solar-induced NO_x enhancements observed with MIPAS/ENVISAT, J. Geophys. Res., 115, D00I11, doi:10.1029/2009JD012569, 2010.

Rodger, C. J., Kavanagh, A. J., Clilverd, M. A., and Marple, S. R.: Comparison between POES energetic electron precipitation observations and riometer absorptions: Implications for determining true precipitation fluxes, J. Geophys. Res.-Space., 118, 7810–7821, doi:10.1002/2013JA019439, 2013.

Rong, P. P., Russell III, J. M., Mlynczak, M. G., Remsberg, E. E., Marshall, B. T., Gordley, L. L., and López-Puertas, M.: Validation of Thermosphere Ionosphere Mesosphere Energetics and Dynamics/Sounding of the Atmosphere using Broadband Emission Radiometry (TIMED/SABER) v1.07 ozone at 9.6 μm in altitude range 15–70 km, J. Geophys. Res., 114, D04306, doi:10.1029/2008JD010073, 2009.

Rusch, D. W., Gerard, J. C., Solomon, S., Crutzen, P. J., and Reid, G. C.: The effect of particle precipitation events on the neutral and ion chemistry of the middle atmosphere – I. Odd nitrogen, Planet. Space Sci., 29, 767–774, doi:10.1016/0032-0633(81)90048-9, 1981.

Sinnhuber, M., Kazeminejad, S., and Wissing, J. M.: Interannual variation of NO_x from the lower thermosphere to the upper stratosphere in the years 1991–2005, J. Geophys. Res., 116, A02312, doi:10.1029/2010JA015825, 2011.

Sinnhuber, M., Nieder, H., and Wieters, N.: Energetic particles precipitation and the chemistry of the mesosphere/lower thermosphere, Surv. Geophys., 33, 1281–1334, doi:10.1007/s10712-012-9201-3, 2012.

Solomon, S., Rusch, D. W., Gerard, J. C., Reid, G. C., and Crutzen, P. J.: The effect of particle precipitation events on the neutral and ion chemistry of the middle atmosphere: II. odd hydrogen, Planet. Space Sci., 29, 885–892, doi:10.1016/0032-0633(81)90078-7, 1981.

Solomon, S., Crutzen, P. J., and Roble, R. G.: Photochemical coupling between the thermosphere and the lower atmosphere 1. odd nitrogen from 50 to 120 km, J. Geophys. Res., 87, 7206–7220, doi:10.1029/JC087iC09p07206, 1982.

Steck, T., von Clarmann, T., Fischer, H., Funke, B., Glatthor, N., Grabowski, U., Höpfner, M., Kellmann, S., Kiefer, M., Linden, A., Milz, M., Stiller, G. P., Wang, D. Y., Allaart, M., Blumenstock, Th., von der Gathen, P., Hansen, G., Hase, F., Hochschild, G., Kopp, G., Kyrö, E., Oelhaf, H., Raffalski, U., Redondas Marrero, A., Remsberg, E., Russell III, J., Stebel, K., Steinbrecht, W., Wetzel, G., Yela, M., and Zhang, G.: Bias determination and precision validation of ozone profiles from MIPAS-Envisat retrieved with the IMK-IAA processor, Atmos. Chem. Phys., 7, 3639–3662, doi:10.5194/acp-7-3639-2007, 2007.

Stiller, G. P., Tsidu, G. M., von Clarmann, T., Glatthor, N., Höpfner, M., Kellmann, S., Linden, A., Ruhnke, R., Fischer, H., López-Puertas, M., Funke, B., and Gil-López, S.: An enhanced HNO_3 second maximum in the Antarctic midwinter upper stratosphere 2003, J. Geophys. Res, 110, D20303, doi:10.1029/2005JD006011, 2005.

Urban, J., Lautié, N., Le Flochmoën, E., Jiménez, C., Eriksson, P., de La Noë, J., Dupuy, E., Ekström, M., El Amraoui, L., Frisk, U., Murtagh, D., Olberg, M., and Ricaud, P.: Odin/SMR limb observations of stratospheric trace gases: Level 2 processing of ClO, N_2O, HNO_3, and O_3, J. Geophys. Res., 110, D14307, doi:10.1029/2004JD005741, 2005.

Von Clarmann, T., Glatthor, N., Grabowski, U., Höpfner, M., Kellmann, S., Kiefer, M., Linden, A., Mengistu Tsidu, G., Milz, M., Steck, T., Stiller, G. P., Wang, D. Y., Fischer, H., Funke, B., Gil-López, S., and López-Puertas, M.: Retrieval of temperature and tangent altitude pointing from limb emission spectra recorded from space by the Michelson Interferometer for Passive Atmospheric Sounding (MIPAS), J. Geophys. Res., 108, 4736, doi:10.1029/2003JD003602, 2003.

von Clarmann, T., Höpfner, M., Kellmann, S., Linden, A., Chauhan, S., Funke, B., Grabowski, U., Glatthor, N., Kiefer, M., Schieferdecker, T., Stiller, G. P., and Versick, S.: Retrieval of temperature, H_2O, O_3, HNO_3, CH_4, N_2O, $ClONO_2$ and ClO from MIPAS reduced resolution nominal mode limb emission measurements, Atmos. Meas. Tech., 2, 159–175, doi:10.5194/amt-2-159-2009, 2009.

Wissing, J. M. and Kallenrode, M. B.: Atmospheric Ionization Module Osnabrück (AIMOS): a 3-D model to determine atmospheric ionization by energetic charged particles from different populations, J. Geophys. Res., 114, A06104, doi:10.1029/2008JA013884, 2009.

Wissing, J. M., Kallenrode, M. B., Wieters, N., Winkler, H., and Sinnhuber, M.: Atmospheric Ionization Module Osnabrück (AIMOS): 2. Total particle inventory in the October–November 2003 event and ozone, J. Geophys. Res., 115, A02308, doi:10.1029/2009JA014419, 2010.

10

The influence of clouds on radical concentrations: observations and modelling studies of HO$_x$ during the Hill Cap Cloud Thuringia (HCCT) campaign in 2010

L. K. Whalley[1,2], D. Stone[2], I. J. George[2,*], S. Mertes[3], D. van Pinxteren[3], A. Tilgner[3], H. Herrmann[3], M. J. Evans[4,5], and D. E. Heard[1,2]

[1]National Centre for Atmospheric Science, University of Leeds, Leeds, LS2 9JT, UK
[2]School of Chemistry, University of Leeds, Leeds, LS2 9JT, UK
[3]Leibniz-Institut für Troposphärenforschung (TROPOS), Permoserstr. 15, 04318 Leipzig, Germany
[4]National Centre for Atmospheric Science, University of York, York, YO10 5DD, UK
[5]Department of Chemistry, University of York, York, YO10 5DD, UK
[*]now at: National Risk Management Research Laboratory, US Environmental Protection Agency, Research Triangle Park, North Carolina 27711, USA

Correspondence to: L. K. Whalley (l.k.whalley@leeds.ac.uk)

Abstract. The potential for chemistry occurring in cloud droplets to impact atmospheric composition has been known for some time. However, the lack of direct observations and uncertainty in the magnitude of these reactions led to this area being overlooked in most chemistry transport models. Here we present observations from Mt Schmücke, Germany, of the HO$_2$ radical made alongside a suite of cloud measurements. HO$_2$ concentrations were depleted in-cloud by up to 90 % with the rate of heterogeneous loss of HO$_2$ to clouds necessary to bring model and measurements into agreement, demonstrating a dependence on droplet surface area and pH. This provides the first observationally derived assessment for the uptake coefficient of HO$_2$ to cloud droplets and was found to be in good agreement with theoretically derived parameterisations. Global model simulations, including this cloud uptake, showed impacts on the oxidising capacity of the troposphere that depended critically on whether the HO$_2$ uptake leads to production of H$_2$O$_2$ or H$_2$O.

1 Introduction

Clouds occupy around 15 % of the volume of the lower troposphere and can impact atmospheric composition through changes in transport, photolysis, wet deposition and in-cloud oxidation of sulfur. Modelling studies have shown that aqueous-phase chemistry can also significantly reduce gaseous HO$_2$ concentrations by heterogeneous uptake and loss into cloud droplets (Jacob, 1996; Tilgner et al., 2005; Huijnen et al., 2014). This chemistry is predicted to reduce OH and O$_3$ concentrations, also due to the reduction in the gas-phase concentration of HO$_2$. This in turn decreases the self-cleansing capacity of the atmosphere and increases the lifetime of many trace gases (Lelieveld and Crutzen, 1990), with impacts for climate and air quality. Aqueous-phase models have been developed which combine multi-phase chemistry with detailed microphysics (Tilgner et al., 2005), but there are limited experimental field data of gas-phase radical concentrations within clouds to corroborate model predictions of heterogeneous loss of radicals to cloud droplets. There have been a number of aircraft campaigns which have measured OH and HO$_2$ radical concentrations within clouds (Mauldin et al., 1997, 1998; Olson et al., 2004; Commane et al., 2010); often, however, simultaneous observations of cloud droplet number and size distributions (or other key gas-phase radical precursors) were not made during these studies, making it difficult to assess the full impact of clouds on radical concentrations. In general therefore climate and air quality models do not consider this impact of clouds on atmospheric composition.

Within the literature, a wide range of uptake coefficients of HO_2 to liquid and aerosol surfaces have been considered to reproduce observed HO_2 concentrations (e.g. Sommariva et al., 2004; Haggerstone et al., 2005; Emmerson et al., 2007; Whalley et al., 2010) with often large uptake coefficients (up to 1 at times) used to reconcile model over-predictions. A wide range of uptake coefficients, not wholly consistent with each other, have been reported from laboratory studies (Abbatt et al., 2012). From measurements conducted in our laboratory, uptake probabilities of HO_2 to sub-micron aerosols were found to be less than 0.02 at room temperature (George et al., 2013) for aqueous aerosols that did not contain significant transition metal ions; similarly low uptake coefficients were derived by Thornton and Abbatt (2005). In contrast, measurements by Taketani et al. (2008) suggest higher uptakes of ~ 0.1 with enhancements observed with increasing relative humidity.

The uptake of HO_2 to aqueous aerosols is driven by its high solubility in water owing to its high Henry's law constant ($= 4.0 \times 10^3$ M atm^{-1} at 298.15 K; Hanson et al., 1992). Once in the aqueous phase, reaction between dissolved HO_2 and its conjugate base, O_2^-, occurs rapidly. Thornton et al. (2008) have demonstrated that the solubility and reactivity of HO_2 are temperature and pH dependent and, if the well-characterised aqueous phase reactions (Sect. 2.3 (R1–R5)) alone are representative of the heterogeneous loss processes, only small uptake coefficients would be expected at room temperature, consistent with the work by George et al. (2013) and Thornton and Abbatt (2005). The enhanced uptake coefficients reported by Taketani et al. (2008) suggest that there may be additional competing mechanisms occurring, however.

Further uncertainties arise in the literature relating to the eventual gas-phase products from these aqueous-phase reactions. The general consensus, until recently, was that these reactions would ultimately produce H_2O_2 (Jacob, 1996), but the significance of the reactions depends critically on whether this is the case or whether, instead, H_2O is produced (Macintyre and Evans, 2011). This is significant, as H_2O_2 can photolyse to return odd hydrogen ($HO_x = OH + HO_2$) to the gas phase, whilst cloud uptake of HO_2 to form H_2O provides a terminal sink for HO_x. Recent work by Mao et al. (2013) postulates that a catalytic mechanism involving the coupling of the transition metal ions Cu(I)/Cu(II) and Fe(II)/Fe(III) may rapidly convert HO_2 to H_2O, rather than H_2O_2 in aqueous aerosols. The concentration and availability of dissolved Fe and Cu in cloud droplets tends to be much lower than in aqueous aerosol (Jacob, 2000), with a large fraction of Cu ions present as organic complexes (Spokes et al., 1996; Nimmo and Fones, 1997), which are far less reactive towards O_2^- and $HO_2(aq)$ than the free ions (Jacob, 2000), and so it is uncertain whether the mechanism put for-

ward by Mao et al. (2013) could be extended to heterogeneous processes occurring within cloud droplets.

To better understand the role of clouds and heterogeneous processes in the oxidative capacity of the troposphere, coordinated gas-phase measurements of OH and HO_2 within clouds together with aerosol–cloud microphysical measurements are needed. The Hill Cap Cloud Thuringia 2010 (HCCT-2010) campaign which took place in 2010 aimed to characterise the interaction of particulate matter and trace gases in orographic clouds. This paper presents the impact of cloud droplets on measured gas-phase OH and HO_2 and uses these observations to assess the proposed aqueous-phase mechanisms and determine the global impact of clouds on the tropospheric oxidising capacity.

2 Experimental

The HCCT-2010 campaign took place at the Thüringer Wald mountain range in central Germany during September and October 2010. The radical measurements were made from the German Weather Service (DWD) and the Federal Environmental Office (UBA) research station located close to the summit of Mt Schmücke (the highest peak in the mountain range, 937 m a.s.l., 10°46′8.5″ E, 50°39′16.5″ N). In October, the UBA station is immersed in cloud for 25 days on average (Herrmann et al., 2005) and, hence, is highly suitable for the study of gas and aerosol interactions with orographic cloud. Two additional experimental sites, approximately 4 km upwind of the summit site at Goldlauter and approximately 3 km downwind of the summit at Gelhberg, were also equipped with a number of instruments which enabled the processing of a single air parcel as it passed through a cloud to be assessed by multi-phase trajectory models such as SPACCIM (SPectral Aerosol Cloud Chemistry Interaction Model; Wolke et al., 2005; see Sect. 2.3). Further details of the locations may be found in Herrmann et al. (2005).

2.1 Radical measurements

OH and HO_2 measurements were made using the fluorescence assay by gas expansion technique (FAGE). Details of the instrumentation can be found in Whalley et al. (2010). A single FAGE fluorescence cell was used for sequential measurements of OH and HO_2. This was operated from the top of a 22 m high tower to co-locate with cloud measurements and ensure that the measurements were performed in full cloud. The cell was held at 1 Torr using a roots blower backed rotary pump system which was housed in an air-conditioned shipping container at the base of the tower (Fig. 1) and was connected to the cell via 30 m of flexible hosing (5 cm outer diameter – OD). A 308 nm tuneable, pulsed laser light was used to electronically excite OH radicals; this was delivered to the cell via a 30 m fibre optic cable (Oz optics) with the laser system (a Nd:YAG pumped Ti:Sapphire, Photonic In-

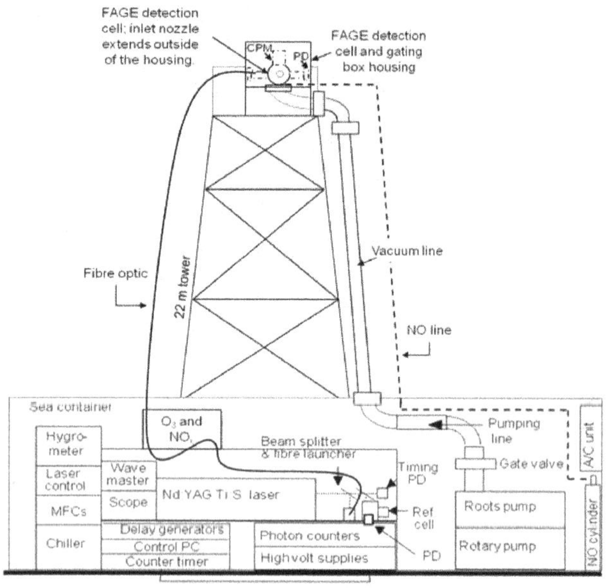

Figure 1. Schematic of the FAGE instrument set-up during the HCCT-2010 campaign. "PD" refers to photodiode, used to normalise the observed HO_2 signal to laser power.

dustries) housed in the shipping container. Fluorescence was detected by a channel photo multiplier (CPM) (Perkin Elmer) and gated photon counting. Data were acquired every second (photon counts from 5000 laser shots), with a data acquisition cycle consisting of 220 s with the laser wavelength tuned to the OH transition (NO was injected after 110 s to rapidly convert HO_2 to OH, to allow the quantification of HO_2) and 110 s tuned away from the OH transition to determine the background signal from laser scattered light.

The sensitivity of the fluorescence cell for OH and HO_2 was determined twice weekly during the measurement period through calibration using VUV photolysis of H_2O vapour in a turbulent flow of zero air (BOC, BTCA air). Calibrations were performed at relevant H_2O vapour concentrations so as to encompass the ambient H_2O vapour concentrations observed. As such, no correction for quenching of the fluorescence signal due to changing conditions was necessary. The impact of H_2O (v) on the sensitivity of this FAGE cell type (as outlined by Commane et al., 2010) has been studied by systematically varying the H_2O concentration from 500 to 10 000 ppmv, and only $\sim 10\%$ reduction in sensitivity over this range for both OH and HO_2 was observed. This reduction is entirely explained by the known quenching of fluorescence by H_2O molecules. The lamp flux was determined by N_2O actinometry (see Commane et al. (2010) for further details); this was carried out before and after the campaign and the values agreed within 21 %; the average flux was used to determine the sensitivity. The limit of detection (LOD) at a signal-to-noise ratio of one to one data acquisition cycle was $\sim 6 \times 10^5$ molec cm^{-3} and $\sim 8.5 \times 10^5$ molecule cm^{-3} for OH and HO_2, respectively.

A number of operational modifications (from the standard University of Leeds ground-based operations; Whalley et al., 2010) were necessary to facilitate measurements of the gas-phase concentrations of the radicals within clouds. As tower measurements were required (a schematic of the measurement set-up is provided in Fig. 1), a single, smaller (4.5 cm (ID) diameter stainless steel cylinder) FAGE fluorescence cell based on the University of Leeds aircraft cell design (Commane et al., 2010) was used for sequential measurements of OH and HO_2. Ambient air was drawn into the cell through a 1 mm diameter pinhole nozzle. The distance between sampling nozzle and radical detection region was 18 cm and NO (10 SCCM, BOC, 99.5 %) was injected ~ 8 cm below the nozzle for titration of HO_2 to OH.

The fluorescence cell was orientated with the nozzle pointing horizontal to the ground in an attempt to minimise water pooling on the nozzle and being sucked into the cell during cloud events. Occasional droplets were ingested by the cell and resulted in an instantaneous large increase in the laser scattered signal. These spiked increases were discrete and short-lived; the data presented here have been filtered to remove these spikes, which were easy to identify.

Tests have been conducted post-campaign to determine the level of HO_2 interference from RO_2 radicals (Fuchs et al., 2011). Under this particular experimental set-up, an equivalent number of ethene-derived RO_2 radicals to HO_2 were found to contribute 46 % to the total HO_2 signal (Whalley et al., 2013). The FAGE instrument was found not to be sensitive to CH_3O_2 and other short-chain alkane-derived RO_2 radicals, but is sensitive to other alkene- and aromatic-derived RO_2 radicals, with similar sensitivities to that for ethene-derived RO_2. The instrument is also sensitive to longer-chain alkane-derived RO_2 radicals ($> C_3$), albeit to a smaller extent, as reported by Whalley et al. (2013). For this rural environment, at this time of year, however, the contribution of alkene- and aromatic-derived RO_2 radicals to the total RO_2 budget is expected to be small, as the parent VOCs for these particular RO_2 types were at low concentrations; isoprene concentrations, for example, were on average just 12.6 pptv. As a consequence of this, the resultant HO_2 interference from RO_2 radicals should also be low.

2.2 Model expression and constraints

An analytical expression has been used to predict the mean diurnal HO_2 concentrations throughout the campaign, both during cloud events and outside of cloud events. This expression was originally developed by Carslaw et al. (1999) for modelling OH, HO_2 and RO_2 radicals in the marine boundary layer and was found to agree with full Master Chemical Mechanism (MCM) model predictions for OH and HO_2 to within 20 % for daytime hours. It has since been extended further by Smith et al. (2006) to include additional HO_2 sinks, such as heterogeneous loss (k_{Loss}). The expression, given in Eq. (3), derives from the solution of simultaneous steady-state expressions for OH and CH_3O_2 (Eq. (1) and Eq. (2) below) and includes any primary sources of HO_2 not coming from radical propagation steps such as formaldehyde photolysis:

$$[OH] = \qquad (1)$$

$$\frac{2fj(O^1D)[O_3] + [HO_2](k_{HO_2+NO}[NO] + k_{HO_2+O_3}[O_3])}{\begin{array}{l} k_{CO+OH}[CO] + k_{H_2+OH}[H_2] + k_{HCHO+OH}[HCHO] \\ + k_{CH_4+OH}[CH_4] + k_{NO_2+OH}[NO_2] + k_{O_3+OH}[O_3] \end{array}}$$

$$[CH_3O_2] = \frac{k_{CH_4+OH}[CH_4][OH]}{k_{CH_3O_2+HO_2}[HO_2] + k_{CH_3O_2+NO}[NO]} \qquad (2)$$

$$\beta[HO_2]^3 + \gamma[HO_2]^2 + \delta[HO_2] + \varepsilon = 0, \qquad (3)$$

where

$$\beta = 2k_{T2}(k_{T3}B + k_{T1}A)$$

$$\begin{aligned} \gamma &= 2k_{T3}k_{T2}J_1 + 2k_{T3}k_{P5}[NO]B + 2k_{T2}k_{P4}[CH_4]B \\ &+ k_T[NO_2]k_{T2}B + 2Ak_{T1}k_{P5}[NO] \end{aligned}$$

$$\begin{aligned} \delta &= 2k_{T3}k_{P5}J_1[NO] + 2k_{T2}k_{P4}J_1[CH_4] + k_T J_1[NO_2]k_{T2} \\ &+ k_T B[NO_2]k_{P5}[NO] - (J_1 + J_2)Ak_{T2} \end{aligned}$$

$$\varepsilon = J_1 k_T[NO_2]k_{P5}[NO] - (J_1 + J_2)Ak_{P5}[NO],$$

where

$$J_1 = P(OH) = 2f[O_3]j(O^1D)$$

(f is the fraction of $O(^1D)$ that reacts with H_2O vapour to form OH, rather than being quenched to $O(^3P)$).

$$J_2 = 2j(HCHO \rightarrow 2HO_2)[HCHO]$$

$$A = k_{CO+OH}[CO] + k_{H_2+OH}[H_2] + k_{HCHO+OH}[HCHO]$$
$$+ k_{CH_4+OH}[CH_4] + k_{NO_2+OH}[NO_2] + k_{O_3+OH}[O_3]$$

$$B = k_{HO_2+NO}[NO] + k_{HO_2+O_3}[O_3] + k_{loss}$$

$$k_T = k_{OH+NO_2}k_{T1} = k_{HO_2+HO_2}k_{T2} = k_{HO_2+CH_3O_2}$$

$$k_{T3} = k_{OH+HO_2}k_{P4} = k_{CH_4+OH}$$

$$k_{P5} = k_{CH_3O_2+NO}$$

Limited CO concentration data are available from the summit site during the project, owing to instrumental problems for the first 2 weeks of measurements. An average CO concentration of 231 ppbv was used in the analytical expression to determine HO_2 concentrations, although additional model runs at + and -1σ of this average concentration (297 ppbv and 165 ppbv respectively) were also made to assess the sensitivity of the predicted HO_2 concentration to this constraint. Similarly, only discrete (non-continuous) measurements of HCHO were made during the project; an average value of 479 pptv was used as a model constraint and further model runs at + and -1σ of this average concentration (818 pptv and 139 pptv respectively) were made.

$j(O^1D)$ was measured from the top of the 22 m tower, alongside the FAGE detection cell, using a 2-π filter radiometer (Bohn et al., 2008) which pointed skywards

Table 1. Details of ancillary measurements used for comparison with radical observations and cubic model constraints.

Measurement	Instrument
Liquid water content	Gerber particle volume monitor
Particle surface area (drops)	Gerber particle volume monitor
Effective drop radius	Gerber particle volume monitor
Temperature	Automatic weather station
Relative humidity	Automatic weather station
$j(O^1D)$	Filter radiometer
Cloud droplet pH	Mettler 405-60 88TE-S7/120
NO_x	Chemiluminescence detector
O_3	TEI 42c, UV absorption
CO	Thermo Electron CO analyser
HCHO	2,4-dinitrophenylhydrazine (DNPH) cartridge samples

throughout the campaign. The photolysis rates of formaldehyde, $j(HCHO)$, have been calculated using the Tropospheric Ultraviolet and Visible (TUV) radiation model (Madronich and Flocke, 1998). The correlation between TUV calculated $j(HCHO)$ with TUV calculated $j(O^1D)$ was determined, allowing these photolysis rates to be scaled to the measured $j(O^1D)$ values to account for the presence of clouds. During cloud events, upward radiation will increase, with the magnitude of this increase dependent on the cloud optical depth (COD) and measurement height (Bohn, 2014). The contribution of upward radiation as a function of COD has been estimated using the TUV model using the methodology outlined by Bohn (2014). This estimated increase in upward radiation has been added to the in-cloud photolysis rates presented in Sect. 3. On average, photolysis rates are enhanced by $\sim 17\%$ during cloud events due to upwelling. A constant value of 1760 ppbv was assumed for CH_4 and a value of 508 ppbv was taken for H_2. O_3 and NO_x measurements were made from the top of the tower using commercial analysers which ran continuously from 16 September (day 3 of the field project). Details of the ancillary measurements used for comparison and model constraints are provided in Table 1. Further details of many of the measurement techniques can be found in the overview paper from an earlier hill cap cloud experiment, the Field Investigations of Budgets and Conversions of Particle Phase Organics in Tropospheric Cloud Processes (FEBUKO) project (Herrmann et al., 2005).

Rate coefficients are taken from the most recent recommendations in the Master Chemical Mechanism (MCMv3.2, http://mcm.leeds.ac.uk/MCM/).

A constant uptake rate for HO_2 (k_{Loss}) of $0.14\,s^{-1}$ to cloud droplets was included during cloud events to reproduce the average HO_2 in-cloud observations. Additional model runs with no uptake during cloud events have also been run for comparison, as have model runs in which the first-order loss to droplets was varied to replicate the HO_2 observations as a function of (i) cloud droplet surface area and (ii) pH (Sect. 3.1).

2.3 Aqueous-phase chemistry

An outline of the aqueous-phase reactions thought to be occurring, and which converts HO_2 to H_2O_2, is given below:

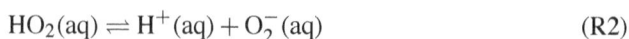

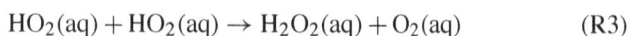

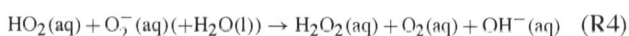

$$HO_2(g) \rightleftharpoons HO_2(aq) \tag{R1}$$

$$HO_2(aq) \rightleftharpoons H^+(aq) + O_2^-(aq) \tag{R2}$$

$$HO_2(aq) + HO_2(aq) \rightarrow H_2O_2(aq) + O_2(aq) \tag{R3}$$

$$HO_2(aq) + O_2^-(aq)(+H_2O(l)) \rightarrow H_2O_2(aq) + O_2(aq) + OH^-(aq) \tag{R4}$$

$$O_2^- + O_3(aq)(+H_2O(l)) \rightarrow OH^-(aq) + OH(aq) + 2O_2 \tag{R5}$$

The equations used to calculate the theoretical increase in γ_{HO_2} with increasing pH, as proposed by Thornton et al. (2008), which have been compared with γ_{HO_2} determined in this work (Sect. 3.1), are given by

$$\frac{1}{\gamma_{HO_2}} = \frac{1}{\alpha_{HO_2}} + \frac{3\omega N_A}{8000(H_{eff}RT)^2 k_{eff}[HO_2(g)]r_P}, \tag{4}$$

where

$$H_{eff} = H_{HO_2}\left[1 + \frac{K_{eq}}{[H^+]}\right] \tag{5}$$

and

$$k_{eff} = \frac{k_3 + \left(\frac{K_{eq}}{[H^+]_{aq}}\right)k_4}{\left(1 + \frac{K_{eq}}{[H^+]_{aq}}\right)^2}. \tag{6}$$

The values used in Eqs. (4)–(6) to calculate γ_{HO_2} are provided in Table 2.

2.4 Trajectory model

In addition to the modelling exercises, outlined in Sect. 2.2 above, an up-to-date chemistry process model, SPACCIM (SPectral Aerosol Cloud Chemistry Interaction Model (Wolke et al., 2005)), has been used to simulate the gas-phase HO_2 radical concentrations along a trajectory during the mountain overflow of an air parcel passing an orographic hill cap cloud to further explore the heterogeneous loss processes occurring during the cloud events encountered. This model combines complex microphysical and detailed multi-phase chemistry, permitting a detailed description of the chemical processing of gases, deliquesced particles and cloud droplets. SPACCIM incorporates the MCMv3.1-CAPRAMv4.0a mechanism (Master Chemical Mechanism (Saunders et al., 2003)/Chemical Aqueous Phase RAdical Mechanism (Tilgner et al., 2013; Braeuer et al., 2015)) with 11 381 gas-phase and 7125 aqueous-phase reactions.

The MCMv3.1-CAPRAM4.0a mechanism incorporates a detailed description of the inorganic and organic multi-phase chemistry, including phase transfer in deliquesced particles and cloud droplets based on time-dependent size-resolved aerosol–cloud spectra. Further details about the SPACCIM model framework and the chemical mechanisms are given elsewhere in the literature (Tilgner et al., 2013; Wolke et al., 2005; Sehili et al., 2005, and references therein).

The measured meteorological data as well as the physical and chemical aerosol and gas-phase data at the upwind site in the village of Goldlauter provided the basis for the time-resolved initialisation of the model. In addition, separate initial box model runs with the MCM mechanism were performed to provide a more comprehensive initialisation of the chemical gas-phase composition at the simulation start. SPACCIM simulations were performed with an air parcel advected along a predefined orography – following the trajectory from the upwind site (Goldlauter) through the hill cap cloud, passing Mt Schmücke (summit site), to the downwind site (Gehlberg). Parcel simulations were performed every 20 min, allowing a time-resolved comparison of the predicted and measured HO_2 data at the summit site.

2.5 Global chemistry transport model

The GEOS-Chem model version 9.1.3 (www.geoschem.org) has been run to assess the global impact of the uptake of HO_2 by cloud droplets. The model was run at $2° \times 2.5°$ global resolution for 2 years. The first year was considered a spin-up and has been ignored. The standard model includes uptake of HO_2 onto aerosols (with an uptake coefficient of 0.2), but the model has been updated in this work to include an uptake of HO_2 onto clouds. This is parameterised as a first-order loss onto clouds in a similar way to that onto aerosols following Schwartz (1984) using the temperature-dependent parameterisation of Thornton et al. (2008) with a cloud pH of 5. The cloud surface area is derived from the cloud liquid water in each model grid box (provided from the meteorological analyses) and cloud droplet radius is taken to be $6\,\mu m$ over continents and $10\,\mu m$ over oceans. Clouds below 258 K are assumed to be ice, and no uptake occurs. The parameterisation takes diffusional limitation in the gas phase into account, but not in the cloud phase. All simulations use the same cloud liquid water fields; thus, the impact of clouds on photolysis, wet deposition and transport is identical in all simulations.

3 Results and discussion

Near-continuous OH and HO_2 measurements were made at the Mt Schmücke site from 13 September to 19 October 2010, during which 35 separate orographic cloud events were encountered which lasted as little as 24 min to more than 2 days in duration. Figure 2 shows the time series of OH, HO_2, $j(O^1D)$, NO, O_3 and liquid water content. OH con-

Table 2. The values used for the calculation of the theoretical uptake coefficient, black triangles, Fig. 5b, as a function of pH; values given at pH $= 5$ here.

Parameter	Value	Comments
T (Temperature)	279 K	Mean HCCT-2010 temperature
H_{HO_2} (Henry's law constant)	1.72×10^4 M atm^{-1}	At 279 K
H_{eff} (Effective Henry's law constant)	8.8×10^4 M atm^{-1}	At 279 K, pH $= 5$
K_{eq} (Equilibrium constant associated with R2)	4.2×10^{-5} M	At 279 K
k_3 (Rate constant for reaction R3)	8.6×10^5 M^{-1} s^{-1}	Bielski et al. (1985)
k_4 (Rate constant for reaction R4)	1.0×10^8 M^{-1} s^{-1}	Bielski et al. (1985)
k_{eff} (Effective second-order rate constant)	1.65×10^7 M^{-1} s^{-1}	At 279 K, pH $= 5$
γ_{HO_2} (Accommodation coefficient)	1	
ω (Mean molecule speed of HO$_2$)	64 000 cm s^{-1}	At 279 K
N_A (Avogadro's number)	6.02×10^{23} mol^{-1}	
R (Universal gas constant)	0.082057 atm L mol^{-1} K^{-1}	
[HO$_2$]	2×10^7 molecule cm^{-3}	
r_p (Particle radius)	6 μm	Mean cloud droplet radius

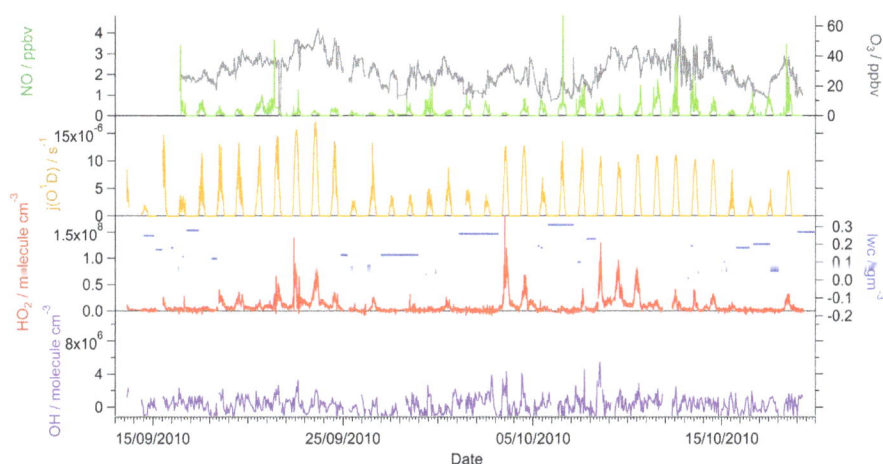

Figure 2. Time series showing the average liquid water content during each cloud episode (blue, horizontal lines), [OH] (purple), [HO$_2$] (red), $j(O^1D)$ (orange), NO (green) and O$_3$ (grey). All data are the average concentrations determined for each FAGE data acquisition cycle, apart from OH concentrations, which are hourly.

centrations were close to or below the limit of detection (LOD) of the instrument for much of the measurement period. A clear diurnal signal was only observable when several days of data were averaged together outside of cloud events (Fig. 3). The peak OH concentration was observed at midday at $\sim 1 \times 10^6$ molecule cm^{-3}. No clear OH diurnal profile was observed during cloud events. HO$_2$ concentrations were variable, depending on whether the site was in cloud or not. The average diurnal peak concentration of HO$_2$ was $\sim 4 \times 10^7$ molecule cm^{-3} outside of cloud events (Fig. 3). A diurnal profile of HO$_2$ was also observed when sampling within clouds with peak concentrations reduced by approximately 90 % on average. The measured rate of ozone photolysis, $j(O^1D)$, varied with time of day and cloud thickness. Daily peak photolysis rates were 8.8×10^{-6} s^{-1}

and 4.1×10^{-6} s^{-1} outside and within clouds, respectively. Clouds thus reduced photolysis rates by ~ 60 %.

Figure 4 shows the dependence of measured HO$_2$ concentration on cloud droplet surface area for all daytime cloud events. The observed HO$_2$ concentration has been divided by the observed $j(O^1D)$ to remove the impact of the changing photolysis rates within the cloud. This ratio has then been normalised to 1 when the droplet surface area was zero and plotted against the cloud droplet surface area. The decrease in the ratio with increasing droplet surface area suggests that in addition to the reduction in HO$_2$ caused by a reduction in the photolysis rates within clouds, there is a further loss process of HO$_2$ that increases with cloud droplet surface area. A similar decrease in the ratio is also observed with increasing liquid water (not shown). From these observations it becomes

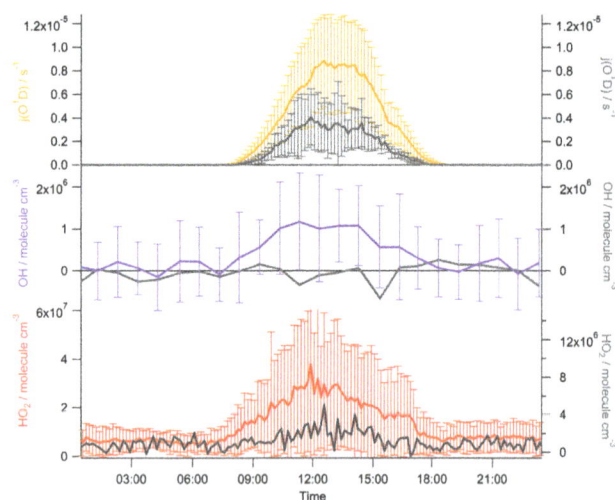

Figure 3. Average diurnal profiles of $j(O^1D$, OH and HO$_2$ in cloud (grey) and out of cloud (coloured). The error bars represent the 1σ variability of the averaged data; only the variability in the out-of-cloud radical data is shown for clarity. Each data point represents 10 min averaged data apart from the OH, for which the hourly averaged data are given.

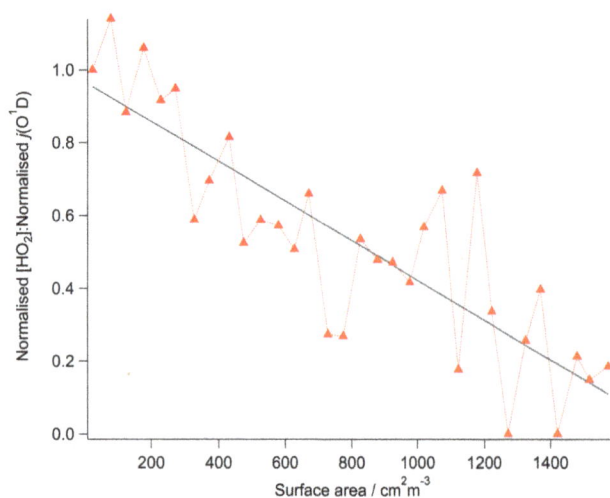

Figure 4. The dependence of the measured HO$_2$ concentration as a function of cloud droplet surface area. To remove the influence of changing photolysis rates, the measured HO$_2$ concentrations have been divided by the correspondingly observed rate of photolysis of ozone ($j(O^1D)$). This ratio has then been normalised to give a value of 1 when the droplet surface area was zero. The systematic decrease in this normalised ratio with increasing droplet surface area suggests that in addition to the reduction in HO$_2$ caused by a reduction in the photolysis rates within clouds, there is a further loss process that increases with cloud droplet surface area. The ratio decreases linearly with increasing droplet surface area up to $1500\,cm^2m^{-3}$, with the line of best fit being ratio $= 1 - 5 \times 10^{-4} \times SA$.

apparent that a heterogeneous process must be occurring in the presence of clouds.

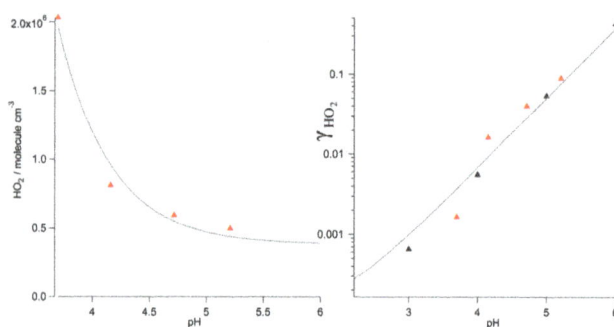

Figure 5. (a) Dependence of the HO$_2$ concentration observed in clouds as a function of cloud pH. All in-cloud HO$_2$ data were averaged into corresponding pH bins (0.6 pH units). The [HO$_2$] decreases exponentially with increasing pH with the line of best fit ([HO$_2$] $= 3.8 \times 10^5 + 5.5 \times 10^9 \exp^{-2.2\,pH}$) displayed by the grey line. Figure 5b: the cloud uptake coefficient estimated by optimising the HO$_2$ concentration calculated from the analytic expression of Carlsaw et al. (1999) compared to the observed HO$_2$ concentration as a function of pH (red triangles). The theoretical expression derived by Thornton et al. (2008) (Eq. 4) using parameters provided in Table 2 is shown as the black triangles with the grey line being a best-fit line for these data ($\gamma_{HO_2} = 2.15 \times 10^{-6} \exp^{2.01\,pH}$).

An insight into the mechanism by which HO$_2$ is lost to clouds is demonstrated by the dependence of the measured HO$_2$ concentration as a function of cloud water pH (Fig. 5a). Throughout the project the pH of the cloud water was recorded every hour and ranged from 3.4 to 5.3. The lowest in-cloud HO$_2$ occurred in clouds with the highest cloud water pH, suggesting that the solubility of HO$_2$ was enhanced at higher pH, as might be expected given that HO$_2$ is a weak acid.

3.1 Determining the uptake coefficient for HO$_2$ to cloud droplets

The analytical expression derived by Carslaw et al. (1999), and given in Eq. (3), has been used to estimate HO$_2$ concentrations both in and out of cloud events (Fig. 6). The expression represents reasonably well the campaign mean diurnal observation of HO$_2$ outside of the cloud events during the daytime (red dashed line and shading). During cloud events, however, the model (black dashed line and shading) overestimates the observed (grey line) HO$_2$ throughout the day. The inclusion of a first-order loss process ($k_{Loss} = 0.14\,s^{-1}$) in the analytical expression is able to bring the observations and calculation into better agreement on average. The cloud droplet surface area was variable during the different cloud events encountered ($1.2 \pm 0.4 \times 10^3\,cm^2m^{-3}$), although no diurnal trend in this parameter was evident. A clear anti-correlation between the observed HO$_2$ concentration and droplet surface area was observed and this correlation could only be reproduced by the analytical expression by increasing k_{Loss} in the model from $2.0 \times 10^{-2}\,s^{-1}$ to $3.5 \times 10^{-1}\,s^{-1}$

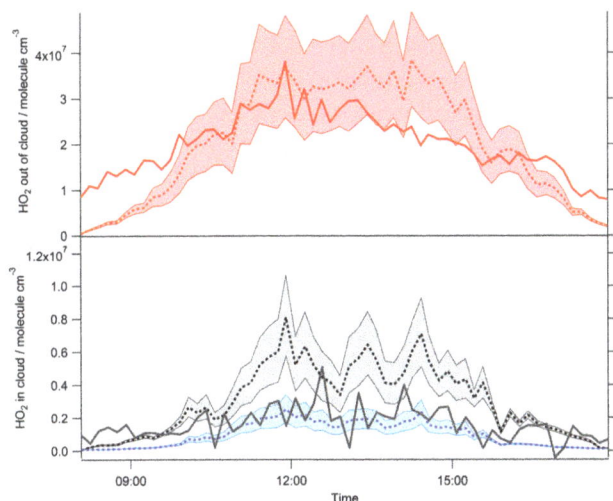

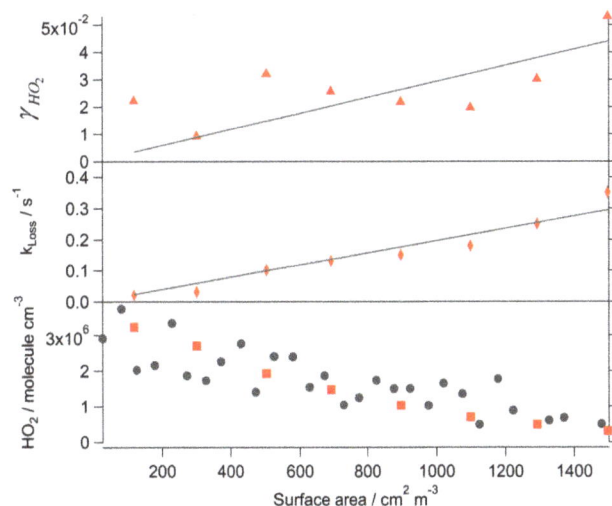

Figure 6. Upper panel. Average measured (solid red line) and simulated (dashed red line) diurnal profile of HO_2 concentrations outside of cloud events. The simulation is based on an expression originally determined by Carslaw et al. (1999) and described further in Sect. 2.2. The shading highlights the sensitivity of the model to $\pm 1\sigma$ changes in the CO and HCHO concentrations used as constraints. Lower panel. Average measured (solid grey line) and modelled (dashed black and blue lines) diurnal profile of HO_2 concentration during cloud events. The model was run without (grey) and with (blue) a loss of HO_2 to cloud droplets equal to a first-order loss rate of $0.14\,s^{-1}$. The shading highlights the sensitivity of the model to $\pm 1\sigma$ changes in the CO and HCHO concentrations used as constraints.

Figure 7. Lower panel. The dependence of the measured HO_2 concentration (grey circles) and modelled HO_2 concentration with a variable first-order loss (red squares) as a function of cloud droplet surface area. Middle panel. The dependence of the first-order loss term used in the model expression to best replicate the observed in-cloud HO_2 as a function cloud droplet surface area, the line of best fit being ($k_{Loss} = 2 \pm 0.1 \times 10^{-4} \times$ SA). Upper panel. The dependence of γ_{HO_2} calculated using Eq. (7) as a function of cloud droplet surface area and constrained with the variable first-order loss term as shown in the middle panel. The line of best fit is ($\gamma_{HO_2} = 2.9 \pm 0.5 \times 10^{-5} \times$ SA).

as the surface area increased from $1.2 \times 10^2\,cm^2\,m^{-3}$ to $1.5 \times 10^3\,cm^2\,m^{-3}$ (Fig. 7).

This first-order loss rate can be converted into an uptake coefficient (γ_{HO_2}) using Eq. (7) (Schwartz, 1984). Using campaign mean values for a cloud surface area (A) of $1.2 \times 10^3\,cm^2\,m^{-3}$, a droplet radius ($r_p$) of $6\,\mu m$, a gasphase diffusion constant for $HO_2 (D_g)$ of $0.25\,cm^2\,s^{-1}$, and a molecular speed of HO_2 (ω) of $64\,000\,cm\,s^{-1}$, gives an uptake coefficient of 0.01; the uptake coefficient as a function of cloud droplet surface area is presented in the upper panel of Fig. 7.

$$k_{loss} = \left(\frac{r_p}{D_g} + \frac{4}{\gamma_{HO_2}\omega} \right)^{-1} A \qquad (7)$$

These derived uptake coefficients are in good agreement with laboratory studies (Abbatt et al., 2012), including recent measurements in our laboratory, which ranged from 0.003 to 0.02, for heterogeneous loss of HO_2 on aqueous $(NH_4)_2SO_4$, NaCl and NH_4NO_3 sub-micron aerosols (George et al., 2013). This methodology provides, for the first time, a direct field assessment of the heterogeneous rate of loss of HO_2.

Repeating this analysis but splitting the observations by cloud pH leads to values of γ_{HO_2} ranging from 1.65×10^{-3}

at a pH of 3.7 to 8.84×10^{-2} at a pH of 5.2 (Fig. 5b). These values are in good agreement with those calculated by Thornton et al. (2008), suggesting that the Thornton mechanism (which is based entirely on the known aqueous-phase chemistry) is in play in real clouds and that it can be used to estimate the heterogeneous loss of HO_2 to cloud surfaces in the troposphere.

SPACCIM simulations (Wolke et al., 2005) have also been carried out, focussing on one particular cloud event which fulfilled the required meteorological and connected flow conditions for the cloud passage experiment (additional simulations relating to the other cloud events encountered during HCCT will be presented in future publications). The modelled and measured HO_2 concentrations at Mt Schmücke during the cloud event, FCE1.1, are presented in Fig. 8. Comparisons between modelled and measured concentrations demonstrate that the simulated HO_2 concentrations are in a similar range as the measurements. The mean simulated HO_2 concentrations of 3.1×10^6 molecule cm^{-3} for FCE1.1 are a factor of 1.4 greater than the HO_2 measurements which were, on average, 2.2×10^6 molecule cm^{-3} during this particular cloud event. A further trajectory model simulation has also been run and compared to measured HO_2 concentrations at Mt Schmücke during a non-cloud event, NCE0.8. Figure 9 reveals that the model is able to reproduce the modelled HO_2 concentrations well and tracks the temporal concentration

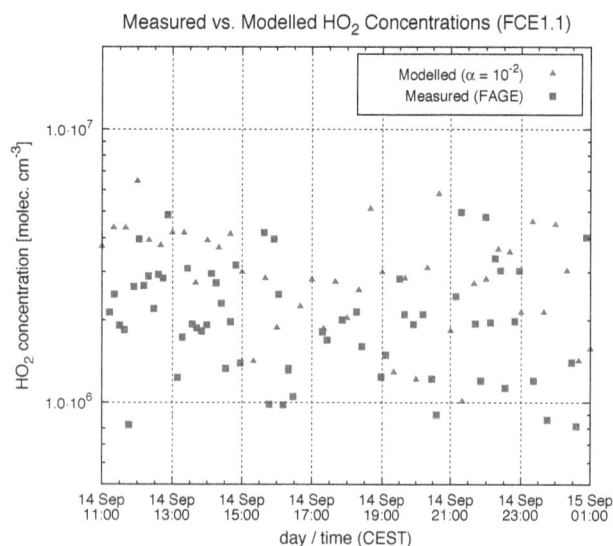

Figure 8. Comparison of the measured (green squares) and modelled (red triangles) gas-phase HO_2 concentrations at the Mt Schmücke site during cloud event FCE1.1 (14 and 15 September 2010, 11:00–01:00 CEST).

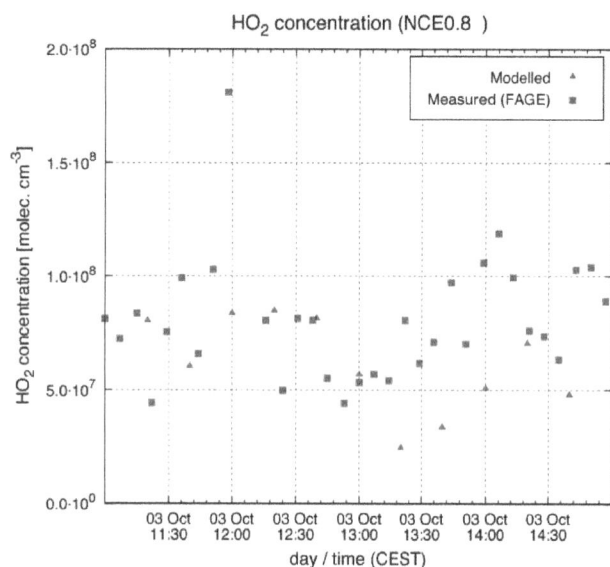

Figure 9. Comparison of the measured (green squares) and modelled (red triangles) gas-phase HO_2 concentrations at the Mt Schmücke site during non-cloud event NCE0.8.

profile throughout this event. The mean predicted HO_2 concentration is just 24 % smaller than the measurements.

The agreement between the trajectory modelled and measured in-cloud HO_2 values confirms the significant reductions in radicals within clouds predicted by complex multiphase box models in the past (Lelieveld and Crutzen, 1990; Tilgner et al., 2005, 2013) and supports the findings presented above. Importantly, the results imply that the phase transfer data for HO_2 used within SPACCIM simulations,

e.g. the applied mass accommodation coefficient ($\alpha_{HO_2} = 10^{-2}$), are appropriate for reproducing the reduced HO_2 concentrations for in-cloud conditions. These applied parameters control the uptake fluxes towards the aqueous phase and, ultimately, the aqueous-phase HO_x levels. Confidence in the values assumed for these parameters is essential to model in-cloud oxidation within the aqueous phase accurately, with the multi-phase chemistry of other important chemical subsystems, such as the S(IV) to S(VI) conversion, the redox cycling of transition metal ions and the processing of organic compounds all heavily dependent upon the values taken.

3.2 Global impact of the uptake of HO_2 onto cloud droplets

The GEOS-Chem (www.geos-chem.org) chemistry transport model has been used to assess the impact of the uptake of HO_2 onto cloud droplets on the global oxidising capacity using the now field-validated mechanism of Thornton et al. (2008). To investigate both the impact of the uptake and whether H_2O_2 is produced, three simulations are run, (i) with no cloud uptake of HO_2, (ii) with cloud uptake (an assumed pH of 5) of HO_2 using the Thornton mechanism to produce H_2O_2, and (iii) with cloud uptake (assumed pH of 5) of HO_2 to produce H_2O. All simulations include HO_2 uptake onto aerosol with γ_{HO_2} of 0.2, which is the standard value used in GEOS-Chem (Martin et al., 2003; Macintyre and Evans, 2011).

Figure 10 shows the annual fractional change in surface HO_2, OH, H_2O_2 and O_3 concentrations with cloud uptake switched on, and with H_2O_2 being either produced or not. Column changes are shown in Fig. 11. Both with and without H_2O_2 production, the impact is most evident in areas with long HO_2 lifetimes, i.e. regions with low NO_x and low HO_2 concentrations, and with significant cloud water densities (see Fig. 12). These are concentrated in the extra-tropics, with up to 25 and 10 % reduction in surface and column concentrations respectively. The impact on the H_2O_2 concentration depends critically on whether H_2O_2 is produced or not within clouds. In the extra-tropics there are up to 30 % increases in surface H_2O_2 if it is produced, with a similar reduction if it is not. The impacts on the surface extra-tropical oxidising capacity (OH) are of the order 10–20 % for both cases, but changes to the column values are only significant in the case where H_2O_2 is not produced. Changes in O_3 concentration are surprisingly small in both simulations. This reflects both the anti-correlation between NO concentrations and HO_2 lifetimes, and the low cloud water densities over the polluted continental regions. The largest fractional changes in HO_2 concentration occur in regions which are not producing O_3. The change in the lifetime due to the HO_2 uptake onto clouds thus has little impact on O_3 production. The large surface impact of the cloud uptake primarily reflects uptake of HO_2 by clouds at the surface (see Fig. 12a) rather than a transported impact of cloud processes from aloft

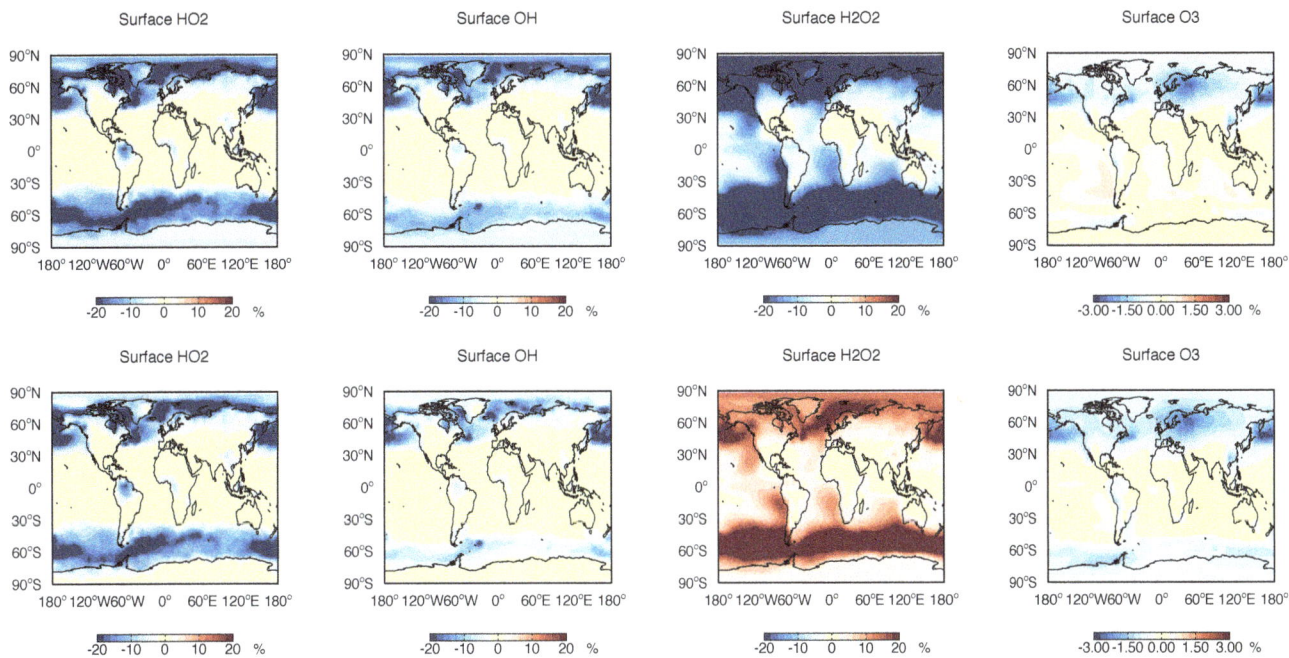

Figure 10. Annually averaged fractional change in surface HO_2, OH, H_2O_2 and O_3 with the inclusion of HO_2 uptake into clouds leading to: (top) the production of H_2O and (bottom) the production of H_2O_2 assuming a cloud pH of 5 and the Thornton et al. (2008) parameterisation.

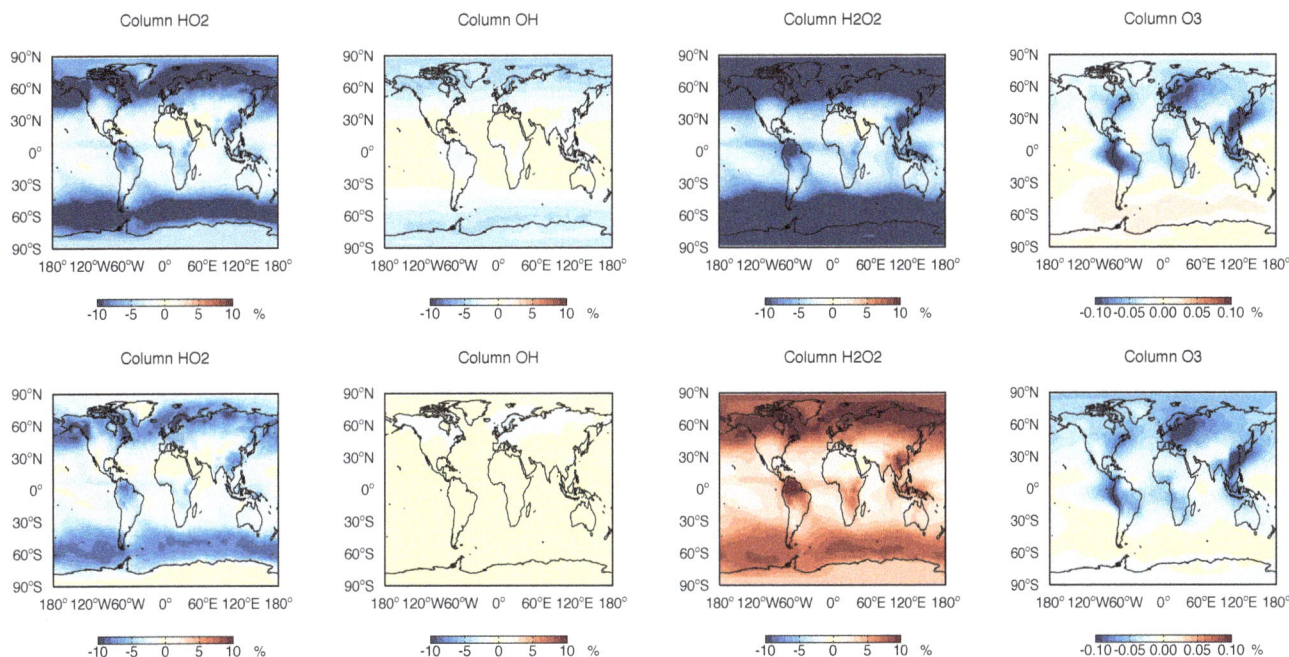

Figure 11. Annually averaged fractional change in column HO_2, OH, H_2O_2 and O_3 with the inclusion of HO_2 uptake into clouds leading to: (top) the production of H_2O and (bottom) the production of H_2O_2 assuming a cloud pH of 5 and the Thornton et al. (2008) parameterisation.

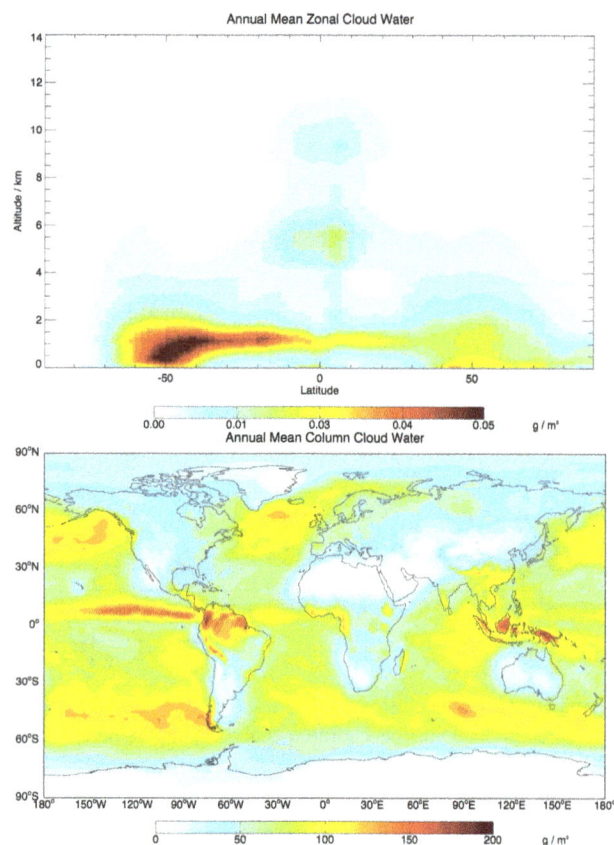

Figure 12. Annually averaged cloud water in the GEOS5 fields as (top) a column total and (bottom) a zonal mean.

downwards. The small impact on O_3 is consistent with results of Liang and Jacob (1997). These simulations make a variety of approximations as outlined in Sect. 2.5. Given the complexity of representing cloud processes occurring over the length scale of metres to hundreds of metres in a comparatively low-resolution global model (hundreds of kilometres), there are significant uncertainties as to the magnitude of these impacts. Further work in higher-resolution cloud-resolving models will be needed to estimate the full impact of these processes. Nevertheless, our observations show that the uptake of HO_2 onto clouds offers a substantial perturbation to the oxidising capacity on the local scale and that this perturbation may propagate into the regional and global scales.

4   Conclusions

We have shown here experimentally for the first time that the uptake of HO_2 onto clouds can have a significant impact on the composition of the atmosphere in a way consistent with theoretical predictions. It seems likely, however, that chemistry occurring within clouds will have other currently unknown impacts on the composition of the atmosphere. Global and regional models need to be developed further to investigate these impacts, with predictive pH an especially important development. The impact of these processes may also change in the future with climate-induced impacts on the hydrological cycle. Further laboratory field studies and modelling are required to help resolve these remaining complex questions.

Acknowledgements. The authors would like to thank Trevor Ingham, John Spence and Matthew Broadbent for help with the development of the FAGE instrument to facilitate tower measurements. Thanks also to Birger Bohn for his useful comments on the in-cloud photolysis rates. HCCT-2010 was partially funded by the German Research Foundation (DFG), grant He 3086/15-1. Stephan Mertes' participation was funded by the DFG, grant ME-3534/1-2. Lisa Whalley, Daniel Stone, Ingrid George, Mathew Evans and Dwayne Heard are grateful to the Natural Environment Research Council for funding.

Edited by: G. McFiggans

References

Abbatt, J. P. D., Lee, A. K. Y., and Thornton, J. A.: Quantifying trace gas uptake to tropospheric aerosol: recent advances and remaining challenges, Chem. Soc. Rev., 41, 6555–6581, doi:10.1039/C2cs35052a, 2012.

Bielski, B. H. J., Cabelli, D. E., Arudi, R. L., and Ross, A. B.: Reactivity of HO_2/O_2 radicals in aqueous solution, J. Phys. Chem. Ref. Data, 14, 1041–1100, doi:10.1063/1.555739, 1985.

Bohn, B., Corlett, G. K., Gillmann, M., Sanghavi, S., Stange, G., Tensing, E., Vrekoussis, M., Bloss, W. J., Clapp, L. J., Kortner, M., Dorn, H.-P., Monks, P. S., Platt, U., Plass-Dülmer, C., Mihalopoulos, N., Heard, D. E., Clemitshaw, K. C., Meixner, F. X., Prevot, A. S. H., and Schmitt, R.: Photolysis frequency measurement techniques: results of a comparison within the ACCENT project, Atmos. Chem. Phys., 8, 5373–5391, doi:10.5194/acp-8-5373-2008, 2008.

Bohn, B.: Interactive comment on "Influence of clouds on the oxidising capacity of the troposphere" by L. K. Whalley et al., Atmos. Chem. Phys. Discuss., 14, C7390–C7394, 2014.

Braeuer, P., Mouchel-Vallon, C., Tilgner, A., Mutzel, A., Böge, O., Rodigast, M., Poulain, L, van Pinxteren, D., Wolke, R., Aumont, B., and Herrmann, H.: Development of a protocol designed for the self-generation of explicit aqueous phase oxidation schemes of organic compounds, Atmos Chem Phys Discuss., in preparation, 2015.

Carslaw, N., Jacobs, P. J., and Pilling, M. J.: Modeling OH, HO_2, and RO_2 radicals in the marine boundary layer 2. Mechanism reduction and uncertainty analysis, J. Geophys. Res.-Atmos., 104, 30257–30273, doi:10.1029/1999jd900782, 1999.

Commane, R., Floquet, C. F. A., Ingham, T., Stone, D., Evans, M. J., and Heard, D. E.: Observations of OH and HO_2 radicals over West Africa, Atmos. Chem. Phys., 10, 8783–8801, doi:10.5194/acp-10-8783-2010, 2010.

Emmerson, K. M., Carslaw, N., Carslaw, D. C., Lee, J. D., Mc-Figgans, G., Bloss, W. J., Gravestock, T., Heard, D. E., Hopkins, J., Ingham, T., Pilling, M. J., Smith, S. C., Jacob, M., and Monks, P. S.: Free radical modelling studies during the UK TORCH Campaign in Summer 2003, Atmos. Chem. Phys., 7, 167–181, doi:10.5194/acp-7-167-2007, 2007.

Fuchs, H., Bohn, B., Hofzumahaus, A., Holland, F., Lu, K. D., Nehr, S., Rohrer, F., and Wahner, A.: Detection of HO_2 by laser-induced fluorescence: calibration and interferences from RO_2 radicals, Atmos. Meas. Tech., 4, 1209–1225, doi:10.5194/amt-4-1209-2011, 2011.

George, I. J., Matthews, P. S. J., Whalley, L. K., Brooks, B., Goddard, A., Romero, M. T. B., and Heard, D. E.: Measurements of uptake coefficients for heterogeneous loss of HO_2 onto submicron inorganic salt aerosols, Phys. Chem. Chem. Phys., 15, 12859–12845, 2013.

Haggerstone, A. L., Carpenter, L. J., Carslaw, N., and Mc-Figgans, G.: Improved model predictions of HO_2 with gas to particle mass transfer rates calculated using aerosol number size distributions, J. Geophys. Res.-Atmos., 110, D04304, doi:10.1029/2004jd005282, 2005.

Hanson, D. R., Burkholder, J. B., Howard, C. J., and Ravishankara, A. R.: Measurement of OH and HO_2 radical uptake coefficients on water and sulfuric-acid surfaces, J. Phys. Chem., 96, 4979–4985, doi:10.1021/J100191a046, 1992.

Herrmann, H., Wolke, R., Muller, K., Bruggemann, E., Gnauk, T., Barzaghi, P., Mertes, S., Lehmann, K., Massling, A., Birmili, W., Wiedensohler, A., Wieprecht, W., Acker, K., Jaeschke, W., Kramberger, H., Svrcina, B., Bachmann, K., Collett, J. L., Galgon, D., Schwirn, K., Nowak, A., van Pinxteren, D., Plewka, A., Chemnitzer, R., Rud, C., Hofmann, D., Tilgner, A., Diehl, K., Heinold, B., Hinneburg, D., Knoth, O., Sehili, A. M., Simmel, M., Wurzler, S., Majdik, Z., Mauersberger, G., and Muller, F.: FEBUKO and MODMEP: field measurements and modelling of aerosol and cloud multiphase processes, Atmos. Environ., 39, 4169–4183, doi:10.1016/j.atmosenv.2005.02.004, 2005.

Huijnen, V., Williams, J. E., and Flemming, J.: Modeling global impacts of heterogeneous loss of HO_2 on cloud droplets, ice particles and aerosols, Atmos. Chem. Phys. Discuss., 14, 8575–8632, doi:10.5194/acpd-14-8575-2014, 2014.

Jacob, D. J.: Chemistry of OH in remote clouds and its role in the production of formic acid and peroxymonosulfate, J. Geophys. Res.-Atmos., D9, 9807–9826, 1986.

Jacob, D. J.: Heterogeneous chemistry and tropospheric ozone, Atmos. Environ., 34, 2131–2159, doi:10.1016/S1352-2310(99)00462-8, 2000.

Lelieveld, J. and Crutzen, P. J.: Influences of cloud photochemical processes on tropospheric ozone, Nature, 343, 227–233, doi:10.1038/343227a0, 1990.

Liang, J. and Jacob, D. J.: Effect of aqueous phase cloud chemistry on tropospheric ozone, J. Geophys. Res.-Atmos., 102, 5993–6001, 1997.

Macintyre, H. L. and Evans, M. J.: Parameterisation and impact of aerosol uptake of HO_2 on a global tropospheric model, Atmos. Chem. Phys., 11, 10965–10974, doi:10.5194/acp-11-10965-2011, 2011.

Madronich, S. and Flocke, S.: The role of solar radiation in atmospheric chemistry, in: Handbook of Environmental Chemistry, edited by: Boule, P., Springer, New York, 1–26, 1998.

Mao, J., Fan, S., Jacob, D. J., and Travis, K. R.: Radical loss in the atmosphere from Cu-Fe redox coupling in aerosols, Atmos. Chem. Phys., 13, 509–519, doi:10.5194/acp-13-509-2013, 2013.

Martin, R. V., Jacob, D. J., Yantosca, R. M., Chin, M., and Ginoux, P.: Global and regional decreases in tropospheric oxidants from photochemical effects of aerosols, J. Geophys. Res.-Atmos., 108, 4097, doi:10.1029/2002JD0022622, 2003.

Mauldin, R. L., Madronich, S., Flocke, S. J., Eisele, F. L., Frost, G. J., and Prevot, A. S. H.: New insights on OH: Measurements around and in clouds, Geophys. Res. Lett., 24, 3033–3036, doi:10.1029/97gl02983, 1997.

Mauldin, R. L., Frost, G. J., Chen, G., Tanner, D. J., Prevot, A. S. H., Davis, D. D., and Eisele, F. L.: OH measurements during the first Aerosol Characterization Experiment (ACE 1): observations and model comparisons, J. Geophys. Res.-Atmos., 103, 16713–16729, doi:10.1029/98jd00882, 1998.

Nimmo, M. and Fones, G. R.: The potential pool of Co, Ni, Cu, Pb and Cd organic complexing ligands in coastal and urban rain waters, Atmos. Environ., 31, 693–702, doi:10.1016/S1352-2310(96)00243-9, 1997.

Olson, J. R., Crawford, J. H., Chen, G., Fried, A., Evans, M. J., Jordan, C. E., Sandholm, S. T., Davis, D. D., Anderson, B. E., Avery, M. A., Barrick, J. D., Blake, D. R., Brune, W. H., Eisele, F. L., Flocke, F., Harder, H., Jacob, D. J., Kondo, Y., Lefer, B. L., Martinez, M., Mauldin, R. L., Sachse, G. W., Shetter, R. E., Singh, H. B., Talbot, R. W., and Tan, D.: Testing fast photochemical theory during TRACE-P based on measurements of OH, HO_2, and CH_2O, J. Geophys. Res.-Atmos., 109, D15s10, doi:10.1029/2003jd004278, 2004.

Saunders, S. M., Jenkin, M. E., Derwent, R. G., and Pilling, M. J.: Protocol for the development of the Master Chemical Mechanism, MCM v3 (Part A): tropospheric degradation of non-aromatic volatile organic compounds, Atmos. Chem. Phys., 3, 161–180, doi:10.5194/acp-3-161-2003, 2003.

Schwartz, S. E.: Gas-phase and aqueous-phase chemistry of HO_2 in liquid water clouds, J. Geophys. Res.-Atmos., 89, 1589–1598, doi:10.1029/Jd089id07p11589, 1984.

Sehili, A. M., Wolke, R., Knoth, O., Simmel, M., Tilgner, A., and Herrmann, H.: Comparison of different model approaches for the simulation of multiphase processes, Atmos. Environ., 39, 4403–4417, doi:10.1016/j.atmosenv.2005.02.039, 2005.

Smith, S. C., Lee, J. D., Bloss, W. J., Johnson, G. P., Ingham, T., and Heard, D. E.: Concentrations of OH and HO_2 radicals during NAMBLEX: measurements and steady state analysis, Atmos. Chem. Phys., 6, 1435–1453, doi:10.5194/acp-6-1435-2006, 2006.

Sommariva, R., Haggerstone, A.-L., Carpenter, L. J., Carslaw, N., Creasey, D. J., Heard, D. E., Lee, J. D., Lewis, A. C., Pilling, M. J., and Zádor, J.: OH and HO_2 chemistry in clean marine air during SOAPEX-2, Atmos. Chem. Phys., 4, 839–856, doi:10.5194/acp-4-839-2004, 2004.

Spokes, L. J., Campos, M. L. A. M., and Jickells, T. D.: The role of organic matter in controlling copper speciation in precipitation, Atmos. Environ., 30, 3959–3966, doi:10.1016/1352-2310(96)00125-2, 1996.

Taketani, F., Kanaya, Y., and Akimoto, H.: Kinetics of heterogeneous reactions of HO$_2$ radical at ambient concentration levels with (NH$_4$)$_2$SO$_4$ and NaCl aerosol particles, J. Phys. Chem. A, 112, 2370–2377, doi:10.1021/Jp0769936, 2008.

Thornton, J. and Abbatt, J. P. D.: Measurements of HO$_2$ uptake to aqueous aerosol: mass accommodation coefficients and net reactive loss, J. Geophys. Res.-Atmos., 110, D08309, doi:10.1029/2004jd005402, 2005.

Thornton, J. A., Jaegle, L., and McNeill, V. F.: Assessing known pathways for HO$_2$ loss in aqueous atmospheric aerosols: regional and global impacts on tropospheric oxidants, J. Geophys. Res.-Atmos., 113, D05303, doi:10.1029/2007jd009236, 2008.

Tilgner, A., Majdik, Z., Sehili, A. M., Simmel, M., Wolke, R., and Herrmann, H.: SPACCIM: simulations of the multiphase chemistry occurring in the FEBUKO hill cap cloud experiments, Atmos. Environ., 39, 4389–4401, doi:10.1016/j.atmosenv.2005.02.028, 2005.

Tilgner, A., Brauer, P., Wolke, R., and Herrmann, H.: Modelling multiphase chemistry in deliquescent aerosols and clouds using CAPRAM3.0i, J. Atmos. Chem., 70, 221–256, doi:10.1007/s10874-013-9267-4, 2013.

Whalley, L. K., Furneaux, K. L., Goddard, A., Lee, J. D., Mahajan, A., Oetjen, H., Read, K. A., Kaaden, N., Carpenter, L. J., Lewis, A. C., Plane, J. M. C., Saltzman, E. S., Wiedensohler, A., and Heard, D. E.: The chemistry of OH and HO$_2$ radicals in the boundary layer over the tropical Atlantic Ocean, Atmos. Chem. Phys., 10, 1555–1576, doi:10.5194/acp-10-1555-2010, 2010.

Whalley, L. K., Blitz, M. A., Desservettaz, M., Seakins, P. W., and Heard, D. E.: Reporting the sensitivity of laser-induced fluorescence instruments used for HO$_2$ detection to an interference from RO$_2$ radicals and introducing a novel approach that enables HO$_2$ and certain RO$_2$ types to be selectively measured, Atmos. Meas. Tech., 6, 3425–3440, doi:10.5194/amt-6-3425-2013, 2013.

Wolke, R., Sehili, A. M., Simmel, M., Knoth, O., Tilgner, A., and Herrmann, H.: SPACCIM: a parcel model with detailed microphysics and complex multiphase chemistry, Atmos. Environ., 39, 4375–4388, doi:10.1016/j.atmosenv.2005.02.038, 2005.

11

Standard climate models radiation codes underestimate black carbon radiative forcing

G. Myhre and B. H. Samset

Center for International Climate and Environmental Research – Oslo (CICERO), Oslo, Norway

Correspondence to: G. Myhre (gunnar.myhre@cicero.oslo.no)

Abstract. Radiative forcing (RF) of black carbon (BC) in the atmosphere is estimated using radiative transfer codes of various complexities. Here we show that the two-stream radiative transfer codes used most in climate models give too strong forward scattering, leading to enhanced absorption at the surface and too weak absorption by BC in the atmosphere. Such calculations are found to underestimate the positive RF of BC by 10 % for global mean, all sky conditions, relative to the more sophisticated multi-stream models. The underestimation occurs primarily for low surface albedo, even though BC is more efficient for absorption of solar radiation over high surface albedo.

1 Introduction

Black carbon (BC) in the atmosphere has been investigated over many decades (Novakov and Rosen, 2013). The first estimate of radiative forcing (RF) of BC on a global scale was provided already 2 decades ago (Haywood and Shine, 1995). However, the diversity in existing estimates of the climate effect of BC is large (Bond et al., 2013; Boucher et al., 2013; Myhre et al., 2013). The causes for the diversity in estimates are many, from emissions (Amann et al., 2013; Cohen and Wang, 2014; Lam et al., 2012; Stohl et al., 2013; Wang et al., 2014b), lifetime and abundance (Hodnebrog et al., 2014; Samset et al., 2014; Wang et al., 2014a) to radiative efficiency (Samset et al., 2013; Zarzycki and Bond, 2010).

When estimating BC RF, the radiative transfer code is a crucial component. Accurate results can be achieved by using multi-stream line-by-line codes. However, these calculations are computationally demanding and are usually not applied for global scale simulations. In present climate mod-els, simplified radiation schemes of various complexity are therefore used and compared against line-by-line results, and each other, as consistency checks.

Several radiation intercomparison exercises have taken place (Boucher et al., 1998; Collins et al., 2006; Ellingson et al., 1991; Forster et al., 2011, 2005; Myhre et al., 2009b; Randles et al., 2013), yielding important suggestions for improvement to the radiative transfer codes. Randles et al. (2013) found that many of the presently used radiative transfer codes underestimate the radiative effect of absorbing aerosols, relative to benchmark multi-stream line-by-line codes. Further, one of the radiative transfer codes was run both as a multi-stream code resembling the benchmark codes, as well as run as a two-stream code resembling the simpler codes used in climate models. These two codes were denoted as numbers 3 and 4, respectively in Randles et al. (2013) and used in the current work. The results indicated that the number of streams in the radiative transfer calculation, i.e. the number of angles through which radiation is allowed to scatter, is crucial for the differences found between the radiation codes. On average, the simpler codes underestimated the radiative effect of BC of the order of 10–15 % relative to the benchmark line-by-line codes. In the present study we further investigate this potential underestimation of BC RF in many of the global climate models, and develop a physical understanding for why it occurs.

2 Models and methods

Simulations in the present paper were performed with a radiative transfer code using the discrete ordinate method (Stamnes et al., 1988). This model has previously been run in idealized experiments with prescribed vertical profiles of aerosol extinction (Randles et al., 2013) and used for global climate simulations (Myhre et al., 2009a). The radiative transfer code was run either in a multi-stream mode (eight streams) or with two streams and the Delta-M method (Wiscombe, 1977). In the global simulations we used meteorological data from ECMWF, and specified aerosol optical properties (Myhre et al., 2009a) and aerosol distribution from the OsloCTM2 chemical transport model (Skeie et al., 2011). To study the impact of the radiation code on global mean RF of BC, input fields and results from OsloCTM2 part of AeroCom Phase II for several aerosol components were used. Here, aerosol BC abundances were specified for 1850 and 2000, and anthropogenic RF defined as the difference between outgoing top-of-atmosphere shortwave radiative flux between these 2 years (Myhre et al., 2013).

3   Results

3.1   Global distribution of underestimated BC RF in models

Figure 1a shows the global mean, clear sky direct effect RF of BC, for a two-stream simulation relative to a simulation with eight streams. As in Randles et al. (2013) we find that the two-stream calculation tends to give lower RF than the eight-stream one. The underestimation in the two-stream simulation is shown here to be largest over ocean, with low surface albedo, whereas over regions with high surface albedo the two-stream more closely reproduces the eight-stream simulation. Under clear sky conditions, the global, annual mean underestimation is 15 % (0.158 vs. 0.187 W m^{-2}) in the two-stream relative to eight-stream simulation (RF (two-stream) divided by RF (eight-stream)).

The albedo of clouds is also affected by the number of streams adopted in the radiative transfer simulations. This makes the top-of-atmosphere reflected solar radiation increase in two-stream calculations, relative to eight-stream simulations. For all sky conditions, the global mean underestimation of RF in the two-stream simulation amounts to 7 %. However, modifying the cloud scattering to get similar top-of-atmosphere solar flux, as in the eight-stream simulation and close to measured fluxes, leads to a 10 % underestimation in the two-stream simulation relative to the eight-stream simulation (0.254 vs. 0.283 W m^{-2}). The largest underestimation is over ocean, and over regions with small cloud cover, as shown in Fig. 1b.

3.2   Underestimation of BC RF as a function of altitude

Global mean RF of BC, as a function of BC located at various altitudes, is shown in Fig. 2. The figure shows results for both two-stream and eight-stream simulations. A similar curve has previously been presented in Samset and Myhre (2011) for eight-stream simulations. The present curve is slightly modified, due to updated ozone and cloud fields. The same ap-

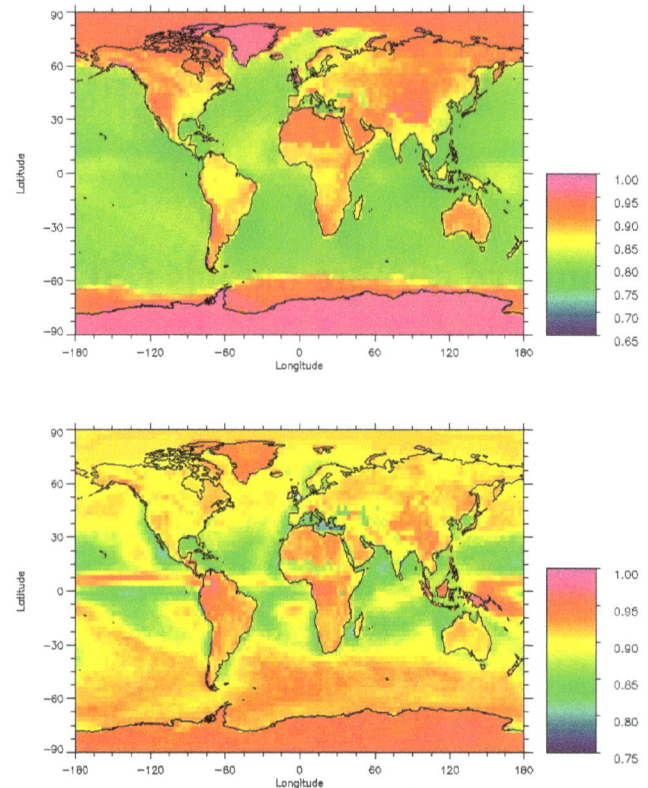

Figure 1. Geographical distribution of ratio between annual mean RF of BC from two-stream simulation relative to eight-stream simulation for clear sky (upper) and all sky (lower).

proach as in Sect. 3.1, to keep cloud scattering and therefore top-of-atmosphere radiative flux for the two-stream simulation equal to the eight-stream simulation, has been applied.

Figure 2a clearly shows the increasing normalized RF (RF exerted per unit aerosol burden) by BC as a function of altitude, due to enhanced effect of absorbing material above scattering components. The underestimation in the two-stream simulation is similar in magnitude for clear sky and all sky conditions, but is, in relative terms, larger for clear sky due to smaller absolute values (Fig. 2b).

For the all sky simulation the underestimation by the two-stream vs. the eight-stream simulation is close to 10 % for BC at all altitudes, except below 900 hPa. Being above scattering components such as clouds increases the absorption by BC, as does the presence of scattering aerosol types, and Rayleigh scattering. Absorption by gases such as ozone and water vapour, as well as absorption by other aerosol types, reduces the absorption by BC. For all sky conditions, Fig. 2 shows a large degree of compensation by scattering and absorption by gases, and other aerosol types than BC. In a model simulation with only BC in the atmosphere, the normalized RF of BC was found to be 1 % higher in two-stream simulations than in eight-stream simulations, showing the importance of

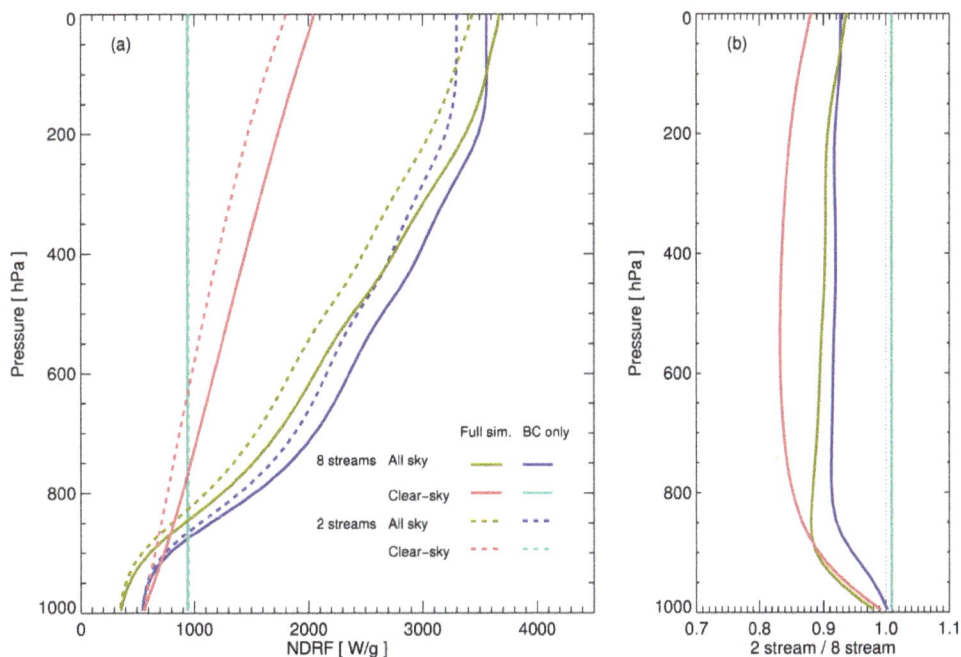

Figure 2. (a) BC RF normalized by abundance, as a function of altitude. Solid lines: eight-stream simulations. Dashed lines: two-stream simulations. Colours represent all sky and clear sky conditions, and whether a full atmospheric simulation including Rayleigh scattering, water vapour and background aerosols was performed ("Full sim"), or if BC was the only radiatively active agent ("BC only"); **(b)** ratio of two-stream to eight-stream simulation results, for the four cases shown in **(a)**.

the other atmospheric components for the correct determination of BC RF.

3.3 Physical description of the underestimation of BC RF

The radiative forcing due to aerosols is known to be a strong function of surface albedo (Haywood and Shine, 1997). This is illustrated in Fig. 3a, where the radiative effect of aerosols with different single scattering albedo has been calculated as a function of surface albedo. We reproduce the well-known characteristics of largest impact of absorbing aerosols over bright surfaces, and of scattering aerosols over dark surfaces.

Figure 3b shows the difference between two-stream and eight-stream calculations, as a function of surface albedo, and for a range of aerosol single scattering albedos. Two-stream and eight-stream results deviate substantially between surface albedos of 0.05 and 0.2. These are surface albedo values where absorbing aerosols have a relatively weak radiative effect. An increasing single scattering albedo gives increasing underestimations of two-stream results (Fig. 3b) and at the same time a decreasing radiative effect (Fig. 3a).

Our interpretation of the cause for the underestimation of two-stream results relative to multi-stream (containing more than two streams) results is lack of sufficient multiple scattering in connection to forward scattering and low surface albedo. Under such conditions the scattering is too strong in the forward direction in two-stream approaches. In addi-

tion the low surface albedo, and thus strong surface absorption, hinders further multiple scattering. Multiple scattering in general enhances the radiative effect of absorbing aerosols.

To illustrate the importance of multiple scattering for the abovementioned underestimation, additional simulations show that purely absorbing aerosols in a non-scattering atmosphere have differences between two-stream and multistream results within only a few percent (less than 2 %), which is the typical deviation as shown in Figure 3b, except for at low surface albedo. The agreement between 8-stream and even higher number of streams such as 16-stream simulations is generally within 1 %, except for very small absolute RF values. Simulations with four streams are generally close to eight-stream simulations. For pure scattering aerosols, two-stream simulations vary with solar zenith angle (see Randles et al., 2013) and surface albedo compared to eight-stream simulations; on a global mean, negative RF for anthropogenic sulphate aerosols is 5 % stronger. The results shown in Fig. 3 are for a solar zenith angle of 30°, but are generally applicable for other solar zenith angles. However, note that the critical single scattering albedo for transitioning from positive to negative radiative effect decreases with increasing solar zenith angle. The underestimation shown in Randles et al. (2013) can also be seen in Figure 3b for a single scattering of 0.75 (close to 0.8 used in the paper) and for a surface albedo of 0.2 at around 10 %.

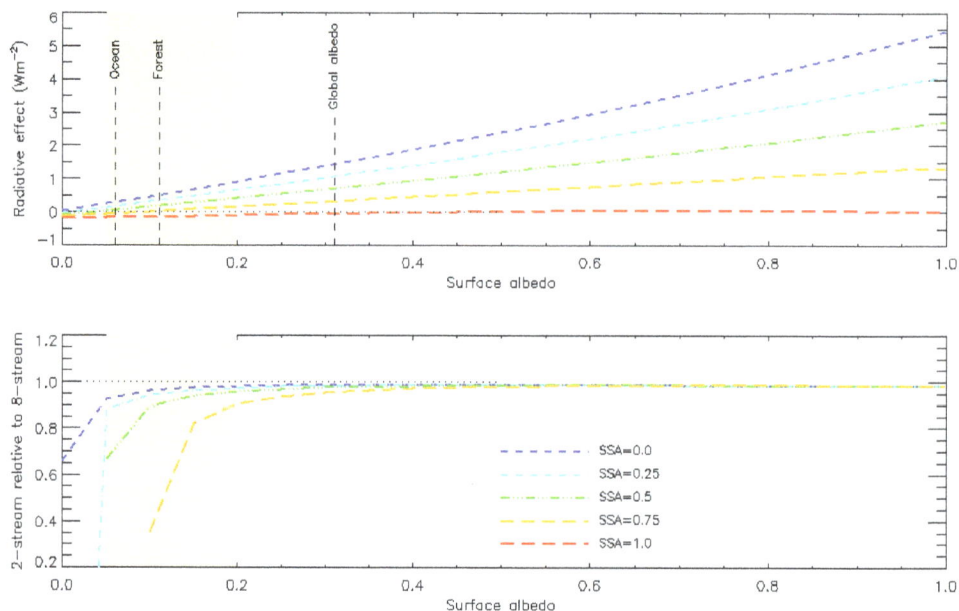

Figure 3. RF as a function of surface albedo for various single scattering albedo (upper), and relative differences between two-stream and multi-stream simulations (lower). In cases where the sign of two-stream and multi-stream simulations for a particular single-scattering albedo differs the results are left out of the lower panel.

4 Conclusions

Two-stream approximations using the Delta-M method, as employed by a majority of present climate models, are found to be relatively accurate for absorbing aerosols. The exception is over areas with low surface albedo. Here, the enhanced forward scattering hinders sufficient multiple scattering, causing an underestimation of the radiative effect of BC. Low albedo occurs in regions with low cloud cover, and low surface albedo such as ocean and snow free forest. In such cases the underestimation relative to more advanced radiation schemes can be of the order of 20–25 %. The underestimation for BC is largest in the presence of scattering components. This also applies to gases with solar absorption. However, under clear sky condition, underestimation of a similar magnitude to BC will only be caused by gases with solar absorption in UV and visible region where Rayleigh scattering is strong. Thus ozone in the lower troposphere is the only gas that is substantially influenced by the number of streams in the radiative transfer simulations. For a global increase in water vapour by 20 % in the lowest 1–2 km of the atmosphere, the difference between two-stream and eight-stream simulations is found to be less than 1 %.

On a global scale, we simulate a 10 % underestimation for RF of BC for all sky conditions, and 15 % for clear sky, for two-stream relative to eight-stream simulations. The clear sky results for selected profiles and solar zenith angles in Randles et al. (2013) showed an average model underestimation between 12 and 15 % compared to benchmark model simulations. The implication of the underestimation is that recent estimates of global mean RF due to BC, e.g., in Myhre et al. (2013) and Bond et al. (2013), where the latter is based on radiative transfer calculations in Schulz et al. (2006), could be up to 10 % too weak, as they are primarily based on models with radiative transfer codes with two-stream simulations. It must however be noted that other issues related to radiative transfer codes may lead to compensation of this underestimation, or additional underestimation. In addition, uncertainties in the abundance of BC, and in its optical properties, are much larger than 10 %. Burden of BC and the normalized RF has a standard deviation of the order of 50 % relative to mean values for the 15 global aerosol models in AeroCom Phase II (Myhre et al., 2013). Even so, considerations for improvements of radiation schemes in global climate models should be made to provide more accurate calculations of present and future radiative forcing due to BC.

Acknowledgements. This work has received funding from the European Union Seventh Framework Programme (FP7/2007–2013) under grant agreement no. 282688 – ECLIPSE as well as from the Research Council of Norway through the projects EarthClim, AEROCOM-P3, and SLAC.

Edited by: Y. Balkanski

References

Amann, M., Klimont, Z., and Wagner, F.: Regional and Global Emissions of Air Pollutants: Recent Trends and Future Scenarios, in: Annual Review of Environment and Resources, Vol. 38, Annual Review of Environment and Resources, edited by: Gadgil, A. and Liverman, D. M., 31–55, 2013.

Bond, T. C., Doherty, S. J., Fahey, D. W., Forster, P. M., Berntsen, T., DeAngelo, B. J., Flanner, M. G., Ghan, S., Karcher, B., Koch, D., Kinne, S., Kondo, Y., Quinn, P. K., Sarofim, M. C., Schultz, M. G., Schulz, M., Venkataraman, C., Zhang, H., Zhang, S., Bellouin, N., Guttikunda, S. K., Hopke, P. K., Jacobson, M. Z., Kaiser, J. W., Klimont, Z., Lohmann, U., Schwarz, J. P., Shindell, D., Storelvmo, T., Warren, S. G., and Zender, C. S.: Bounding the role of black carbon in the climate system: A scientific assessment, J. Geophys. Res.-Atmos., 118, 5380–5552, 2013.

Boucher, O., Schwartz, S. E., Ackerman, T. P., Anderson, T. L., Bergstrom, B., Bonnel, B., Chylek, P., Dahlback, A., Fouquart, Y., Fu, Q., Halthore, R. N., Haywood, J. M., Iversen, T., Kato, S., Kinne, S., Kirkevag, A., Knapp, K. R., Lacis, A., Laszlo, I., Mishchenko, M. I., Nemesure, S., Ramaswamy, V., Roberts, D. L., Russell, P., Schlesinger, M. E., Stephens, G. L., Wagener, R., Wang, M., Wong, J., and Yang, F.: Intercomparison of models representing direct shortwave radiative forcing by sulfate aerosols, J. Geophys. Res.-Atmos., 103, 16979–16998, 1998.

Boucher, O., Randall, D., Artaxo, P., Bretherton, C., Feingold, G., Forster, P., Kerminen, V.-M., Kondo, Y., Liao, H., Lohmann, U., Rasch, P., Satheesh, S. K., Sherwood, S., Stevens, B., and Zhang, X.-Y.: Clouds and Aerosols, in: Climate Change 2013: The Physical Science Basis. Contribution of Working Group I to the Fifth Assessment Report of the Intergovernmental Panel on Climate Change, edited by: Stocker, T. F., Qin, D., Plattner, G.-K., Tignor, M., Allen, S. K., Boschung, J., Nauels, A., Xia, Y., Bex, V., Midgley, P. M., Cambridge University Press, Cambridge, United Kingdom and New York, NY, USA, 571–657, 2013.

Cohen, J. B. and Wang, C.: Estimating global black carbon emissions using a top-down Kalman Filter approach, J. Geophys. Res.-Atmos., 119, 307–323, 2014.

Collins, W. D., Ramaswamy, V., Schwarzkopf, M. D., Sun, Y., Portmann, R. W., Fu, Q., Casanova, S. E. B., Dufresne, J. L., Fillmore, D. W., Forster, P. M. D., Galin, V. Y., Gohar, L. K., Ingram, W. J., Kratz, D. P., Lefebvre, M. P., Li, J., Marquet, P., Oinas, V., Tsushima, Y., Uchiyama, T., and Zhong, W. Y.: Radiative forcing by well-mixed greenhouse gases: Estimates from climate models in the Intergovernmental Panel on Climate Change (IPCC) Fourth Assessment Report (AR4), J. Geophys. Res.-Atmos., 111, D14317, doi:10.1029/2005JD006713, 2006.

Ellingson, R. G., Ellis, J., and Fels, S.: The intercomparison of radiation codes used in climate models – long-wave results, J. Geophys. Res.-Atmos., 96, 8929–8953, 1991.

Forster, P. M. D., Burkholder, J. B., Clerbaux, C., Coheur, P. F., Dutta, M., Gohar, L. K., Hurley, M. D., Myhre, G., Portmann, R. W., Shine, K. P., Wallington, T. J., and Wuebbles, D.: Resolution of the uncertainties in the radiative forcing of HFC-134a, J. Quant. Spectrosc. Ra., 93, 447–460, 2005.

Forster, P. M. D., Fomichev, V. I., Rozanov, E., Cagnazzo, C., Jonsson, A. I., Langematz, U., Fomin, B., Iacono, M. J., Mayer, B., Mlawer, E., Myhre, G., Portmann, R. W., Akiyoshi, H., Falaleeva, V., Gillett, N., Karpechko, A., Li, J. N., Lemennais, P., Morgenstern, O., Oberlander, S., Sigmond, M., and Shibata, K.: Evaluation of radiation scheme performance within chemistry climate models, J. Geophys. Res.-Atmos., 116, D10302, doi:10.1029/2010JD015361, 2011.

Haywood, J. M. and Shine, K. P.: The effect of anthropogenic sulfate and soot aerosol on the clear-sky planetary radiation budget, Geophys. Res. Lett., 22, 603–606, 1995.

Haywood, J. M. and Shine, K. P.: Multi-spectral calculations of the direct radiative forcing of tropospheric sulphate and soot aerosols using a column model, Q. J. Roy. Meteor. Soc., 123, 1907–1930, 1997.

Hodnebrog, Ø., Myhre, G., and Samset, B. H.: How shorter black carbon lifetime alters its climate effect, Nat. Commun., 5, 5065, doi:10.1038/ncomms6065, 2014.

Lam, N. L., Chen, Y., Weyant, C., Venkataraman, C., Sadavarte, P., Johnson, M. A., Smith, K. R., Brem, B. T., Arineitwe, J., Ellis, J. E., and Bond, T. C.: Household Light Makes Global Heat: High Black Carbon Emissions From Kerosene Wick Lamps, Environ. Sci. Technol., 46, 13531–13538, 2012.

Myhre, G., Berglen, T. F., Johnsrud, M., Hoyle, C. R., Berntsen, T. K., Christopher, S. A., Fahey, D. W., Isaksen, I. S. A., Jones, T. A., Kahn, R. A., Loeb, N., Quinn, P., Remer, L., Schwarz, J. P., and Yttri, K. E.: Modelled radiative forcing of the direct aerosol effect with multi-observation evaluation, Atmos. Chem. Phys., 9, 1365–1392, doi:10.5194/acp-9-1365-2009, 2009a.

Myhre, G., Kvalevag, M., Radel, G., Cook, J., Shine, K. P., Clark, H., Karcher, F., Markowicz, K., Kardas, A., Wolkenberg, P., Balkanski, Y., Ponater, M., Forster, P., Rap, A., and de Leon, R. R.: Intercomparison of radiative forcing calculations of stratospheric water vapour and contrails, Meteorol. Z., 18, 585–596, 2009b.

Myhre, G., Samset, B. H., Schulz, M., Balkanski, Y., Bauer, S., Berntsen, T. K., Bian, H., Bellouin, N., Chin, M., Diehl, T., Easter, R. C., Feichter, J., Ghan, S. J., Hauglustaine, D., Iversen, T., Kinne, S., Kirkevåg, A., Lamarque, J.-F., Lin, G., Liu, X., Lund, M. T., Luo, G., Ma, X., van Noije, T., Penner, J. E., Rasch, P. J., Ruiz, A., Seland, Ø., Skeie, R. B., Stier, P., Takemura, T., Tsigaridis, K., Wang, P., Wang, Z., Xu, L., Yu, H., Yu, F., Yoon, J.-H., Zhang, K., Zhang, H., and Zhou, C.: Radiative forcing of the direct aerosol effect from AeroCom Phase II simulations, Atmos. Chem. Phys., 13, 1853–1877, doi:10.5194/acp-13-1853-2013, 2013.

Novakov, T. and Rosen, H.: The Black Carbon Story: Early History and New Perspectives, Ambio, 42, 840–851, 2013.

Randles, C. A., Kinne, S., Myhre, G., Schulz, M., Stier, P., Fischer, J., Doppler, L., Highwood, E., Ryder, C., Harris, B., Huttunen, J., Ma, Y., Pinker, R. T., Mayer, B., Neubauer, D., Hitzenberger, R., Oreopoulos, L., Lee, D., Pitari, G., Di Genova, G., Quaas, J., Rose, F. G., Kato, S., Rumbold, S. T., Vardavas, I., Hatzianastassiou, N., Matsoukas, C., Yu, H., Zhang, F., Zhang, H., and Lu, P.: Intercomparison of shortwave radiative transfer schemes in global aerosol modeling: results from the AeroCom Radiative Transfer Experiment, Atmos. Chem. Phys., 13, 2347–2379, doi:10.5194/acp-13-2347-2013, 2013.

Samset, B. H. and Myhre, G.: Vertical dependence of black carbon, sulphate and biomass burning aerosol radiative forcing, Geophys. Res. Lett., 38, L24802, doi:10.1029/2011gl049697, 2011.

Samset, B. H., Myhre, G., Schulz, M., Balkanski, Y., Bauer, S., Berntsen, T. K., Bian, H., Bellouin, N., Diehl, T., Easter, R. C., Ghan, S. J., Iversen, T., Kinne, S., Kirkevåg, A., Lamarque, J.-

F., Lin, G., Liu, X., Penner, J. E., Seland, Ø., Skeie, R. B., Stier, P., Takemura, T., Tsigaridis, K., and Zhang, K.: Black carbon vertical profiles strongly affect its radiative forcing uncertainty, Atmos. Chem. Phys., 13, 2423–2434, doi:10.5194/acp-13-2423-2013, 2013.

Samset, B. H., Myhre, G., Herber, A., Kondo, Y., Li, S.-M., Moteki, N., Koike, M., Oshima, N., Schwarz, J. P., Balkanski, Y., Bauer, S. E., Bellouin, N., Berntsen, T. K., Bian, H., Chin, M., Diehl, T., Easter, R. C., Ghan, S. J., Iversen, T., Kirkevåg, A., Lamarque, J.-F., Lin, G., Liu, X., Penner, J. E., Schulz, M., Seland, Ø., Skeie, R. B., Stier, P., Takemura, T., Tsigaridis, K., and Zhang, K.: Modelled black carbon radiative forcing and atmospheric lifetime in AeroCom Phase II constrained by aircraft observations, Atmos. Chem. Phys., 14, 12465–12477, doi:10.5194/acp-14-12465-2014, 2014.

Schulz, M., Textor, C., Kinne, S., Balkanski, Y., Bauer, S., Berntsen, T., Berglen, T., Boucher, O., Dentener, F., Guibert, S., Isaksen, I. S. A., Iversen, T., Koch, D., Kirkevåg, A., Liu, X., Montanaro, V., Myhre, G., Penner, J. E., Pitari, G., Reddy, S., Seland, Ø., Stier, P., and Takemura, T.: Radiative forcing by aerosols as derived from the AeroCom present-day and pre-industrial simulations, Atmos. Chem. Phys., 6, 5225–5246, doi:10.5194/acp-6-5225-2006, 2006.

Skeie, R. B., Berntsen, T. K., Myhre, G., Tanaka, K., Kvalevåg, M. M., and Hoyle, C. R.: Anthropogenic radiative forcing time series from pre-industrial times until 2010, Atmos. Chem. Phys., 11, 11827–11857, doi:10.5194/acp-11-11827-2011, 2011.

Stamnes, K., Tsay, S. C., Wiscombe, W., and Jayaweera, K.: Numerically Stable Algorithm For Discrete-Ordinate-Method Radiative-Transfer In Multiple-Scattering And Emitting Layered Media, Appl. Optics, 27, 2502–2509, 1988.

Stohl, A., Klimont, Z., Eckhardt, S., Kupiainen, K., Shevchenko, V. P., Kopeikin, V. M., and Novigatsky, A. N.: Black carbon in the Arctic: the underestimated role of gas flaring and residential combustion emissions, Atmos. Chem. Phys., 13, 8833–8855, doi:10.5194/acp-13-8833-2013, 2013.

Wang, Q. Q., Jacob, D. J., Spackman, J. R., Perring, A. E., Schwarz, J. P., Moteki, N., Marais, E. A., Ge, C., Wang, J., and Barrett, S. R. H.: Global budget and radiative forcing of black carbon aerosol: Constraints from pole-to-pole (HIPPO) observations across the Pacific, J. Geophys. Res.-Atmos., 119, 195–206, 2014a.

Wang, R., Tao, S., Balkanski, Y., Ciais, P., Boucher, O., Liu, J. F., Piao, S. L., Shen, H. Z., Vuolo, M. R., Valari, M., Chen, H., Chen, Y. C., Cozic, A., Huang, Y., Li, B. G., Li, W., Shen, G. F., Wang, B., and Zhang, Y. Y.: Exposure to ambient black carbon derived from a unique inventory and high-resolution model, P. Natl. Acad. Sci. USA, 111, 2459–2463, 2014b.

Wiscombe, W. J.: Delta-M method – rapid yet accurate radiative flux calculations for strongly asymmetric phase functions, J. Atmos. Sci., 34, 1408–1422, 1977.

Zarzycki, C. M. and Bond, T. C.: How much can the vertical distribution of black carbon affect its global direct radiative forcing?, Geophys. Res. Lett., 37, L20807, doi:10.1029/2010gl044555, 2010.

12

Hygroscopic properties of NaCl and NaNO$_3$ mixture particles as reacted inorganic sea-salt aerosol surrogates

D. Gupta, H. Kim, G. Park, X. Li, H.-J. Eom, and C.-U. Ro

Department of Chemistry, Inha University, Incheon, 402-751, South Korea

Correspondence to: C.-U. Ro (curo@inha.ac.kr)

Abstract. NaCl in fresh sea-salt aerosol (SSA) particles can partially or fully react with atmospheric NO$_x$/HNO$_3$, so internally mixed NaCl and NaNO$_3$ aerosol particles can co-exist over a wide range of mixing ratios. Laboratory-generated, micrometer-sized NaCl and NaNO$_3$ mixture particles at 10 mixing ratios (mole fractions of NaCl (X_{NaCl} = 0.1 to 0.9) were examined systematically to observe their hygroscopic behavior, derive experimental phase diagrams for deliquescence and efflorescence, and understand the efflorescence mechanism. During the humidifying process, aerosol particles with the eutonic composition ($X_{NaCl} = 0.38$) showed only one phase transition at their mutual deliquescence relative humidity (MDRH) of 67.9 (± 0.5) %. On the other hand, particles with other mixing ratios showed two distinct deliquescence transitions; i.e., the eutonic component dissolved at MDRH, and the remainder in the solid phase dissolved completely at their DRHs depending on the mixing ratios, resulting in a phase diagram composed of four different phases, as predicted thermodynamically. During the dehydration process, NaCl-rich particles ($X_{NaCl} > 0.38$) showed a two stage efflorescence transition: the first stage was purely driven by the homogeneous nucleation of NaCl and the second stage at the mutual efflorescence RH (MERH) of the eutonic components, with values in the range of 30.0–35.5 %. Interestingly, aerosol particles with the eutonic composition ($X_{NaCl} = 0.38$) also showed two-stage efflorescence, with NaCl crystallizing first followed by heterogeneous nucleation of the remaining NaNO$_3$ on the NaCl seeds. NaNO$_3$-rich particles ($X_{NaCl} \leq 0.3$) underwent single-stage efflorescence transitions at ERHs progressively lower than the MERH because of the homogeneous nucleation of NaCl and the almost simultaneous heterogeneous nucleation of NaNO$_3$ on the NaCl seeds. SEM/EDX elemental mapping indicated that the effloresced NaCl–NaNO$_3$ particles at all mixing ratios were composed of a homogeneously crystallized NaCl moiety in the center, surrounded either by the eutonic component (for $X_{NaCl} > 0.38$) or NaNO$_3$ (for $X_{NaCl} \leq 0.38$). During the humidifying or dehydration process, the amount of eutonic composed part drives particle/droplet growth or shrinkage at the MDRH or MERH (second ERH), respectively, and the amount of pure salts (NaCl or NaNO$_3$ in NaCl- or NaNO$_3$-rich particles, respectively) drives the second DRHs or first ERHs, respectively. Therefore, their behavior can be a precursor to the optical properties and direct radiative forcing for these atmospherically relevant mixture particles representing the coarse, reacted inorganic SSAs. In addition, the NaCl–NaNO$_3$ mixture aerosol particles can maintain an aqueous phase over a wider RH range than pure NaCl particles as SSA surrogate, making their heterogeneous chemistry more probable.

1 Introduction

Atmospheric aerosols play important roles in global climate change, directly by scattering or absorbing incoming solar radiation and indirectly by serving as cloud condensation nuclei (Pandis et al., 1995; Satheesh and Moorthy, 2005). The radiative effects depend on the chemical composition and sizes of the atmospheric aerosol particles. The optical properties and the chemical reactivity of the atmospheric aerosols also depend on their mixing states and different aerosol phases (Martin, 2000). Studies of the hygroscopic properties of inorganic salt particles as aerosol surrogates can provide better insights to several of these important aerosol properties, such as (1) alteration of aerodynamic properties;

(2) cloud-droplet nucleation efficiency; (3) optical properties; and (4) physicochemical changes, through complex heterogeneous chemical reactions with atmospheric gas-phase species (Wang and Martin, 2007; Haywood and Boucher, 2000; ten Brink, 1998; Krueger et al., 2003).

Sea-salt or sea-spray aerosols (SSAs) comprise a large proportion of the atmospheric particulate mass (25–50 %) (Finlayson–Pitts and Pitts, 2000). Thus far, many studies have examined the hygroscopic behavior of both airborne and laboratory-generated SSAs (Tang et al., 1997; Wise et al., 2007, 2009; Prather et al., 2013), but the hygroscopicity of the SSAs is not completely understood (Meskhidze et al., 2013). NaCl in the nascent SSAs can react quickly (within a few minutes to an hour of residence in air) with the atmospheric NO_x/HNO_3 (ten Brink, 1998; Saul et al., 2006; Liu et al., 2007). This can lead to the formation of partially or fully reacted particles containing NaCl and $NaNO_3$ over a range of mixing ratios. Indeed, studies of individual marine aerosols have clearly shown the existence of fully or partially reacted SSA particles, and a significant portion of these particles were reported to be mixtures of sodium chloride, nitrate, and/or sulfate (Gard et al., 1998; Ro et al., 2001; Laskin et al., 2003; Ault et al., 2014). Moreover, the further reactive uptake of N_2O_5 was reported to be dependent on the chloride to nitrate ratio of the reacted SSAs and their phases (Ryder et al., 2014). The primary and secondary organics, biogenic particulates, sea-salt sulfates (ss-SO_4^{2-}), non-sea-salt sulfates (nss-SO_4^{2-}), etc., add greater complexity to these SSAs (O'Dowd and de Leeuw, 2007; Keene et al., 2007; Prather et al., 2013; Beardsley et al., 2013; Ault et al., 2013b). A detailed knowledge of the hygroscopic properties, mixing states, and the spatial distribution of the chemical components in NaCl–$NaNO_3$ mixture particles, as partially or fully reacted SSA surrogates, can serve as a good preliminary step to a better understanding of the complex chemical/physical mixing states, hygroscopic behavior, and reactivity of ambient SSAs.

Many studies have examined the hygroscopic properties of two-component inorganic salt particles as inorganic aerosol surrogates (Cohen et al., 1987a; Tang and Munkelwitz, 1993, 1994a, b; Tang et al., 1978; Ge et al., 1996, 1998; Chang and Lee, 2002). Stepwise phase transitions generally occur for particles composed of two inorganic salts during the humidifying process (Wexler and Seinfeld, 1991). On the other hand, particles with the eutonic composition deliquesce completely at the mutual deliquescence relative humidity (MDRH), resulting in a single phase transition. For two-component inorganic hygroscopic salt particles, the first transition generally occurs at their MDRH, and the aqueous phase resulting from the partial deliquescence has the eutonic composition. The partially dissolved particles keep absorbing water with further increases in the RH and the residual solid component completely dissolves when the RH reaches their DRH, which depends on the composition of the particle. As the humidifying processes of in-

organic salts are governed by thermodynamics, a range of thermodynamic models have been developed to predict the deliquescence behavior or the ionic activity coefficients of two-component aerosol particles (Tang, 1976; Ansari and Pandis, 1999; Clegg et al., 1998; Wexler and Clegg, 2002; Zuend et al., 2008, 2011). The Extended Atmospheric Inorganics Model (E-AIM) predicts the physical state and chemical compositions of aerosols containing several atmospherically relevant inorganic ionic species and/or organic species (http://www.aim.env.uea.ac.uk/aim/aim.php). The Aerosol Inorganic-Organic Mixtures Functional groups Activity Coefficient (AIOMFAC) model allows calculations of the activity coefficients in organic and/or inorganic mixtures from simple binary solutions to complex multicomponent systems (http://www.aiomfac.caltech.edu). On the other hand, there have been few systematic, experimental hygroscopic studies to support the theoretical models for mixed salt particles.

During the dehydration process, where the RH is decreased from high to low, the concentration of single salts in the aqueous droplets becomes dense and the inorganic single salts can be finally crystallized at their efflorescence RH (ERH). The ERH is sometimes significantly lower than the DRH. For example, pure NaCl particles have a DRH of ~ 75 and ERH of ~ 45–47 % (Martin, 2000). From a thermodynamic point of view, or as observed in bulk ternary systems, aqueous droplets with double salts should show step-wise efflorescence transitions: a component in the aqueous droplets precipitates first at their ERH and then the aqueous phase of the eutonic composition effloresces at their mutual ERH (MERH), which should be lower than either ERHs of the pure salts. Therefore, effloresced mixed particles may form a heterogeneous, core-shell crystal structure owing to the stepwise crystallization process (Ge et al., 1996). On the other hand, aqueous droplets with a eutonic composition are expected to crystallize simultaneously, resulting in a homogeneous crystal structure. Efflorescence, however, is a kinetic or rate-driven process that requires a sufficient activation energy to overcome the kinetic barrier (Martin, 2000). This kinetic or critical-nucleation barrier in turn depends on a range of factors, such as the mixing states of the chemical components, micro-physical states, supersaturation levels, vapor pressure, interfacial tension, viscosity, inter-ionic forces, and solute-water and solute-solute interactions (Cohen et al., 1987b). Therefore, the ERHs of single or multi-component salts are difficult to predict theoretically (Seinfeld and Pandis, 2006). A theoretical model for the efflorescence behavior of the NaCl–Na_2SO_4 mixed system was reported, where the efflorescence was considered to be driven primarily by the homogeneous nucleation of the more supersaturated salt, resulting in a sufficiently large seed for the subsequent heterogeneous nucleation of the other salt (Gao et al., 2007). However, there is no general model that covers the efflorescence of multi-component particles, because it depends on many complicated parameters and varies with the salt characteristics. Therefore, the best way to understand the efflorescence behavior of aerosols is through experimental measurements (Seinfeld and Pandis, 2006). For example, a recent experimental study of the two component NaCl–KCl mixture particles showed that during the dehydration process, the aqueous droplets of various mixing ratios underwent single step efflorescence (Li et al., 2014). Based on the experimentally obtained efflorescence phase diagram and X-ray elemental maps of the effloresced NaCl–KCl mixture particles at various mixing ratios, it was suggested that the more supersaturated salt nucleated homogeneously to crystallize in the center and the other salt underwent heterogeneous crystallization on the former almost simultaneously in the time-scale of the measurements. Full phase diagrams, covering the entire range of mixing ratios, are needed to fully understand the hygroscopic behavior of multi-component aerosol particles (Martin, 2000).

Up until now, there have been only a few studies on the hygroscopic properties of mixed NaCl–NaNO$_3$ aerosol system. Tang and Munkelwitz (1994a) examined the temperature dependent deliquescence behavior of equimolar mixed NaCl–NaNO$_3$ particles using a single-particle levitation technique. Ge et al. (1998) also studied the deliquescence behavior of mixed NaCl–NaNO$_3$ particles with a NaCl mole fraction of 0.2, 0.378 (eutonic), and 0.8 using rapid single-particle mass spectrometry (RSMS), where only MDRH was measured experimentally, and it was claimed that the second stage DRHs agreed with the thermodynamic predictions of AIM. Although the crystallization process was not studied experimentally, a core-shell type of heterogeneous morphology in the NaCl–NaNO$_3$ particles was proposed based on the two-step deliquescence phase transitions observed during the humidifying process (Ge at al., 1998; Hoffmann et al, 2004) and by measuring the secondary electron yields from the nebulized mixture particles (Ziemann and McMurry, 1997). On the other hand, as observed for the NaCl–KCl mixture particles, particles that exhibit two-step deliquescence phase transitions do not always have a core-shell type (Li et al., 2014). Moreover, it was reported that pure NaNO$_3$ droplets do not crystallize easily during the dehydration process, and at very low RHs they appear to exist in an amorphous form (Hoffmann et al, 2004; Gibson et al., 2006; Kim et al., 2012). Therefore, the mixing states in NaCl–NaNO$_3$ particles can be understood only when the efflorescence phenomena for these mixture particles are elucidated. In this study, the hygroscopic properties and microstructure of mixed NaCl–NaNO$_3$ particles at various mixing ratios were examined extensively by optical microscopy and scanning electron microscopy/energy dispersive X-ray spectroscopy (SEM/EDX). The phase transitions of the mixed NaCl–NaNO$_3$ aerosol particles were observed by monitoring the size changes of the particles on the optical images as a function of the RH. SEM/EDX mapping was used to investigate the compositional distribution in the effloresced particles. This paper describes the hygroscopic behavior of the

NaCl–NaNO$_3$ binary aerosol particles as reacted SSA surrogates at 10 different mixing ratios for the first time.

2 Experimental section

2.1 Preparation of mixed NaCl–NaNO$_3$ particles

Mixed NaCl–NaNO$_3$ particles were generated by the nebulization of mixed aqueous solutions. Pure solutions (1.0 M) of NaCl and NaNO$_3$ (NaCl, > 99.9 % purity, Aldrich; NaNO$_3$, 99.9 % purity, Aldrich) were prepared, and the desired solution was made by mixing the two solutions volumetrically. A single jet atomizer (HCT4810) was used to generate aerosol particles to be deposited on transmission electron microscopy (TEM) grids (200 mesh Cu coated with Formvar stabilized with carbon, Ted Pella, Inc.), which behave as hydrophobic substrates (Eom et al., 2014). The aqueous aerosol particles were dried by passing through a silica packed diffusion dryer (HCT4920) with a residence time of ~ 2 s. The size of the dry particles ranged from 1 to 10 μm.

In this study, NaCl–NaNO$_3$ particles with 10 different mixing ratios were investigated; i.e., 9 compositions with NaCl mole fractions of 0.1–0.9 ($X_{NaCl} = 0.1, 0.2, 0.3, 0.4, 0.5, 0.6, 0.7, 0.8$, and 0.9, where X_{NaCl} represents the mole fraction of NaCl.) and a eutonic composition ($X_{NaCl} = 0.38$, which was calculated from the ionic activity products predicted by the AIOMFAC model). Based on the mixing ratio of two salts, the mixed NaCl–NaNO$_3$ particles were divided into three categories: (1) eutonic particles ($X_{NaCl} = 0.38$), (2) NaCl-rich particles containing larger NaCl fraction than the eutonic composition ($X_{NaCl} > 0.38$), and (3) NaNO$_3$-rich particles containing larger NaNO$_3$ fraction than the eutonic composition ($X_{NaCl} < 0.38$).

2.2 Hygroscopic property measurement

The hygroscopic properties of the particles were investigated using a "see-through" inertia impactor apparatus equipped with an optical microscope. The experimental set-up is described in detail elsewhere (Ahn et al., 2010). Briefly, the apparatus is composed of three parts: (a) see-through impactor, (b) optical microscope and (c) humidity controlling system. A TEM grid on which aerosol particles were deposited was mounted on the impaction plate in the see-through impactor. The RH inside the impactor was controlled by mixing dry and wet (saturated with water vapor) N$_2$ gases. The wet N$_2$ gas was obtained by bubbling through deionized water reservoirs. The flow rates of the dry and wet N$_2$ gases were controlled by mass flow controllers to obtain the desired RH in the range of ~ 3–93 %, which was monitored using a digital hygrometer (Testo 645). The digital hygrometer was calibrated using a dew-point hygrometer (M2 Plus-RH, GE), providing RH readings with ±0.5 % reproducibility. To achieve a steady state for condensing or evaporating water, each humidity condition was sustained for at least 2 min. The particles on the impaction plate were observed through a nozzle throat using an optical microscope (Olympus, BX51M). Images of the particles were recorded continuously using a digital camera (Canon EOS 5D, full frame, Canon EF f/3.5 L macro USM lens) during the humidifying (by increasing RH from $\sim$3 to $\sim$93 %) and dehydration (by decreasing RH from $\sim$93 to $\sim$3 %) experiments. The image size was 4368×2912 pixels, and the image recording condition was set to ISO200. The exposure time was 0.4 s, and the depth of focus (DOF) was F/3.5. All hygroscopic experiments were conducted at room temperature ($T = 22 \pm 1$ °C).

The change in particle size with the variation of RH was monitored by measuring the particle areas in the optical images. The particle images were processed using image analysis software (Matrox, Inspector v9.0). The size of the imaging pixel was calibrated using 10 μm Olympus scale bars. Particles with $D_p > 0.5 \mu$m could be analyzed using the present system (Ahn et al., 2010; Eom et al., 2014).

2.3 SEM/EDX measurement

After the hygroscopicity measurements of the individual particles, SEM/EDX was performed for the effloresced particles to determine the morphology and spatial distribution of the chemical elements (Ahn et al., 2010; Li et al., 2014). The measurements were carried out using a Jeol JSM-6390 SEM equipped with an Oxford Link SATW ultrathin window EDX detector. The resolution of the detector was 133 eV for the Mn Kα X-rays. The X-ray spectra and elemental maps were recorded under the control of Oxford INCA Energy software. A 10 kV accelerating voltage and 0.2 nA beam current was used and the typical measuring times were 10 min for elemental mapping.

3 Results and discussion

3.1 Hygroscopic behavior of pure NaCl and NaNO$_3$ particles

Aerosol particles generated from a pure NaCl aqueous solution showed typical hysteresis curves with DRH $= 75.5$ (±0.5) % and ERH $= 47.6$–46.3 %, and these values were consistent with the reported values (Wise et al., 2007; Tang et al., 1997). The DRH and ERH of the single-component NaCl aerosol particles are denoted as those of the "pure NaCl limit". The dry-deposited NaNO$_3$ powder particles exhibited typical hygroscopic curves with definite phase transitions at DRH $= 74.0$ (±0.5) %, and ERH $= 45.7$–26.7 %, which are similar to the values reported by Tang and Munkelwitz (1994b). Hereafter, the DRH of the "pure NaNO$_3$ limit" is defined as 74.0 %. On the other hand, most wet deposited aerosol particles (>90 %) generated by nebulization from a NaNO$_3$ aqueous solution grew continuously and shrank without any phase transition during the humidifying and dehydration processes, which have also been reported (Gysel

et al., 2002; McInnes et al., 1996; Lee et al., 2000; Hoffman et al., 2004). However, a few wet deposited particles showed ERH in the range, 25.8–18.9 %. This discrepancy was attributed to the different nucleation mechanisms, i.e. homogeneous and heterogeneous nucleation, for pure and impure (seed containing) NaNO$_3$ particles, respectively. Detailed discussions can be found elsewhere (Kim et al., 2012). Nebulized NaNO$_3$ particles may exist as amorphous particles with no visible ERHs (Hoffmann et al., 2004; Kim et al., 2012).

3.2 Hygroscopic behavior of mixed NaCl–NaNO$_3$ particles

To describe the measurement procedure for observing the hygroscopic behavior of individual aerosol particles during humidifying and dehydration processes, Fig. 1 presents representative optical images of NaCl–NaNO$_3$ particles with a NaCl-rich composition, i.e., $X_{\text{NaCl}} = 0.8$, taken at various RHs. Images (a–e) and (f–j) were recorded when the RH was first increased ($\uparrow$) from $\sim$3 to 90 % (humidifying process), and then decreased ($\downarrow$) from $\sim$90 to 3 % (dehydration process), respectively. As the optical images of the particles were recorded using a digital camera, the data for 10–15 particles in each image field was obtained. During the humidifying process, the sizes and shapes of the particles did not change until all the particles absorbed moisture and showed a first partial deliquescence transition at RH $= 67.9$ (±0.5) %, which is the DRH of the eutonic composed part (i.e., $X_{\text{NaCl}} = 0.38$) (see Fig. 2c). Upon further increases in RH, the particles absorbed more moisture and grew in size until a second deliquescence transition occurred at RH $= 73.7$ %, where all the particles were fully converted to homogeneous aqueous droplets. After the second deliquescence transition, liquid droplets underwent hygroscopic growth with increasing RH due to the condensation of water vapor. During the dehydration process, the aqueous droplets decreased gradually in size, until the first efflorescence transitions were observed over the range of RH $= 45.2$–44.7 % for different droplets in the image field, where they became partially crystallized. Upon further decreases in RH, they underwent a second and final efflorescence transition over the range of RH $= 33.5$–30.0 %. All particles in the image field were finally transformed into solids at RH $= 30.0$ %, below which no further decrease in size was observed. All the particles in the image field showed two-stage deliquescence transitions at specific RHs because the deliquescence transitions are prompt. On the other hand, the two-stage efflorescence transitions occurred over a range of RH because the efflorescence driven by the nucleation kinetics is a stochastic process (Martin, 2000; Krieger et al., 2012). As the projected optical image of a particle placed on the substrate was monitored during the hygroscopic measurements, the shape and size of the effloresced particles did not appear the same as the orig-

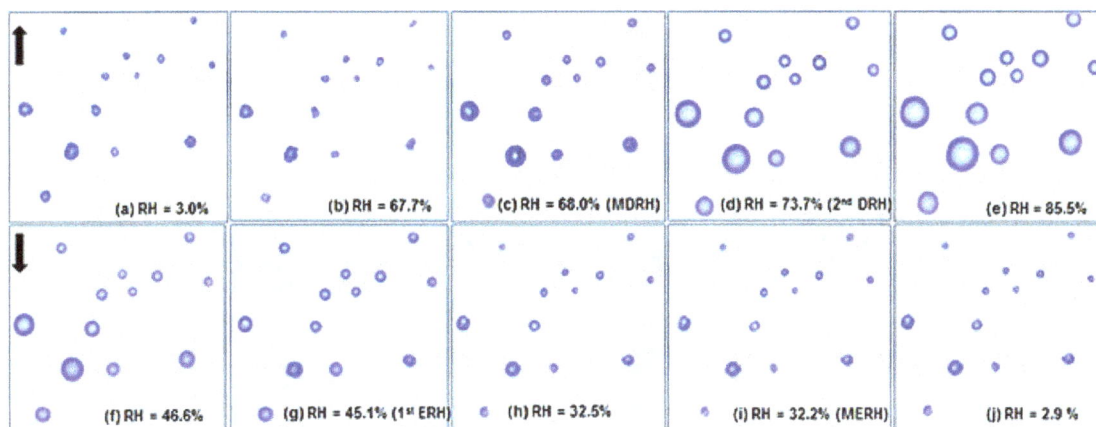

Figure 1. Optical images of NaCl–NaNO$_3$ particles with a mole fraction of $X_{NaCl} = 0.8$, obtained during humidifying ($\uparrow$) and dehydration ($\downarrow$) processes for the same image field.

inal dry particle due to the rearrangement of a solid particle when it crystallizes during the dehydration process.

Figure 2 presents the humidifying and dehydration curves for mixed NaCl–NaNO$_3$ particles at different mixing ratios. The humidifying and dehydration curves are represented as the area ratio (A/A_0: left-hand axis), which was obtained by dividing the 2-D projected particle area at a given RH (A) by that before starting the humidifying process (A_0). The hygroscopic behavior of the mixed NaCl–NaNO$_3$ particles differed according to the categories, i.e. eutonic, NaCl-rich, and NaNO$_3$-rich particles, which are discussed in the next sections.

3.2.1 Eutonic particles ($X_{NaCl} = 0.38$)

Figure 2c shows the 2-D projected area ratio plot as a function of the RH obtained during humidifying and dehydration processes for a representative eutonic particle. During the humidifying process, a single phase transition from solid particles to liquid droplets was observed at RH = 67.1–67.9 %. After deliquescence, the size of the liquid droplet grew gradually and continuously with further increases in RH. As the eutonic particles deliquesced at RH = 67.9 ($\pm$0.5) %, they approached the MDRH of the mixed NaCl–NaNO$_3$ particles. The measured MDRH is consistent with the value calculated from the ionic activity products predicted by the AIOMFAC model and other experimental values (Tang and Munkelwitz, 1994a). During the dehydration process, however, the eutonic droplets showed two-stage efflorescence transitions. The particle (Fig. 2c) decreased gradually in size due to water evaporation and the particle size decreased sharply at the first efflorescence transition at RH = 37.3–36.6 %. The particle was then observed to undergo a second efflorescence transition at RH = 35.3–34.4 %. All eutonic droplets in the optical image field showed first and second ERHs over the range of RH = 37.7–35.7 and 35.4–33.4 %, respectively.

In general, the effloresced particle areas are different from the original ones; i.e., the A/A_0 values deviate from unity (see Fig. 2) due to the rearrangement of the particles during recrystallization. Because only the top-view 2-D images are obtained from particles sitting on the substrate, the morphology of the particles may not appear the same after recrystallization unless they are perfectly spherical.

3.2.2 NaCl-rich particles ($X_{NaCl} > 0.38$)

Figure 2e presents the 2-D-area ratio plot as a function of the RH of a NaCl-rich particle ($X_{NaCl} = 0.8$). During the humidifying process, the particle size remained constant until RH = $\sim$ 65.5 %, where a slight decrease in size was observed due to water adsorption in the lattice imperfections of the solid salts in the particle and structural rearrangement inside the crystal lattice. A first deliquescence transition was observed from RH = 67.7 to 68.0 %, where its size increased sharply. With further increases in RH, it grew gradually until RH = 73.7 %, at which point the second transition occurred. Thereafter, with further increases in RH, the particle grew continuously. The first phase transition at RH = 68.0 %, i.e., MDRH (RH = 67.9 ($\pm$0.5) %) of the NaCl–NaNO$_3$ system, was assigned to the deliquescence of the eutonic component in the particle. At the MDRH, the particle consisted of a mixed phase of liquid droplets (eutonic solution) and NaCl solid inclusion. The solid inclusions in the partially deliquesced mixed inorganic salt particles were also observed by environmental transmission electron microscopy (ETEM) (Freney et al., 2009, 2010). Above the MDRH, with the increase in RH, water vapor kept being absorbed, and NaCl solid dissolved thoroughly at RH = 73.7 %, which is the second DRH of the particles with $X_{NaCl} = 0.8$.

All other NaCl-rich particles with different compositions (e.g., $X_{NaCl} = 0.5$ and 0.9 in Fig. 2d and f, respectively) also exhibited two-stage phase transitions during the humidifying process: the first transition at MDRH (RH = 67.9 ($\pm$0.5) %)

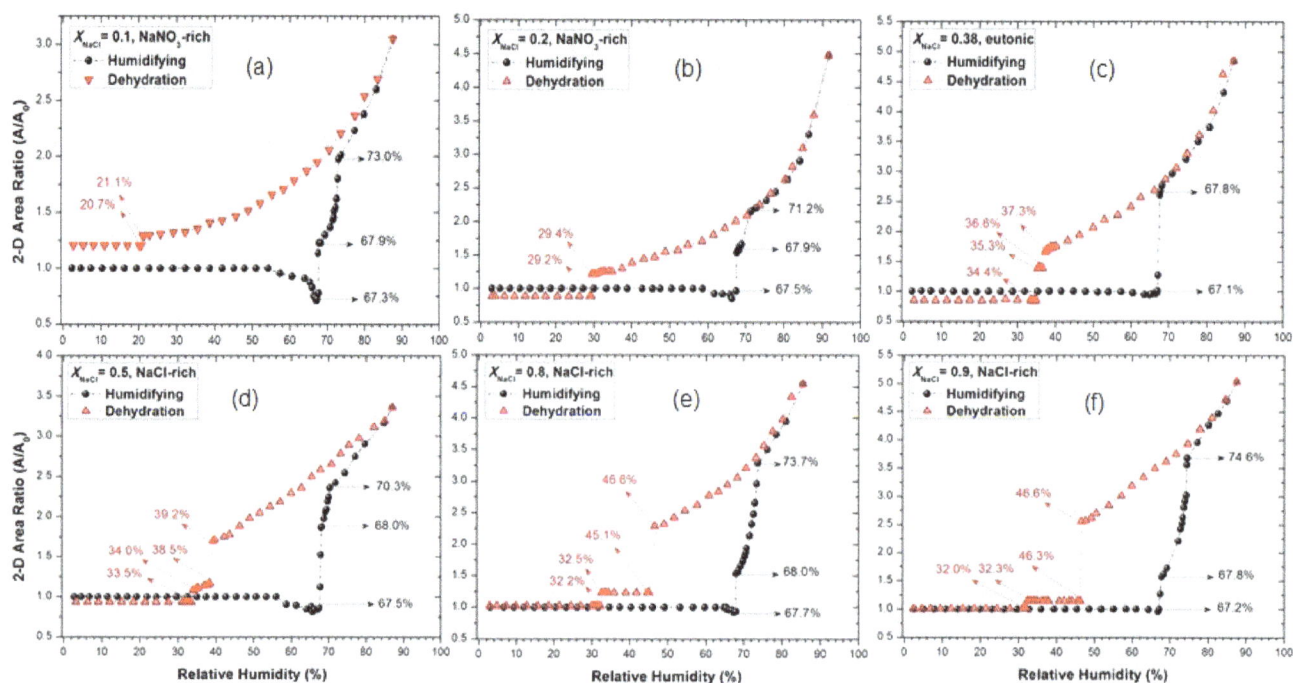

Figure 2. Plots of 2-D projected area ratio as a function of the relative humidity (humidifying process: closed circles; dehydration process: open triangles) for different mole fractions of NaCl–NaNO$_3$, expressed in X_{NaCl} values of **(a)** 0.1 (NaNO$_3$-rich), **(b)** 0.2 (NaNO$_3$-rich), **(c)** 0.38 (eutonic), **(d)** 0.5 (equimolar/NaCl-rich), **(e)** 0.8 (NaCl-rich), and **(f)** 0.9 (NaCl-rich). The transition relative humidity in both humidifying and dehydration processes are marked with arrows.

due to deliquescence of the eutonic component and the second one at their DRHs owing to complete deliquescence of the particles. The MDRH is independent of the particle composition. On the other hand, the second DRHs are dependent on the compositions and shift toward the pure NaCl limit (DRH = 75.5 ($\pm$0.5) %) with increasing NaCl mole fraction. Figure 3 plots the measured DRHs for the NaCl-rich particles with various compositions as a function of the NaCl mole fraction, showing that the experimental DRH values are in good agreement with the values calculated from the AIOM-FAC model.

During the dehydration process (Fig. 2e), a representative NaCl-rich particle with the composition of $X_{NaCl} = 0.8$ shows two-stage phase transition. The liquid droplet gradually decreased in size with decreasing RH and became supersaturated by NaCl below 73.7 % RH (DRH for X_{NaCl} = 0.8). With the further decreases in RH, the droplet size decreased sharply at RH = 46.6–45.1 % due to the crystallization of NaCl in the droplet. At RH = 45.1 %, the first ERH for X_{NaCl} = 0.8, the particle was composed of a mixed phase of eutonic solution and NaCl solid. With further decreases in RH, the eutonic component in the particle precipitated at RH = 32.2 %, which is the MERH of the aerosol particle, resulting in the formation of a completely effloresced solid particle. The measured first ERH and MERH for the particles with a composition of X_{NaCl} = 0.8 varied among the particles and were in the range, RH = 45.2–44.7 % and

RH = 33.5–30.0 %, respectively. All other NaCl-rich particles with different compositions also exhibited two-stage transitions during the dehydration process: the first transition at their ERH, which is specific to their compositions, owing to the homogeneous nucleation and crystallization of NaCl, and the second transition at the MERH due to precipitation of the eutonic component.

As the NaCl-rich particles show two-stage phase transitions during the dehydration process, they may form a core-shell type structure where the crystalline NaCl occupies the center and the eutonic solid is present at the surface. Detailed microstructures of the effloresced NaCl-rich particles will be discussed later.

3.2.3 NaNO$_3$-rich particles ($X_{NaCl} < 0.38$)

All NaNO$_3$-rich particles also showed two-stage transitions during the humidifying process. For example, in Fig. 2b, a NaNO$_3$-rich particle with the composition of $X_{NaCl} = 0.2$ showed the first and second transition at RH = 67.5–67.9 % (MDRH) and RH = 71.2 %, respectively. At the MDRH, the deliquesced component formed a eutonic aqueous solution and the un-deliquesced NaNO$_3$ solid was still included in the eutonic solution. Above MDRH, the aerosol particle absorbed water continuously with a further increase in the RH, and it deliquesced completely at RH = 71.2 %, which is the second DRH of the particle of X_{NaCl} = 0.2. Above DRH, the

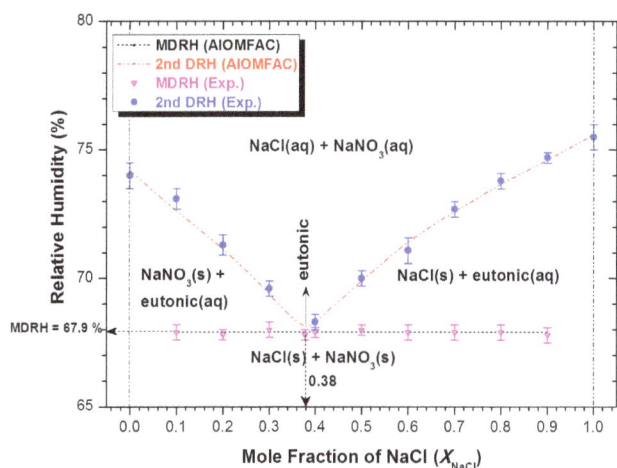

Figure 3. Measured first DRH or MDRH (open triangles), second DRH (closed circles) values, calculated MDRH (dotted line), and the second DRHs (dash-dotted curve) from the AIOMFAC, plotted as a function of the mole fraction of NaCl in NaCl–NaNO₃ mixture particles. The phase notations shown in brackets are s = solid; and aq = aqueous.

size of the liquid droplet increased with increasing RH due to the condensation of water vapor. Other NaNO₃-rich particles (e.g., $X_{NaCl} = 0.1$ in Fig. 2a) also exhibited two-stage phase transitions during the humidifying process: the first transition at MDRH due to the deliquescence of the eutonic component (independent of chemical composition) and the second one at their DRH due to the complete deliquescence of the particles. In all NaNO₃-rich particles, the second DRH was dependent on the mixing ratio of the particles and shifted toward the pure NaNO₃ limit (DRH of NaNO₃ = 74.0 ($\pm$0.5) %) with increasing NaNO₃ mole fraction (Fig. 3).

The hygroscopic behavior of the NaNO₃-rich particles during the dehydration process was different from that of NaCl-rich particles. The NaCl-rich particles showed two-stage transitions, whereas the NaNO₃-rich particles showed a single-stage transition. For example, a droplet of $X_{NaCl} = 0.2$ decreased continuously in size with decreasing RH until it showed an efflorescence transition at RH = 29.4–29.2 % (Fig. 2b). Similar to the NaCl-rich case, the NaNO₃-rich particles would be expected to exhibit two-stage transitions during the dehydration process, i.e., the first transition accompanying the precipitation of solid NaNO₃, and the second one due to the efflorescence of the eutonic component. However, all the NaNO₃-rich particles with $X_{NaCl} = 0.1, 0.2$, and 0.3 showed single-stage efflorescence transitions during the dehydration process (Figs. 2a, b, and 4), suggesting that all the components in the NaNO₃-rich particles crystallized (almost) simultaneously. The supersaturated NaNO₃ in the droplets did not appear to crystallize until the NaCl crystallized and acted as heterogeneous nuclei for the almost simultaneous solidification of NaNO₃. To confirm this assumption, particles with small fractions of NaCl, such as

particles with $X_{NaCl} = 0.01, 0.03$, and 0.05, were investigated. The particles with $X_{NaCl} = 0.05$ showed two-stage and single-stage transitions during the humidifying and dehydration processes, respectively, which is similar to that observed for the NaNO₃-rich particles with a composition of $X_{NaCl} \geq 0.1$. On the other hand, the particles with $X_{NaCl} = 0.01$ and 0.03 underwent continuous hygroscopic growth and shrinkage without phase transitions during the humidifying and dehydration processes, respectively, which is similar to that of aerosol particles generated from an aqueous single-component NaNO₃ solution. Pure nebulized NaNO₃ did not show clear efflorescence transitions apparently because of its amorphous nature (Hoffman et al., 2004; Gibson et al., 2006; Kim et al., 2012). Therefore, a sufficient quantity of heterogeneous nuclei (NaCl in this case) is needed to induce the crystallization of NaNO₃. A similar observation was reported for the crystallization of NH₄NO₃ and NH₄HSO₄ aerosol particles (Schlenker and Martin, 2005). Both the pure one-component salts did not effloresce, even at RH = 1 %, and their crystallization could be promoted with the addition of some fraction of inclusions that could serve as good heterogeneous nuclei.

3.3 Deliquescence phase diagram of mixed NaCl–NaNO₃ particles

Figure 3 presents the measured MDRHs (the first DRHs) and second DRHs of the NaCl–NaNO₃ mixture particles with different mole fractions along with the measured DRHs of the pure NaCl and NaNO₃ particles. As shown in Fig. 3, a clearly demarked phase diagram depicting their deliquescence behavior was obtained experimentally as follows:

1. NaCl(s) + NaNO₃(s) phase in Fig. 3: both NaCl and NaNO₃ are mixed as solids below the MDRH at all mole fractions;

2. NaCl(s) + eutonic(aq) phase: a mixed phase of solid NaCl and aqueous eutonic components between the MDRH and second DRHs for $X_{NaCl} > 0.38$;

3. NaNO₃(s) + eutonic(aq) phase: a mixed phase of solid NaNO₃ and aqueous eutonic components between the MDRH and second DRHs for $X_{NaCl} < 0.38$; and

4. NaCl(aq) + NaNO₃(aq) phase: both NaCl and NaNO₃ are mixed in the aqueous phase above the second DRHs at all mole fractions.

This MDRH and second DRHs obtained experimentally agrees well with the values calculated from the ionic activity products of constituents predicted by the AIOMFAC model, as shown in Fig. 3 (dotted line for MDRH and dash-dotted curve for the second DRHs). Although many theoretical models have been developed to predict the hygroscopic behavior of a mixed sodium chloride and nitrate system (Tang, 1976; Ansari and Pandis, 1999; Clegg et al., 1998;

Wexler and Clegg, 2002), only a few experimental results have been reported. Tang and Munkelwitz (1994a) reported an MDRH = 68.0 (±0.4) % at 25° C. Ge et al. (1998) examined the deliquescence behavior of mixed NaCl-NaNO$_3$ particles with a NaCl mole fraction of 0.2, 0.38 (eutonic), and 0.8, using RSMS. They claimed that the DRHs were generally consistent with the AIM thermodynamic model predictions. On the other hand, the measured DRH values were not reported except for the MDRH of 67.0 %, which was lower than the 67.9 % obtained in the present study, because the RSMS appeared to detect only the start of the transition similar to the present system where mutual deliquescence for $X_{NaCl} = 0.38$ began at 67.1 % (Fig. 2b), whereas the actual MDRH was at the end of the transition at 67.9 % when the eutonic particle dissolved completely.

All the mixed NaCl–NaNO$_3$ particles showed the first phase transition at the MDRH regardless of the mixing ratio of the two salts. Thermodynamically, as the phase transition of mixed-salts is governed by the water activity at the eutonic point, the MDRH of the mixed-salt particles is independent of the initial composition of the mixture. Other inorganic mixed particles, such as NaCl–KCl, Na$_2$SO$_4$-NaNO$_3$ and NH$_4$Cl–NaCl particles, also exhibited a mutual deliquescence transition at RHs, which were independent of the initial dry-salt compositions, and the phase diagram followed the typical pattern for these two-component inorganic salt + water ternary systems (Tang and Munkelwitz, 1994a; Li et al., 2014; Kelly et al., 2008). For the NaCl-rich particles of $X_{NaCl} > 0.38$, which contain more NaCl than the eutonic composition, the second DRH value approached the DRH of the pure NaCl salt as the NaCl concentration was increased. For the NaNO$_3$-rich particles of $X_{NaCl} < 0.38$, the DRH approached that of pure NaNO$_3$ as the NaNO$_3$ mole fraction was increased (Fig. 3). This suggests that the second-stage deliquescence is purely driven by the solid salt remaining after the first deliquescence of the eutonic composition.

3.4 Efflorescence phase diagram of mixed NaCl–NaNO$_3$ particles

Figure 4 shows the measured ERHs and MERHs for the mixed NaCl–NaNO$_3$ particles with various mixing ratios as a function of the NaCl mole fraction. Unlike the deliquescence phase diagram, which showed four systematic phases, the efflorescence phase diagram is composed of three distinct phases:

1. NaCl(aq) + NaNO$_3$(aq) phase: both NaCl and NaNO$_3$ are mixed in the aqueous phase above the first ERHs at all mixing ratios;

2. NaCl(s) + eutonic(aq) phase: a mixed phase of solid NaCl and aqueous eutonic components between the first ERH and second ERH (MERH) for $X_{NaCl} \geq 0.38$; and

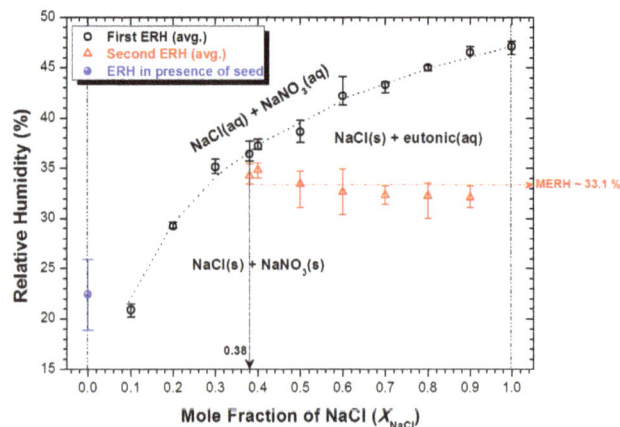

Figure 4. Measured first-ERH values (open circles) and second-ERH values (open triangles) as a function of the mole fraction of NaCl in NaCl–NaNO$_3$ mixture particles as well as ERH (closed circle) for wet deposited NaNO$_3$ particles containing seeds. The phase notations shown in brackets are s = solid; and aq = aqueous.

3. NaCl(s) + NaNO$_3$(s) phase: both NaCl and NaNO$_3$ are mixed as solids below the second ERH (MERH) for $X_{NaCl} \geq 0.38$ and below the first ERHs for $X_{NaCl} < 0.38$.

This experimental phase diagram for efflorescence is reported for the first time, for which no theoretical predictions or other experimental reports exist to the best of the authors' knowledge.

The ERH of NaCl-rich droplets ($X_{NaCl} > 0.38$) shifted toward the pure NaCl limit (RH = 47.6–46.3 %) when the NaCl content was increased (see Fig. 4). The measured MERH was observed at a relatively wide range of RH = 30.0–35.5 %. The droplets with the eutonic composition ($X_{NaCl} = 0.38$) showed the first and second ERHs over a range of 37.7–35.7 and 35.4–33.4 %, respectively. The NaNO$_3$-rich droplets ($X_{NaCl} < 0.38$) had only one-stage efflorescence transition, and the decreasing trend of their ERHs as the NaCl mole fraction was decreased clearly follows that for the NaCl-rich and eutonic droplets (Fig. 4). This suggests that the first efflorescence of the NaCl–NaNO$_3$ mixture droplets at all mixing ratios is driven by the homogeneous nucleation of NaCl. For the NaCl-rich droplets, crystallized NaCl acts as a seed for further precipitation of the remaining metastable eutonic aqueous part. For the eutonic droplets, NaCl is crystallized homogeneously at the first ERH and the remaining NaNO$_3$ solidify heterogeneously on the NaCl seeds at the second ERH. For NaNO$_3$-rich droplets, due to the low NaCl content, the homogeneous nucleation rate of NaCl decreases with decreasing NaCl mole fraction, and the NaNO$_3$ appears to undergo almost simultaneous heterogeneous crystallization (precipitation) on the NaCl seeds under the time scale of measurements (i.e., within 2 min of the equilibrating time for recording images during the first observed efflorescence transition).

3.5 Spatial distribution of effloresced NaCl–NaNO₃ solid particles

To examine the morphology and spatial distribution of the chemical components in NaCl–NaNO₃ particles at various mixing ratios, SEM/EDX were performed for the effloresced solid particles formed after the humidifying and dehydration cycles. Figure 5a shows the secondary electron image (SEI) and elemental X-ray mapping images of a NaCl-rich ($X_{NaCl} = 0.8$) particle. The elemental X-ray maps suggest that Cl (from NaCl) is concentrated in the central part, whereas O (from NaNO₃) is more concentrated at the edges. This suggests that NaCl nucleates homogeneously to crystallize in the center at the first ERH, whereas NaCl and NaNO₃ from the eutonic phase crystallized on these central NaCl seeds and precipitated on the edges at the second ERH (MERH).

For a typical particle with a eutonic composition ($X_{NaCl} = 0.38$), the elemental map of Cl suggests that NaCl is again more concentrated at the central part, like NaCl-rich particles, and O from NaNO₃ is concentrated around this NaCl core (Fig. 5b), suggesting that NaCl is nucleated homogeneously to crystallize in the center at the first ERH, whereas the remaining NaNO₃ precipitated at the second ERH. The thermodynamically expected, homogeneously mixed eutonic particles were not observed, and the spatial distribution was rather ruled by the two-stage efflorescence transitions.

In the case of NaNO₃-rich particles, as shown in Fig. 5c, the Cl component was localized in the core region, and Na and O are distributed over the entire particle. This suggests that even for NaNO₃-rich particles, NaCl is nucleated homogeneously to crystallize in the core at the single observed ERHs, whereas the NaNO₃ precipitate simultaneously on this core within the time scale of the measurements. Although the NaNO₃-rich droplet was supersaturated with NaNO₃ with decreasing RH during the dehydration process, NaNO₃ could not crystallize easily even at the MERH of the NaCl-rich mixtures. Therefore, RHs lower than MERH (see Fig. 4 for the lower ERHs in the case of $X_{NaCl} \leq 0.3$ than MERH) were necessary for the homogeneous crystallization of NaCl followed by the almost simultaneous, induced heterogeneous crystallization of NaNO₃.

Up until now, the binary inorganic salt aerosols, except the eutonic composition, are generally believed to form a core-shell type heterogeneous solid of a pure salt core surrounded by the eutonic component. The formation of the core-shell type had been reported for a range of binary mixed aerosol particles, such as NaCl–KCl, KCl–KI, (NH₄)₂SO₄-NH₄NO₃ system (Ge et al., 1996). Based on the secondary electron yield measurements of the NaCl–NaNO₃ mixture particles, it was claimed that individual particles exist in core-shell form with the richer salt (NaCl in NaCl-rich or NaNO₃ in

NaNO₃-rich), and the eutonic components occupy the core and shell, respectively (Ziemann and McMurry, 1997). However, the study was conducted on particles nebulized from aqueous solutions that had not gone through the proper dehydration process. Hoffmann et al. (2004) predicted a core-shell type structure based on their observations of NaCl inclusions in the partially deliquesced particles only. On the other hand, a previous report on NaCl–KCl mixture particles showed that NaCl and KCl were crystallized as separate phases and not necessarily in the core-shell configuration (Li et al., 2014). In addition, a eutonic solid shell in the dry particle was found to be unnecessary for the exhibition of the two-stage deliquescence transitions during the humidifying process at all mixing ratios. The efflorescence phase diagram (Fig. 4) and the X-ray maps (Fig. 5) showed that effloresced NaCl–NaNO₃ particles of all mixing ratios had NaCl crystallized homogeneously in the center, surrounded either by the eutonic component for $X_{NaCl} > 0.38$ or NaNO₃ for $X_{NaCl} \leq 0.38$. These micro-structures and spatial distributions of chemical components have obvious atmospheric implications (Ziemann and McMurry, 1997).

4 Atmospheric implications

The particle/droplet size variations with RH for the different mixing states (Fig. 2), the four and three distinct phases observed during the humidifying (Fig. 3) and dehydration (Fig. 4) processes, respectively, have important atmospheric implications in terms of radiative forcing (Baynard et al., 2006; Ma et al., 2008), cloud nucleation efficiency (Petters et al., 2007; Wex et al., 2008), atmospheric chemistry (Ault et al., 2013a; Ryder et al., 2014; Wang and Laskin, 2014), and gas adsorption/desorption (gas-particle partitioning) (Woods et al., 2012). Some of these implications related to the observed experimental results are discussed.

4.1 Particle size change at different phase transitions

In this study, the measured NaCl–NaNO₃ mixture particles, $1–10\,\mu m$ in size and their size variations according to the RH change, are atmospherically relevant, because inorganics are dominant in the super-micron or coarse size fraction of SSAs (Keene et al., 2007; Ault et al., 2013b; Prather et al., 2013). In Fig. 6a, 2-D diameter ratios ($d/d_x = \sqrt{(A/A_x)}$), where x is the particle state at RHs before the transitions), which represent the size increase due to mutual deliquescence of the eutonic part at MDRH (compared to the rearranged solid particle) as well as to deliquescence of the pure salts at the second DRHs (compared to the partially deliquesced particle at MDRH), were plotted as a function of the NaCl mole fractions. As shown in Figs. 2 and 6a, the particle size variations during the humidifying process depends on the mixing ratio of the two-component NaCl–NaNO₃ particles as the extent of the increase in particle size due to the

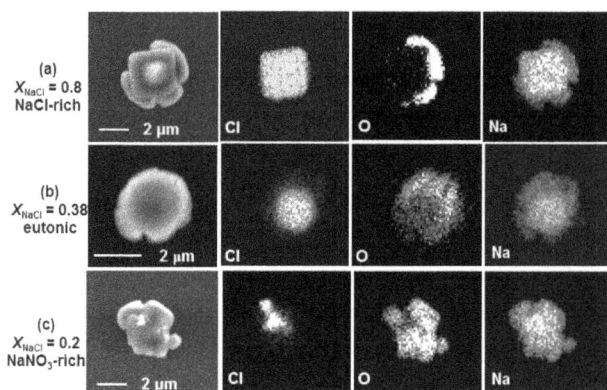

Figure 5. Secondary electron images (SEIs) and elemental X-Ray maps for Cl (from NaCl), O (from NaNO$_3$), and Na of the effloresced NaCl–NaNO$_3$ mixture particles with compositions of (a) $X_{NaCl} = 0.8$ (NaCl-rich); (b) $X_{NaCl} = 0.38$ (eutonic); and (c) $X_{NaCl} = 0.2$ (NaNO$_3$-rich).

water uptake is associated with the chemical composition of the particles. For single-component NaCl and NaNO$_3$ particles, their 2-D diameters increase ~ 2.0 and ~ 1.5 fold, respectively, when they deliquesced at their DRHs (Fig. 6a). The difference was attributed to the different solubility of the salts. Because NaCl is less soluble than NaNO$_3$ (their solubility is 36.0/100 g and 91.2/100 g, respectively) (Lide, 2002), NaCl requires more water than NaNO$_3$ to form a saturated solution, resulting in a larger increase in size when solid NaCl particles deliquesce. Therefore, for the mixed NaCl–NaNO$_3$ particles, the extent of the particle size increases at the MDRH or the second DRH is associated with the mixing ratios of the two salts. At the MDRH, where only the eutonic component deliquesces, the extent of the size increase depends on the fraction of the eutonic component in the particles. Therefore, at the MDRH, particles with a eutonic composition ($X_{NaCl} = 0.38$) show the largest increase in size ($d/d_x = \sim 1.7$ as shown in Fig. 6a). For NaCl-rich particles ($X_{NaCl} > 0.38$), the size increase at the MDRH decreases with increasing NaCl fraction (Figs. 2d–f and 6a) because the eutonic fraction, which depends on the NaNO$_3$ content in the particle, decreases with increasing NaCl mole fraction. In contrast, for NaNO$_3$-rich particles ($X_{NaCl} < 0.38$), the size at the MDRH increases with increasing NaCl fraction (Figs. 2a, b, and 6a), because the eutonic fraction, which depends on the NaCl content, increases with increasing NaCl mole fraction. The particle size change at the RHs between MDRH and the second DRH during the humidifying process depends on the residual solid components after the mutual deliquescence. For NaCl-rich particles, the particles with a larger NaCl fraction grow more at the RHs between MDRH and their second DRHs (Figs. 2d–f and 6a) because the partially deliquesced particles have a larger residual NaCl solid fraction. In contrast, for the NaNO$_3$-rich particles, the particles

with the larger NaCl fraction grow less (Figs. 2a, b, and 6a), because they have a smaller residual NaCl solid fraction.

At the first ERH, where efflorescence occurs by the homogeneous nucleation of NaCl, the NaCl-rich and eutonic droplets ($X_{NaCl} \geq 0.38$) show a larger decrease in size for droplets of a larger NaCl fraction (Fig. 6b). For the NaNO$_3$-rich droplets, the simultaneous heterogeneous precipitation of the NaNO$_3$ on the homogeneously crystallized NaCl seeds occurs at their ERH, and the droplets of higher NaNO$_3$ content show a larger decrease in size. The decrease in size at the second efflorescence transition for aerosol particles of $X_{NaCl} \geq 0.38$ should be governed by the eutonic content in the aerosol particles; i.e., the more eutonic content, the greater the size decrease (Fig. 6b).

Aerosol particles with mixing ratios of $X_{NaCl} \geq 0.38$ did not show a noticeable change in the 2-D area when the RH was decreased from their first ERH to the start of the second ERH (MERH) (see Figs. 1g–i and 2c–f). At this RH range, the aerosol particles are expected to gradually decrease in size due to the evaporation of water in the eutonic solution. As discussed above, however, less soluble NaCl solid particles require more water than NaNO$_3$ to form a saturated solution, resulting in a larger increase in size when NaCl particles deliquesce (Fig. 6a). On the other hand, when aqueous NaCl droplets crystallize, more water in the droplets evaporates than when NaNO$_3$ is present in the droplets. For NaCl-rich droplets, because a large portion of water in the droplet might be removed when NaCl crystallizes at their first ERH, only a small amount of water remains in the eutonic solution, which evaporates gradually until complete efflorescence occurs, even though shrinkage was not observed well on their optical images. However, in this RH range, the particle boundary appeared to be clearer when it became more concentrated with decreasing RH. Indeed, the change was clear when viewed directly through the optical microscope (unlike the digital images) during the dehydration measurements. The size variations of aerosol particles with different mixing ratios according to the RH change can help predict their aerodynamic properties and hence their residence time in ambient air. These variations, however, also depend on the original size of the particles, possibly more for sub-micron particles (Hu et al., 2010).

4.2 Hygroscopic growth and cloud droplet nucleation

Cloud droplet nucleation can begin when the air is supersaturated with water vapor (i.e., RH is above 100 %) and the hygroscopic growth at high RHs are correlated with the cloud condensation nuclei (CCN) activity (Petters et al., 2007; Wex et al., 2008). Under these experimental conditions, the highest working RH was ~ 93 %. On the other hand, as the DRHs of pure NaCl and NaNO$_3$ are 75.5 (± 0.5) % and 74.0 (± 0.5) %, respectively, the hygroscopic growth begins at less than ~ 76 % RH for particles of all mixing ratios. In Fig. 7, the growth factors in terms of the 2-D diameter ratios (d_{95}/d_0

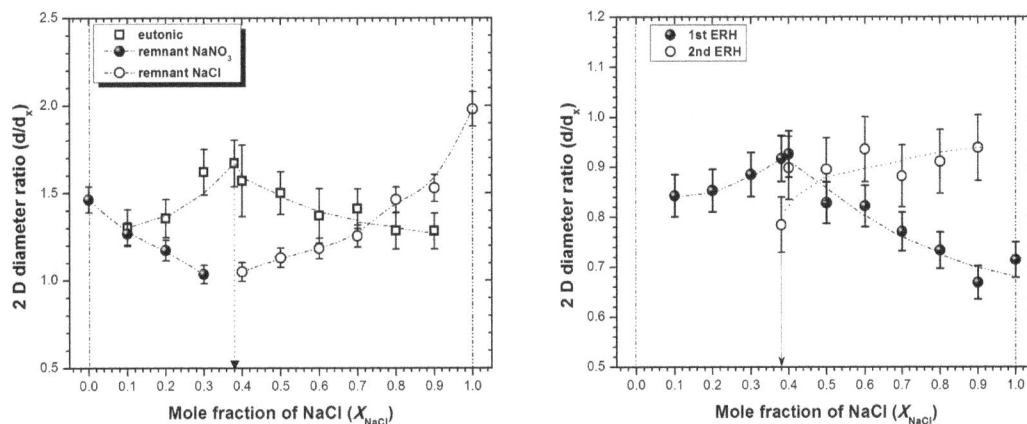

Figure 6. (a) 2-D diameter ratios (d/d_x) of the particles/aerosols plotted as a function of the NaCl mole fractions (X_{NaCl}), which represent the size increase due to the mutual deliquescence of the eutonic part at MDRH ("d") compared to that of the solid particle at the start of the first transition ("d_x") (eutonic: open squares); and due to the complete deliquescence of the pure salts (in the figure, NaNO3: closed circles and NaCl: open circles) at their second DRHs ("d") compared to that of the partially deliquesced particle at MDRH ("d_x") and **(b)** 2-D diameter ratios (d/d_x) of droplets/aerosols plotted as a function of the NaCl mole fractions (X_{NaCl}), which represent the size decrease due to the efflorescence of NaCl ("d") compared to that of the liquid droplets at the start of the first efflorescence ("d_x") (first ERH: closed circles); and due to the mutual efflorescence of the eutonic compositions (second ERH: open circles) ("d") compared to that of the partially effloresced particles ("d_x").

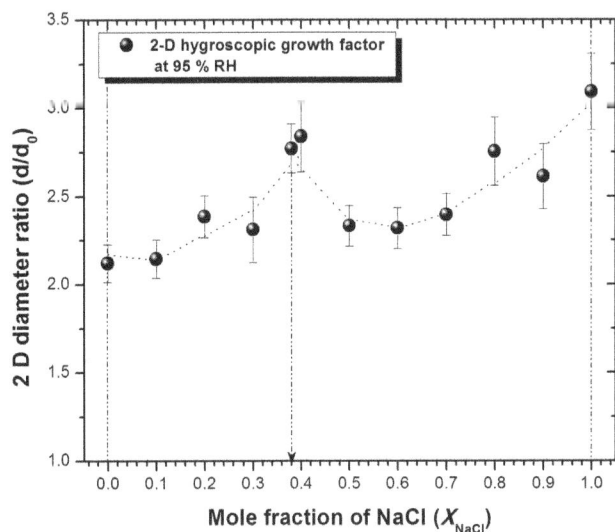

Figure 7. Hygroscopic growth factors in terms of the 2-D diameter ratios (d_{95}/d_0 where d_{95} = the diameter at ∼ 95 %, determined by extrapolation, and d_0 = dry diameter at lowest or ∼ 3 % RH) are plotted as a function of the mole fraction of NaCl for the NaCl–NaNO3 mixture particles.

where d_{95} = the diameter at RH = 95 %, determined by extrapolation, and d_0 = dry diameter at ∼ 3 % RH) are plotted as a function of the mole fraction of NaCl for the NaCl–NaNO3 mixture particles to indicate the cloud droplet nucleation efficiency of these particles at various mixing ratios. NaCl, eutonic, and NaNO3 particles, which have single deliquescence transitions (one DRH), appear to show

relatively higher hygroscopic growth (in the following order of NaCl > eutonic > NaNO3) than particles with other mixing ratios, having two-stage deliquescence (MDRH and the second DRH). The hygroscopic growth first decreases with decreasing eutonic content for NaCl-rich particles with $0.38 < X_{NaCl} ≤ 0.6$ and then begins to increase with increasing NaCl content for particles with $0.6 < X_{NaCl} ≤ 1.0$. Similar behavior was observed for NaNO3-rich particles. Therefore, for NaCl–NaNO3 mixture particles, the CCN efficiency should also follow the same trend. This can serve as a precursor to the hygroscopic growth and CCN activity of more complex, reacted SSAs containing other salts and organics (King et al., 2012). However, an estimation of the full CCN activity and cloud droplet number concentration require further experiments on a wider dry particle size range under both un- and super-saturated conditions (Fuentes et al., 2011; Mochida et al., 2011).

4.3 Atmospheric chemistry

The knowledge of mixing states, phases, and spatial distribution of chemicals in NaCl–NaNO3 mixture particles, as reacted SSA surrogates, at various RHs is expected to help better understand the complexity of real ambient SSAs, their hygroscopic properties, aqueous phase chemistry, etc. For SSAs comprising NaCl–NaNO3 mixture particles generated from a partial or full reaction with NOx/HNO3, their aqueous surface regions are crucial for atmospheric heterogeneous chemistry because the aqueous phases of NaCl, NaNO3, or the eutonic composed part at different RHs are expected to be available for further reactions with gas phase species, such as N2O5 (Ault et al., 2013a; Ryder et al., 2014) or organics

(Wang and Laskin, 2014), and/or for facile gas-particle partitioning (Woods et al., 2012).

SSAs are generated by sea-spray action, so that they are initially aqueous droplets when they become airborne. Although real ambient SSAs are complex mixtures containing NaCl, other inorganic salts, and organics (O'Dowd and de Leeuw, 2007; Keene et al., 2007; Prather et al., 2013; Beardsley et al., 2013; Ault et al., 2013b), NaCl particles have been studied as genuine SSA surrogates in many laboratories (Martin, 2000; Krieger et al., 2012). Based on the experimental phase diagrams of deliquescence and efflorescence for the NaCl-NaNO$_3$ system (Figs. 3 and 4), once airborne as droplets, the genuine SSA surrogate would remain in the aqueous phase unless the droplets experience an ambient RH below their ERH of 47.6–46.3 % (pure NaCl). Their aqueous phase can facilitate heterogeneous reactions with gaseous species, such as NO$_x$/HNO$_3$, to become NaCl–NaNO$_3$ mixture droplets. When mixture droplets are formed, their full efflorescence occurs at lower RHs (i.e., at MERH of ~ 33.1 % for the mixture droplets of $X_{NaCl} \geq 0.38$ or at ERH of ~ 35–21 % for those of $X_{NaCl} < 0.38$; see Fig. 4) than ERH for pure NaCl, so that their chance for a further gas-particle interaction would be larger with less probability to fully effloresce. Even when the mixed particles become solids due to the full efflorescence below either their MDRH or ERH, they can deliquesce at a lower DRH (i.e., the NaCl–NaNO$_3$ mixed particles with various mixing ratios partially deliquesce at a MDRH of ~ 67.9 %, whereas pure NaCl deliquesces at a DRH of ~ 75.5 %; see Fig. 3). In other words, NaCl–NaNO$_3$ mixture aerosol particles can maintain an aqueous phase over a wider RH range than the genuine SSA surrogate, making their heterogeneous chemistry more probable. Therefore, NaCl-rich particles with higher Cl$^-$/NO$_3^-$ can show higher uptake of N$_2$O$_5$ at ambient RHs (usually >60 %) in the marine boundary layer (Ryder et al., 2014). On the other hand, a high uptake of HNO$_3$ due to surface adsorbed moisture on crystalline NaCl particles has also been reported (Saul et al., 2006; Liu et al., 2007). The ionic activities of aqueous Na$^+$/Cl$^-$/NO$_3^-$ at high to very low water activities during hydration and dehydration, indicating the chemical reactivity of these ions, can be calculated using the AIOMFAC model.

5   Conclusions

The hygroscopic behavior and microstructures of mixed NaCl–NaNO$_3$ particles at various mixing ratios collected on TEM grids were investigated by the use of optical microscopy and SEM/EDX. The DRHs and ERHs of the mixed NaCl–NaNO$_3$ particles in the micrometer size range at room temperature were determined by monitoring the change in the particle area in 2-D optical images with the RH variation of ~ 3–93 %. During the humidifying process, the particles with a eutonic composition ($X_{NaCl} = 0.38$) exhibited single-stage deliquescence behavior at MDRH $= 67.9$ (± 0.5) %.

The mixed NaCl–NaNO$_3$ particles, except for the eutonic composition, showed two-stage phase transitions: the first transition at MDRH, which is independent of the chemical compositions, and the second transition at the DRHs, which depends on the mixing ratios of the two salts. For the NaCl-rich particles ($X_{NaCl} > 0.38$), the increase in the mole fraction of NaCl shifted the DRH toward the pure NaCl limit (DRH $= 75.5$ %). For NaNO$_3$-rich particles ($X_{NaCl} < 0.38$), the increase in the mole fraction of NaNO$_3$ shifted the DRH toward the pure NaNO$_3$ limit (DRH $= 74.0$ %). The measured DRH values agreed well with those calculated from the thermodynamic AIOMFAC model.

The dehydration behavior of the mixed NaCl–NaNO$_3$ particles depend on the mixing ratios of the two salts. NaCl-rich particles ($X_{NaCl} > 0.38$) showed two-stage efflorescence transitions: the first stage, which is driven purely by the homogeneous nucleation of NaCl, and the second stage at MERH. Interestingly, the eutonic composed particles ($X_{NaCl} = 0.38$) also showed two-stage efflorescence with NaCl crystallizing first followed by the heterogeneous nucleation of NaNO$_3$ on NaCl seeds. NaNO$_3$-rich particles ($X_{NaCl} \leq 0.3$) showed single-stage efflorescence transitions at ERHs progressively lower than MERH due to the slower homogeneous nucleation of NaCl and the simultaneous heterogeneous efflorescence of NaNO$_3$ on the NaCl seeds. SEM/EDX elemental mapping shows that effloresced NaCl–NaNO$_3$ particles of all mixing ratios had NaCl crystallized homogeneously in the center, surrounded either by the eutonic component for $X_{NaCl} > 0.38$ or NaNO$_3$ for $X_{NaCl} \leq 0.38$. A pure salt core and eutonic composed solid shell in the dry particle was not necessary for exhibiting the two-stage deliquescence transitions in all mixing ratios.

During humidifying or dehydration processes, the amount of the eutonic composed part drives the particle/droplet growth or shrinkage at the MDRH or MERH (second ERH), respectively, and the amount of pure salts (NaCl or NaNO$_3$ in NaCl- or NaNO$_3$-rich particles, respectively) does at the second DRHs or first ERHs, respectively. Therefore, their behavior can be a precursor to the optical properties and direct radiative forcing for these atmospherically relevant mixture particles representing the coarse, reacted inorganic SSAs. On the other hand, the hygroscopic growth at a RH of 95 % for the various mixing ratios of these mixture particles can be a precursor to the cloud droplet nucleation efficiency and indirect radiative forcing. In addition, NaCl–NaNO$_3$ mixture aerosol particles can maintain an aqueous phase over a wider RH range than the genuine SSA surrogate (i.e., pure NaCl particles), making their heterogeneous chemistry more likely.

Acknowledgements. This study was supported by Basic Science Research Program through the National Research Foundation of Korea (NRF) funded by the Ministry of Education, Science and Technology (2012R1A2A1A05026329), by Metrology Research Center funded by the Korea Research Institute of Standards and Science (KRISS – 2013 – 13011055), and by Inha University.

Edited by: M. Ammann

References

Ahn, K.-H., Kim, S.-M., Jung, H.-J., Lee, M.-J., Eom, H.-J., Maskey, S., and Ro, C.-U.: Combined use of optical and electron microscopic techniques for the measurement of hygroscopic property, chemical composition, and morphology of individual aerosol particles, Anal. Chem., 82, 7999–8009, doi:10.1021/ac101432y, 2010.

Ansari, A. S. and Pandis, S. N.: Prediction of multicomponent inorganic atmospheric aerosol behavior, Atmos. Environ., 33, 745–757, doi:1016/S1352-2310(98)00221-0, 1999.

Ault, A. P., Guasco, T. L., Ryder, O. S., Baltrusaitis, J., Cuadra-Rodriguez, L. A., Collins, D. B., Ruppel, M. J., Bertram, T. H., Prather, K. A., and Grassian, V. H.: Inside versus Outside: Ion Redistribution in Nitric Acid Reacted Sea Spray Aerosol Particles as Determined by Single Particle Analysis, J. Am. Chem. Soc., 135, 14528–14531, doi:10.1021/ja407117x, 2013a.

Ault, A. P., Moffet, R. C., Baltrusaitis, J., Collins, D. B., Ruppel, M. J., Cuadra-Rodriguez, L. A., Zhao, D., Guasco, T. L., Ebben, C. J., Geiger, F. M., Bertram, T. H., Prather, K. A., and Grassian, V. H.: Size-Dependent Changes in Sea Spray Aerosol Composition and Properties with Different Seawater Conditions, Environ. Sci. Technol., 47, 5603–5612, doi:10.1021/es400416g, 2013b.

Ault, A. P., Guasco, T. L., Baltrusaitis, J., Ryder, O. S., Trueblood, J. V., Collins, D. B., Ruppel, M. J., Cuadra-Rodriguez, L. A., Prather, K. A., and Grassian, V. H.: Heterogeneous Reactivity of Nitric Acid with Nascent Sea Spray Aerosol: Large Differences Observed between and within Individual Particles, J. Phys. Chem. Lett., 5, 2493–2500, doi:10.1021/jz5008802, 2014.

Baynard, T., Garland, R. M., Ravishankara, A. R., Tolbert, M. A., and Lovejoy, E. R.: Key factors influencing the relative humidity dependence of aerosol light scattering, Geophys. Res. Lett., 33, L06813, doi:10.1029/2005gl024898, 2006.

Beardsley, R., Jang, M., Ori, B., Im, Y., Delcomyn, C. A., and Witherspoon, N.: Role of sea salt aerosols in the formation of aromatic secondary organic aerosol: yields and hygroscopic properties, Environ. Chem., 10, 167–177, doi:10.1071/EN13016, 2013.

Chang, S.-Y. and Lee, C.-T.: Applying GC-TCD to investigate the hygroscopic characteristics of mixed aerosols, Atmos. Environ., 36, 1521–1530, doi:10.1016/S1352-2310(01)00546-5, 2002.

Clegg, S. L., Brimblecombe, P., and Wexler, A. S.: A thermodynamic model of the system $H^+-NH_4^+-Na^+-SO_4^{2-}-NO_3^--Cl^--H_2O$ at 298.15 K, J. Phys. Chem. A, 102, 2155–2171, doi:10.1021/jp973043j, 1998.

Cohen, M. D., Flagan, R. C., and Seinfeld, J. H.: Studies of concentrated electrolyte solutions using the electrodynamic balance. 2. Water activities for mixed-electrolyte solutions, J. Phys. Chem., 91, 4575–4582, doi:10.1021/j100301a030, 1987a.

Cohen, M. D., Flagan, R. C., and Seinfeld, J. H.: Studies of concentrated electrolyte solutions using the electrodynamic balance. 3. Solute nucleation, J. Phys. Chem., 91, 4583–4590, doi:10.1021/j100301a031, 1987b.

Eom, H.-J., Gupta, D., Li, X., Jung, H.-J., Kim, H., and Ro, C.-U.: Influence of Collecting Substrates on the Characterization of Hygroscopic Properties of Inorganic Aerosol Particles, Anal. Chem., 86, 2648–2656, doi:10.1021/ac4042075, 2014.

Finlayson-Pitts, B. J. and Pitts, J. N. J.: Chemistry of the upper and lower atmosphere theory, experiments, and applications, Academic Press, San Diego, 2000.

Freney, E. J., Martin, S. T., and Buseck, P. R.: Deliquescence and Efflorescence of Potassium Salts Relevant to Biomass-Burning Aerosol Particles, Aerosol Sci. Technol., 43, 799–807, 2009.

Freney, E. J., Adachi, K., and Buseck, P. R.: Internally mixed atmospheric aerosol particles: Hygroscopic growth and light scattering, J. Geophys. Res., 115, D19210, doi:10.1029/2009JD013558, 2010.

Fuentes, E., Coe, H., Green, D., and McFiggans, G.: On the impacts of phytoplankton-derived organic matter on the properties of the primary marine aerosol – Part 2: Composition, hygroscopicity and cloud condensation activity, Atmos. Chem. Phys., 11, 2585–2602, doi:10.5194/acp-11-2585-2011, 2011.

Gao, Y., Yu, L. E., and Chen, S. B.: Efflorescence Relative Humidity of Mixed Sodium Chloride and Sodium Sulfate Particles, J. Phys. Chem. A, 111, 10660–10666, doi:10.1021/jp073186y, 2007.

Gard, E. E., Kleeman, M. J., Gross, D. S., Hughes, L. S., Allen, J. O., Morrical, B. D., Fergenson, D. P., Dienes, T., Galli, M. E., Johnson, R. J., Cass, G. R., and Prather, K. A.: Direct Observation of Heterogeneous Chemistry in the Atmosphere Science, Science, 279, 1184–1187, doi:10.1126/science.279.5354.1184, 1998.

Ge, Z., Wexler, A. S., and Johnston, M. V.: Multicomponent Aerosol Crystallization, J. Colloid Interf. Sci., 183, 68–77, doi:10.1006/jcis.1996.0519, 1996.

Ge, Z., Wexler, A. S., and Johnston, M. V.: Deliquescence Behavior of Multicomponent Aerosols, J. Phys. Chem. A, 102, 173–180, doi:10.1021/jp972396f, 1998.

Gibson, E. R., Hudson, P. K., and Grassian, V. H.: Physicochemical Properties of Nitrate Aerosols: Implications for the Atmosphere, J. Phys. Chem. A, 110, 11785–11799, doi:10.1021/jp063821k, 2006.

Gysel, M., Weingartner, E., and Baltensperger, U.: Hygroscopicity of aerosol particles at low temperatures. 2. Theoretical and experimental hygroscopic properties of laboratory generated aerosols, Environ. Sci. Technol., 36, 63–68, doi:10.1021/es010055g, 2002.

Haywood, J. and Boucher, O.: Estimates of the direct and indirect radiative forcing due to tropospheric aerosols: A review, Rev. Geophys., 38, 513–543, doi:10.1029/1999rg000078, 2000.

Hoffman, R. C., Laskin, A., and Finlayson-Pitts, B. J.: Sodium nitrate particles: physical and chemical properties during hydration and dehydration, and implications for aged sea salt aerosols, J. Aerosol Sci., 35, 869–887, doi:10.1016/j.jaerosci.2004.02.003, 2004.

Hu, D., Qiao, L., Chen, J., Ye, X., Yang, X., Cheng, T., and Fang, W.: Hygroscopicity of Inorganic Aerosols: Size and Relative Humidity Effects on the Growth Factor, Aerosol Air Qual. Res., 10, 255–264, doi:10.4209/aaqr.2009.12.0076, 2010.

Keene, W. C., Maring, H., Maben, J. R., Kieber, D. J., Pszenny, A. A. P., Dahl, E. E., Izaguirre, M. A., Davis, A. J., Long, M. S., Zhou, X., Smoydzin, L., and Sander, R.: Chemical and physical characteristics of nascent aerosols produced by bursting bubbles at a model air-sea interface, J. Geophys. Res.-Atmos., 112, D21202, doi:10.1029/2007jd008464, 2007.

Kelly, J. T., Wexler, A. S., Chan, C. K., and Chan, M. N.: Aerosol thermodynamics of potassium salts, double salts, and water content near the eutectic, Atmos. Environ., 42, 3717–3728, doi:10.1016/j.atmosenv.2008.01.001, 2008.

Kim, H.-K., Lee, M.-J., Jung, H.-J., Eom, H.-J., Maskey, S., Ahn, K.-H., and Ro, C.-U.: Hygroscopic behavior of wet dispersed and dry deposited NaNO$_3$ particles, Atmos. Environ., 60, 68–75, doi:10.1016/j.atmosenv.2012.06.011, 2012.

King, S. M., Butcher, A. C., Rosenoern, T., Coz, E., Lieke, K. I., de Leeuw, G., Nilsson, E. D., and Bilde, M.: Investigating Primary Marine Aerosol Properties: CCN Activity of Sea Salt and Mixed Inorganic–Organic Particles, Environ. Sci. Technol., 46, 10405–10412, doi:10.1021/es300574u, 2012.

Krieger, U. K., Marcolli, C., and Reid, J. P.: Exploring the complexity of aerosol particle properties and processes using single particle techniques, Chem. Soc. Rev., 41, 6631–6662, doi:10.1039/c2cs35082c, 2012.

Krueger, B. J., Grassian, V. H., Iedema, M. J., Cowin, J. P., and Laskin, A.: Probing Heterogeneous Chemistry of Individual Atmospheric Particles Using Scanning Electron Microscopy and Energy-Dispersive X-ray Analysis, Anal. Chem., 75, 5170–5179, doi:10.1021/ac034455t, 2003.

Laskin, A., Gaspar, D. J., Wang, W., Hunt, S. W., Cowin, J. P., Colson, S. D., and Finlayson-Pitts, B. J.: Reactions at Interfaces As a Source of Sulfate Formation in Sea-Salt Particles, Science, 301, 340–344, doi:10.1126/science.1085374, 2003.

Lee C.-T. and Hsu, W.-C.: The Measurement of Liquid Water Mass Associated with Collected Hygroscopic Particles, J. Aerosol Sci., 31, 189–197, doi:10.1016/S0021-8502(99)00048-8, 2000.

Li, X., Gupta, D., Eom, H.-J., Kim, H., and Ro, C.-U.: Deliquescence and efflorescence behavior of individual NaCl and KCl mixture aerosol particles, Atmos. Environ., 82, 36–43, doi:10.1016/j.atmosenv.2013.10.011, 2014.

Lide, D. R. (Ed.): Handbook of Chemistry and Physics Eighty third ed., CRC Press, Boca Raton, Florida, 2002.

Liu, Y., Cain, J. P., Wang, H., and Laskin, A.: Kinetic Study of Heterogeneous Reaction of Deliquesced NaCl Particles with Gaseous HNO$_3$ Using Particle-on-Substrate Stagnation Flow Reactor Approach, J. Phys. Chem. A, 111, 10026–10043, doi:10.1021/jp072005p, 2007.

Ma, X., von Salzen, K., and Li, J.: Modelling sea salt aerosol and its direct and indirect effects on climate, Atmos. Chem. Phys., 8, 1311–1327, doi:10.5194/acp-8-1311-2008, 2008.

Martin, S. T.: Phase Transitions of Aqueous Atmospheric Particles, Chem. Rev., 100, 3403–3454, doi:10.1021/cr990034t, 2000.

McInnes, L. M., Quinn, P. K., Covert, D. S., and Anderson, T. L.: Gravimetric analysis, ionic composition, and associated water mass of the marine aerosol, Atmos. Environ., 30, 869–884, doi:10.1016/1352-2310(95)00354-1, 1996.

Meskhidze, N., Petters, M. D., Tsigaridis, K., Bates, T., O'Dowd, C., Reid, J., Lewis, E. R., Gantt, B., Anguelova, M. D., Bhave, P. V., Bird, J., Callaghan, A. H., Ceburnis, D., Chang, R., Clarke, A., de Leeuw, G., Deane, G., DeMott, P. J., Elliot, S., Facchini,

M. C., Fairall, C. W., Hawkins, L., Hu, Y., Hudson, J. G., Johnson, M. S., Kaku, K. C., Keene, W. C., Kieber, D. J., Long, M. S., Mårtensson, M., Modini, R. L., Osburn, C. L., Prather, K. A., Pszenny, A., Rinaldi, M., Russell, L. M., Salter, M., Sayer, A. M., Smirnov, A., Suda, S. R., Toth, T. D., Worsnop, D. R., Wozniak, A., and Zorn, S. R.: Production mechanisms, number concentration, size distribution, chemical composition, and optical properties of sea spray aerosols, Atmos. Sci. Lett., 14, 207–213, doi:10.1002/asl2.441, 2013.

Mochida, M., Nishita-Hara, C., Furutani, H., Miyazaki, Y., Jung, J., Kawamura, K., and Uematsu, M.: Hygroscopicity and cloud condensation nucleus activity of marine aerosol particles over the western North Pacific, J. Geophys. Res.-Atmos., 116, D06204, doi:10.1029/2010jd014759, 2011.

O'Dowd, C. D. and de Leeuw, G.: Marine aerosol production: a review of the current knowledge, Philos. T. R. Soc. A, 365, 1753–1774, doi:10.1098/rsta.2007.2043, 2007.

Pandis, S. N., Wexler, A. S., and Seinfeld, J. H.: Dynamics of tropospheric aerosols (Review), J. Phys. Chem., 99, 9646–9659, 1995.

Petters, M. D. and Kreidenweis, S. M.: A single parameter representation of hygroscopic growth and cloud condensation nucleus activity, Atmos. Chem. Phys., 7, 1961–1971, doi:10.5194/acp-7-1961-2007, 2007.

Prather, K. A., Bertram, T. H., Grassian, V. H., Deane, G. B., Stokes, M. D., DeMott, P. J., Aluwihare, L. I., Palenik, B. P., Azam, F., Seinfeld, J. H., Moffet, R. C., Molina, M. J., Cappa, C. D., Geiger, F. M., Roberts, G. C., Russell, L. M., Ault, A. P., Baltrusaitis, J., Collins, D. B., Corrigan, C. E., Cuadra-Rodriguez, L. A., Ebben, C. J., Forestieri, S. D., Guasco, T. L., Hersey, S. P., Kim, M. J., Lambert, W. F., Modini, R. L., Mui, W., Pedler, B. E., Ruppel, M. J., Ryder, O. S., Schoepp, N. G., Sullivan, R. C., and Zhao, D.: Bringing the ocean into the laboratory to probe the chemical complexity of sea spray aerosol, P. Natl. Acad. Sci., 110, 7550–7555, doi:10.1073/pnas.1300262110, 2013.

Ro, C.-U., Oh, K.-Y., Kim, H., Kim, Y. P., Lee, C. B., Kim, K.-H., Osan, J., de Hoog, J., Worobiec, A., and Van Grieken, R.: Single Particle Analysis of Aerosols at Cheju Island, Korea, Using Low-Z Electron Probe X-ray Microanalysis: A Direct Proof of Nitrate Formation from Sea-Salts, Environ. Sci. Technol., 35, 4487–4494, doi:10.1021/es0155231, 2001.

Ryder, O. S., Ault, A. P., Cahill, J. F., Guasco, T. L., Riedel, T. P., Cuadra-Rodriguez, L. A., Gaston, C. J., Fitzgerald, E., Lee, C., Prather, K. A., and Bertram, T. H.: On the Role of Particle Inorganic Mixing State in the Reactive Uptake of N$_2$O$_5$ to Ambient Aerosol Particles, Environ. Sci. Technol., 48, 1618–1627, doi:10.1021/es4042622, 2014.

Satheesh, S. K. and Moorthy, K. K.: Radiative effects of natural aerosols: A review, Atmos. Environ., 39, 2089–2110, doi:10.1016/j.atmosenv.2004.12.029, 2005.

Saul, T. D., Tolocka, M. P., and Johnston, M. V.: Reactive Uptake of Nitric Acid onto Sodium Chloride Aerosols Across a Wide Range of Relative Humidities, J. Phys. Chem. A, 110, 7614–7620, doi:10.1021/jp060639a, 2006.

Schlenker, J. C. and Martin, S. T.: Crystallization Pathways of Sulfate-Nitrate-Ammonium Aerosol Particles, J. Phys. Chem. A, 109, 9980–9985, 2005.

Seinfeld, J. H. and Pandis, S. N.: Atmospheric chemistry and physics : from air pollution to climate change, 2nd ed., J. Wiley, Hoboken NJ, 1203 pp., 2006.

Tang, I. N.: Phase Transformation and Growth of Aerosol Particles Composed of Mixed Salts, J. Aerosol Sci., 7, 361–371, doi:10.1016/0021-8502(76)90022-7, 1976.

Tang, I. N. and Munkelwitz, H. R.: Composition and temperature dependence of the deliquescence properties of hygroscopic aerosols, Atmos. Environ., 27A, 467–473, doi:10.1016/0960-1686(93)90204-C, 1993.

Tang, I. N. and Munkelwitz, H. R.: Aerosol Phase Transformation and Growth in the Atmosphere, J. Appl. Meteorol., 33, 791–796, doi:10.1175/1520-0450(1994)033<0791:APTAGI>2.0.CO;2, 1994a.

Tang, I. N. and Munkelwitz, H. R.: Water activities, densities, and refractive indices of aqueous sulfates and sodium nitrate droplets of atmospheric importance, J. Geophys. Res., 99, 18801–18808, doi:10.1029/94jd01345, 1994b.

Tang, I. N., Munkelwitz, H. R., and Davis, J. G.: Aerosol Growth Studies – IV Phase Transformation of Mixed Salt Aerosols in a Moist Atmosphere, J. Aerosol Sci., 9, 505–511, doi:10.1016/0021-8502(78)90015-0, 1978.

Tang, I. N., Tridico, A. C., and Fung, K. H.: Thermodynamics and optical properties of sea salt aerosols, J. Geophys. Res., 102, 23269–23275, doi:10.1029/97jd01806, 1997.

ten Brink, H. M.: Reactive uptake of HNO3 and H2SO4 in sea-salt (NaCl) particles, J. Aerosol Sci., 29, 57–64, doi:10.1016/S0021-8502(97)00460-6, 1998.

Wang, B. and Laskin, A.: Reactions between water-soluble organic acids and nitrates in atmospheric aerosols: Recycling of nitric acid and formation of organic salts, J. Geophys. Res.-Atmos., 119, 2013JD021169, doi:10.1002/2013jd021169, 2014.

Wang, J. and Martin, S. T.: Satellite characterization of urban aerosols: Importance of including hygroscopicity and mixing state in the retrieval algorithms, J. Geophys. Res.-Atmos., 112, D17203, doi:10.1029/2006jd008078, 2007.

Wex, H., Stratmann, F., Hennig, T., Hartmann, S., Niedermeier, D., Nilsson, E., Ocskay, R., Rose, D., Salma, I., and Ziese, M.: Connecting hygroscopic growth at high humidities to cloud activation for different particle types, Environ. Res. Lett., 3, 035004, doi:10.1088/1748-9326/3/3/035004, 2008.

Wexler, A. S. and Clegg, S. L.: Atmospheric aerosol models for systems including the ions H+, NH4+, Na+, SO4^2−, NO3−, Cl−, Br−, and H2O, J. Geophys. Res., 107, D14, 4207, doi:10.1029/2001JD000451, 2002.

Wexler, A. S. and Seinfeld, J. H.: Second-generation inorganic aerosol model, Atmos. Environ., 25A, 2731–2748, doi:10.1016/0960-1686(91)90203-J, 1991.

Wise, M. E., Semeniuk, T. A., Bruintjes, R., Martin, S. T., Russell, L. M., and Buseck, P. R.: Hygroscopic behavior of NaCl-bearing natural aerosol particles using environmental transmission electron microscopy, J. Geophys. Res., 112, D10224, doi:10.1029/2006JD007678, 2007.

Wise, M. E., Freney, E. J., Tyree, C. A., Allen, J. O., Martin S. T., and Russell L. M., and Buseck, P. R.: Hygroscopic behavior and liquid-layer composition of aerosol particles generated from natural and artificial seawater, J. Geophys. Res., 114, D03201, doi:10.1029/2008JD010449, 2009.

Woods, E., Yi, C., Gerson, J. R., and Zaman, R. A.: Uptake of Pyrene by NaCl, NaNO3, and MgCl2 Aerosol Particles, J. Phys. Chem. A, 116, 4137–4143, doi:10.1021/jp3014145, 2012.

Ziemann, P. J. and McMurry, P. H.: Spatial Distribution of Chemical Components in Aerosol Particles as Determined from Secondary Electron Yield Measurements: Implications for Mechanisms of Multicomponent Aerosol Crystallization, J. Colloid Interf. Sci., 193, 250–258, doi:10.1006/jcis.1997.5075, 1997.

Zuend, A., Marcolli, C., Luo, B. P., and Peter, T.: A thermodynamic model of mixed organic-inorganic aerosols to predict activity coefficients, Atmos. Chem. Phys., 8, 4559–4593, doi:10.5194/acp-8-4559-2008, 2008.

Zuend, A., Marcolli, C., Booth, A. M., Lienhard, D. M., Soonsin, V., Krieger, U. K., Topping, D. O., McFiggans, G., Peter, T., and Seinfeld, J. H.: New and extended parameterization of the thermodynamic model AIOMFAC: calculation of activity coefficients for organic-inorganic mixtures containing carboxyl, hydroxyl, carbonyl, ether, ester, alkenyl, alkyl, and aromatic functional groups, Atmos. Chem. Phys., 11, 9155–9206, doi:10.5194/acp-11-9155-2011, 2011.

Permissions

All chapters in this book were first published in ACP, by Copernicus Publications; hereby published with permission under the Creative Commons Attribution License or equivalent. Every chapter published in this book has been scrutinized by our experts. Their significance has been extensively debated. The topics covered herein carry significant findings which will fuel the growth of the discipline. They may even be implemented as practical applications or may be referred to as a beginning point for another development.

The contributors of this book come from diverse backgrounds, making this book a truly international effort. This book will bring forth new frontiers with its revolutionizing research information and detailed analysis of the nascent developments around the world.

We would like to thank all the contributing authors for lending their expertise to make the book truly unique. They have played a crucial role in the development of this book. Without their invaluable contributions this book wouldn't have been possible. They have made vital efforts to compile up to date information on the varied aspects of this subject to make this book a valuable addition to the collection of many professionals and students.

This book was conceptualized with the vision of imparting up-to-date information and advanced data in this field. To ensure the same, a matchless editorial board was set up. Every individual on the board went through rigorous rounds of assessment to prove their worth. After which they invested a large part of their time researching and compiling the most relevant data for our readers.

The editorial board has been involved in producing this book since its inception. They have spent rigorous hours researching and exploring the diverse topics which have resulted in the successful publishing of this book. They have passed on their knowledge of decades through this book. To expedite this challenging task, the publisher supported the team at every step. A small team of assistant editors was also appointed to further simplify the editing procedure and attain best results for the readers.

Apart from the editorial board, the designing team has also invested a significant amount of their time in understanding the subject and creating the most relevant covers. They scrutinized every image to scout for the most suitable representation of the subject and create an appropriate cover for the book.

The publishing team has been an ardent support to the editorial, designing and production team. Their endless efforts to recruit the best for this project, has resulted in the accomplishment of this book. They are a veteran in the field of academics and their pool of knowledge is as vast as their experience in printing. Their expertise and guidance has proved useful at every step. Their uncompromising quality standards have made this book an exceptional effort. Their encouragement from time to time has been an inspiration for everyone.

The publisher and the editorial board hope that this book will prove to be a valuable piece of knowledge for researchers, students, practitioners and scholars across the globe.

List of Contributors

H. Lyamani
Andalusian Institute for Earth System Research (IISTA-CEAMA), 18006, Granada, Spain
Department of Applied Physic, University of Granada, 18071, Granada, Spain

A. Valenzuela
Andalusian Institute for Earth System Research (IISTA-CEAMA), 18006, Granada, Spain
Department of Applied Physic, University of Granada, 18071, Granada, Spain

D. Perez-Ramirez
Mesoscale Atmospheric Processes Laboratory, NASA Goddard Space Flight Center, Greenbelt, Maryland 20771, USA
Goddard Earth Sciences Technology and Research, Universities Space Research Association (GESTAR/USRA), Columbia, Maryland, USA

C. Toledano
Atmospheric Optics Group (GOA), University of Valladolid (UVA), 47071, Valladolid, Spain

M. J. Granados-Muñoz
Andalusian Institute for Earth System Research (IISTA-CEAMA), 18006, Granada, Spain
Department of Applied Physic, University of Granada, 18071, Granada, Spain

F. J. Olmo
Andalusian Institute for Earth System Research (IISTA-CEAMA), 18006, Granada, Spain
Department of Applied Physic, University of Granada, 18071, Granada, Spain

L. Alados-Arboledas
Andalusian Institute for Earth System Research (IISTA-CEAMA), 18006, Granada, Spain
Department of Applied Physic, University of Granada, 18071, Granada, Spain

K. Ding
School of Atmospheric Sciences, Nanjing University, Nanjing, Jiangsu 210093, China
Collaborative Innovation Center of Climate Change, Jiangsu 210093, China

J. Liu
School of Atmospheric Sciences, Nanjing University, Nanjing, Jiangsu 210093, China
University of Toronto, Toronto, Ontario, M5S 3G3, Canada
Collaborative Innovation Center of Climate Change, Jiangsu 210093, China

A. Ding
School of Atmospheric Sciences, Nanjing University, Nanjing, Jiangsu 210093, China
Collaborative Innovation Center of Climate Change, Jiangsu 210093, China

Q. Liu
School of Atmospheric Sciences, Nanjing University, Nanjing, Jiangsu 210093, China
Collaborative Innovation Center of Climate Change, Jiangsu 210093, China

T. L. Zhao
Nanjing University of Information Science and Technology, Nanjing, Jiangsu 210044, China

J. Shi
Institute of Remote Sensing Applications, Chinese Academy of Sciences, Beijing 100101, China

Y. Han
School of Atmospheric Sciences, Nanjing University, Nanjing, Jiangsu 210093, China

H. Wang
International Institute for Earth System Sciences, Nanjing University, Nanjing, Jiangsu 210093, China

F. Jiang
International Institute for Earth System Sciences, Nanjing University, Nanjing, Jiangsu 210093, China

S. Dai
State Key Laboratory of Organic Geochemistry and Guangdong Key Laboratory of Environmental Resources Utilization and Protection, Guangzhou Institute of Geochemistry, Chinese Academy of Sciences, Guangzhou 510640, China
University of Chinese Academy of Sciences, Beijing 100049, China

X. Bi
State Key Laboratory of Organic Geochemistry and Guangdong Key Laboratory of Environmental Resources Utilization and Protection, Guangzhou Institute of Geochemistry, Chinese Academy of Sciences, Guangzhou 510640, China

L. Y. Chan
State Key Laboratory of Organic Geochemistry and Guangdong Key Laboratory of Environmental Resources Utilization and Protection, Guangzhou Institute of Geochemistry, Chinese Academy of Sciences, Guangzhou 510640, China

J. He
State Key Laboratory of Organic Geochemistry and
Guangdong Key Laboratory of Environmental Resources
Utilization and Protection, Guangzhou Institute of
Geochemistry, Chinese Academy of Sciences, Guangzhou
510640, China
Institute of Atmospheric Environmental Safety and
Pollution Control, Jinan University, Guangzhou 510632,
China

B. Wang
3Institute of Atmospheric Environmental Safety and
Pollution Control, Jinan University, Guangzhou 510632,
China

X. Wang
State Key Laboratory of Organic Geochemistry and
Guangdong Key Laboratory of Environmental Resources
Utilization and Protection, Guangzhou Institute of
Geochemistry, Chinese Academy of Sciences, Guangzhou
510640, China

P. Peng
State Key Laboratory of Organic Geochemistry and
Guangdong Key Laboratory of Environmental Resources
Utilization and Protection, Guangzhou Institute of
Geochemistry, Chinese Academy of Sciences, Guangzhou
510640, China

G. Sheng
State Key Laboratory of Organic Geochemistry and
Guangdong Key Laboratory of Environmental Resources
Utilization and Protection, Guangzhou Institute of
Geochemistry, Chinese Academy of Sciences, Guangzhou
510640, China

J. Fu
State Key Laboratory of Organic Geochemistry and
Guangdong Key Laboratory of Environmental Resources
Utilization and Protection, Guangzhou Institute of
Geochemistry, Chinese Academy of Sciences, Guangzhou
510640, China

M. Pramitha
National Atmospheric Research Laboratory (NARL),
Gadanki, India

M. Venkat Ratnam
National Atmospheric Research Laboratory (NARL),
Gadanki, India

A. Taori
National Atmospheric Research Laboratory (NARL),
Gadanki, India

B. V. Krishna Murthy
B1, CEBROS, Chennai, India

D. Pallamraju
Physical Research Laboratory (PRL), Ahmadabad, India

S. Vijaya Bhaskar Rao
Department of Physics, Sri Venkateswara University,
Tirupati, India

P. Nabat
Météo-France, CNRM-GAME, Centre national de
recherches météorologiques, UMR3589, Toulouse, France

S. Somot
Météo-France, CNRM-GAME, Centre national de
recherches météorologiques, UMR3589, Toulouse, France

M. Mallet
Laboratoire d'Aérologie, Toulouse, France

M. Michou
Météo-France, CNRM-GAME, Centre national de
recherches météorologiques, UMR3589, Toulouse, France

F. Sevault
Météo-France, CNRM-GAME, Centre national de
recherches météorologiques, UMR3589, Toulouse, France

F. Driouech
Direction de la Météorologie Nationale, Casablanca,
Morocco

D. Meloni
Laboratory for Earth Observations and Analyses, ENEA,
Rome, Italy

A. di Sarra
Laboratory for Earth Observations and Analyses, ENEA,
Rome, Italy

C. Di Biagio
Laboratoire interuniversitaire des systèmes
atmosphériques (LISA), UMR7583 – CNRS, Créteil, France

P. Formenti
Laboratoire interuniversitaire des systèmes
atmosphériques (LISA), UMR7583 – CNRS, Créteil, France

M. Sicard
Universitat Politechnica de Catalunya, Barcelona, Spain

J.-F. Léon
Laboratoire d'Aérologie, Toulouse, France

M.-N. Bouin
Météo-France, CMM, Centre de Météorologie Marine,
Brest, France

K. W. Dawson
Marine, Earth, and Atmospheric Science, North Carolina
State University, Raleigh, NC, USA

N. Meskhidze
Marine, Earth, and Atmospheric Science, North Carolina State University, Raleigh, NC, USA

D. Josset
Science Systems and Applications, Inc./NASA Langley Research Center, Hampton, VA, USA

S. Gassó
GESTAR/Morgan State University, Goddard Space Flight Center, Greenbelt, MD, USA
Naval Research Laboratory, Stennis Space Center, Mississippi, USA

H. Vogelmann
Karlsruhe Institute of Technology, IMK-IFU, Garmisch-Partenkirchen, Germany

R. Sussmann
Karlsruhe Institute of Technology, IMK-IFU, Garmisch-Partenkirchen, Germany

T. Trickl
Karlsruhe Institute of Technology, IMK-IFU, Garmisch-Partenkirchen, Germany

A. Reichert
Karlsruhe Institute of Technology, IMK-IFU, Garmisch-Partenkirchen, Germany

F. Slemr
Max-Planck-Institute for Chemistry, Hahn-Meitner-Weg 1, 55128 Mainz, Germany

H. Angot
Université Grenoble Alpes, LGGE, 38041 Grenoble, France

A. Dommergue
Université Grenoble Alpes, LGGE, 38041 Grenoble, France
CNRS, LGGE, 38041 Grenoble, France

O. Magand
CNRS, LGGE, 38041 Grenoble, France

M. Barret
Université Grenoble Alpes, LGGE, 38041 Grenoble, France
CNRS, LGGE, 38041 Grenoble, France

A. Weigelt
Helmholtz-Zentrum Geesthacht (HZG), Institute of Coastal Research, Max-Planck-Strasse 1, 21502 Geesthacht, Germany

R. Ebinghaus
Helmholtz-Zentrum Geesthacht (HZG), Institute of Coastal Research, Max-Planck-Strasse 1, 21502 Geesthacht, Germany

E.-G. Brunke
5South African Weather Service c/o CSIR, P.O. Box 320, Stellenbosch 7599, South Africa

K. A. Pfaffhuber
Norwegian Institute for Air Research (NILU), P.O. Box 100, 2027 Kjeller, Norway

G. Edwards
Macquarie University, Environmental Science, Sydney, NSW, Australia

D. Howard
Macquarie University, Environmental Science, Sydney, NSW, Australia

J. Powell
CSIRO Ocean and Atmosphere Flagship Research, Aspendale, VIC, Australia

M. Keywood
CSIRO Ocean and Atmosphere Flagship Research, Aspendale, VIC, Australia

F. Wang
Centre for Earth Observation Science, Department of Environment and Geography, University of Manitoba, Winnipeg, MB, R3T 2N2, Canada

T. Fytterer
Institute for Meteorology and Climate Research, Karlsruhe Institute of Technology, Eggenstein-Leopoldshafen, Germany

M. G. Mlynczak
Atmospheric Sciences Division, NASA Langley Research Center, Hampton, VA, USA

H. Nieder
Institute for Meteorology and Climate Research, Karlsruhe Institute of Technology, Eggenstein-Leopoldshafen, Germany

K. Pérot
Department of Earth and Space Sciences, Chalmers University of Technology, Göteborg, Sweden
deceased, 14 August 2014

M. Sinnhuber
Institute for Meteorology and Climate Research, Karlsruhe Institute of Technology, Eggenstein-Leopoldshafen, Germany

G. Stiller
Institute for Meteorology and Climate Research, Karlsruhe Institute of Technology, Eggenstein-Leopoldshafen, Germany

J. Urban
Department of Earth and Space Sciences, Chalmers University of Technology, Göteborg, Sweden
deceased, 14 August 2014

L. K. Whalley
National Centre for Atmospheric Science, University of Leeds, Leeds, LS2 9JT, UK
School of Chemistry, University of Leeds, Leeds, LS2 9JT, UK

D. Stone
School of Chemistry, University of Leeds, Leeds, LS2 9JT, UK

I. J. George
School of Chemistry, University of Leeds, Leeds, LS2 9JT, UK

S. Mertes
Leibniz-Institut für Troposphärenforschung (TROPOS), Permoserstr. 15, 04318 Leipzig, Germany

D. van Pinxteren
Leibniz-Institut für Troposphärenforschung (TROPOS), Permoserstr. 15, 04318 Leipzig, Germany

A. Tilgner
Leibniz-Institut für Troposphärenforschung (TROPOS), Permoserstr. 15, 04318 Leipzig, Germany

H. Herrmann
Leibniz-Institut für Troposphärenforschung (TROPOS), Permoserstr. 15, 04318 Leipzig, Germany

M. J. Evans
National Centre for Atmospheric Science, University of York, York, YO10 5DD, UK
Department of Chemistry, University of York, York, YO10 5DD, UK
now at: National Risk Management Research Laboratory, US Environmental Protection Agency, Research Triangle Park, North Carolina 27711, USA

D. E. Heard
National Centre for Atmospheric Science, University of Leeds, Leeds, LS2 9JT, UK
School of Chemistry, University of Leeds, Leeds, LS2 9JT, UK

G. Myhre
Center for International Climate and Environmental Research – Oslo (CICERO), Oslo, Norway

B. H. Samset
Center for International Climate and Environmental Research – Oslo (CICERO), Oslo, Norway

D. Gupta
Department of Chemistry, Inha University, Incheon, 402-751, South Korea

H. Kim
Department of Chemistry, Inha University, Incheon, 402-751, South Korea

G. Park
Department of Chemistry, Inha University, Incheon, 402-751, South Korea

X. Li
Department of Chemistry, Inha University, Incheon, 402-751, South Korea

H.-J. Eom
Department of Chemistry, Inha University, Incheon, 402-751, South Korea

C.-U. Ro
Department of Chemistry, Inha University, Incheon, 402-751, South Korea

www.ingramcontent.com/pod-product-compliance
Lightning Source LLC
Chambersburg PA
CBHW080300230326

41458CB00097B/5243